PLENUM PRESS HANDBOOKS OF HIGH-TEMPERATURE MATERIALS

No. 1 – MATERIALS INDEX

PLENUM PRESS HANDBOOKS OF
HIGH-TEMPERATURE MATERIALS

No. 1 — MATERIALS INDEX
Peter T. B. Shaffer

No. 2 — PROPERTIES INDEX
G. V. Samsonov

PLENUM PRESS HANDBOOKS OF
HIGH-TEMPERATURE MATERIALS

No. 1
MATERIALS INDEX

by
Peter T. B. Shaffer

with a foreword by
Henry H. Hausner

SPRINGER SCIENCE+BUSINESS MEDIA, LLC 1963

ISBN 978-1-4899-6200-3   ISBN 978-1-4899-6405-2 (eBook)
DOI 10.1007/978-1-4899-6405-2

Library of Congress Catalog Card Number : 64-17206

This handbook was originally published by The Carborundum
Company under the title

PROPERTIES OF HIGH TEMPERATURE MATERIALS
(Borides, Carbides, Elements, Nitrides, Oxides, and Silicides)

FOREWORD

Technological progress at the present time depends to a large extent on the development of materials which withstand high temperatures. Work on these materials is going on in many industrial and government laboratories. The characteristics of these high-temperature materials can be found in a great variety of books, articles, and reports, but they have never been collected in a single volume.

This volume contains data on the properties of more than 520 different materials, such as carbides, borides, nitrides, sulfides, phosphides, silicides, oxides, as well as mixed oxides and mixed carbides. The data are taken from more than 690 references, with the large majority from the U. S. Government report literature. The data concern general, chemical, electrical, mechanical, nuclear, optical, structural, and thermal properties. The references are listed by number as they appear in the tables, as well as alphabetically by name of author. The extensive table of contents on pages vii to xx makes it easy to find the various materials in the tables.

Dr. Shaffer, a graduate of Massachusetts Institute of Technology and The State University of Pennsylvania, is presently Senior Research Associate in the Physics Department of the Carborundum Company, where he has been employed for a number of years. He is a specialist in high-temperature materials and high-temperature test equipment design.

Henry H. Hausner
Adj. Professor, Polytechnic
Institute of Brooklyn
Consulting Engineer,
New York, N. Y.

INDEX

I BORIDES

II CARBIDES

III MIXED CARBIDES

IV ELEMENTS

V NITRIDES

VI OXIDES

VII MIXED OXIDES

$Al_2O_3 \cdot BaO$	388	$9\,Al_2O_3 \cdot 2\,B_2O_3$	413
$Al_2O_3 \cdot BeO$	389	$2\,B_2O_3 \cdot 9\,Al_2O_3$	413
$Al_2O_3 \cdot CaO$	390	$BaO \cdot Al_2O_3$	388
$Al_2O_3 \cdot CoO$	391	$BaO \cdot ZrO_2$	414
$Al_2O_3 \cdot Co_2O_3$	392	$6\,BaO \cdot Ta_2O_5$	415
$Al_2O_3 \cdot FeO$	393	$BeO \cdot Al_2O_3$	389
$Al_2O_3 \cdot Li_2O$	394	$BeO \cdot Cr_2O_3$	416
$Al_2O_3 \cdot MgO$	395 - 397	$BeO \cdot UO_2$	417
$Al_2O_3 \cdot MnO$	398	$2\,BeO \cdot SiO_2$	418
$Al_2O_3 \cdot NiO$	399	$2\,BeO \cdot UO_2$	419
$Al_2O_3 \cdot SiO_2$	400	$3\,BeO \cdot 2\,ZrO_2$	420
$Al_2O_3 \cdot SrO$	401	$CaO \cdot Al_2O_3$	390
$Al_2O_3 \cdot TiO_2$	402	$CaO \cdot Cr_2O_3$	421
$Al_2O_3 \cdot ZnO$	403	$CaO \cdot HfO_2$	422
$Al_2O_3 \cdot ZrO_2$	404	$CaO \cdot SiO_2$	423
$2\,Al_2O_3 \cdot 2\,MgO \cdot 5\,SiO_2$	405	$CaO \cdot TiO_2$	424
$3\,Al_2O_3 \cdot 2\,CeO_2$	406	$CaO \cdot ZrO_2$	425
$3\,Al_2O_3 \cdot 2\,SiO_2$	407 - 408	$2\,CaO \cdot SiO_2$	426
$3\,Al_2O_3 \cdot 2\,SnO_2$	409	$3\,CaO \cdot 5\,Al_2O_3$	411
$3\,Al_2O_3 \cdot 2\,ThO_2$	410	$3\,CaO \cdot SiO_2$	427
$5\,Al_2O_3 \cdot 3\,CaO$	411	$3\,CaO \cdot TiO_2$	428
$5\,Al_2O_3 \cdot Li_2O$	412	$2\,CeO_2 \cdot 3\,Al_2O_3$	406

VII MIXED OXIDES (Cont.)

$CoO \cdot Al_2O_3$	391	$MgO \cdot Al_2O_3$	395 – 397
$Co_2O_3 \cdot Al_2O_3$	392	$MgO \cdot Cr_2O_3$	430
$Cr_2O_3 \cdot BeO$	416	$MgO \cdot 4\,Cr_2O_3$	435
$Cr_2O_3 \cdot CaO$	421	$MgO \cdot Fe_2O_3$	437
$Cr_2O_3 \cdot FeO$	429	$MgO \cdot La_2O_3$	441
$Cr_2O_3 \cdot MgO$	430	$MgO \cdot SiO_2$	442
$Cr_2O_3 \cdot MnO$	431	$MgO \cdot TiO_2$	443
$Cr_2O_3 \cdot ZnO$	432	$MgO \cdot 2\,TiO_2$	444
$Cr_2O_3 \cdot ZrO_2$	433	$MgO \cdot ZrO_2$	445
$2\,Cr_2O_3 \cdot Fe_2O_3$	434	$2\,MgO \cdot 2\,Al_2O_3 \cdot 5\,SiO_2$	405
$4\,Cr_2O_3 \cdot MgO$	435	$2\,MgO \cdot SiO_2$	446
$FeO \cdot Al_2O_3$	393	$2\,MgO \cdot TiO_2$	447
$FeO \cdot Cr_2O_3$	429	$3\,MgO \cdot 2\,UO_2$	448
$2\,FeO \cdot SiO_2$	436	$MnO \cdot Al_2O_3$	398
$Fe_2O_3 \cdot MgO$	437	$MnO \cdot Cr_2O_3$	431
$Fe_2O_3 \cdot ZnO$	438	$NiO \cdot Al_2O_3$	399
$HfO_2 \cdot CaO$	422	$SiO_2 \cdot Al_2O_3$	400
$HfO_2 \cdot SiO_2$	439	$SiO_2 \cdot 2\,BeO$	418
$HfO_2 \cdot SrO$	440	$SiO_2 \cdot CaO$	423
$La_2O_3 \cdot MgO$	441	$SiO_2 \cdot 2\,CaO$	426
$Li_2O \cdot Al_2O_3$	394	$SiO_2 \cdot 3\,CaO$	427
$Li_2O \cdot 5\,Al_2O_3$	412	$SiO_2 \cdot 2\,FeO$	436

VIII SILICIDES

INTRODUCTION

Library searches in conjunction with several projects involving high temperature materials have been made and many property data were collected. In addition, several data summaries of varying degrees of comprehensiveness are at the author's disposal. As knowledge of the availability of this collection of data became known, more and more of the author's fellow researchers referred to it. It was felt, therefore, that a compilation of these data, if made available for others, would save many hours of searching through the literature. Since the references were, in most cases, available this compilation would also provide a ready source of reference to original papers in various fields to which the reader could refer for specific details, procedures, etc.

The task of cataloguing both the primary and secondary references became completely unmanageable. The author, therefore, took the liberty of omitting secondary references to such summaries as "Refractory Hard Metals" by Schwartzkopf and Kieffer; "Thermophysical Properties of Solid Materials" by Goldsmith et al. and others, except in those cases where the references to original papers were not completely clear.

The very nature of such a data summary as this makes it impossible to attain a high degree of completeness. The author wishes to state that no attempt to obtain all the data for any specific compound has been made. This compilation of data covers only those data which were at the author's disposal and did not involve any specific literature searches for the sake of this data compilation.

FORMAT

Property data are divided into main classes and sub classes as follows:

A. GENERAL

Formula

Compound Name

Formula weight

Formula volume

Melting point

Boiling point

Vapor pressure

Evaporation rate

X-ray density

Pycnometric density

B. CHEMICAL

Theoretical analysis

Synthesis

Reactivity and temperature limit of usefulness

C. ELECTRICAL

Resistivity

Critical temperature

Temperature coefficient of resistivity

C. ELECTRICAL (cont.)

      Thermal EMF

      Dielectric constant

      Dissipation factor

      Thermionic work function

      Magnetic susceptibility

      Critical field

D. MECHANICAL

      Strength: Bending
                    Tensile
                    Compressive

      Hardness: Mohs
                    Vickers
                    Knoop
                    Rockwell

      Elastic moduli: Young's
                            Torsion
                            Shear

      Poisson's ratio

      Creep rate

E. NUCLEAR

      Thermal neutron capture cross section

      Radiation damage

F. OPTICAL

      Color

      Form

F. OPTICAL (cont.)

      Refractive index

      Optical sign

G. STRUCTURE

H. THERMAL

      Conductivity

      Expansion

      Specific heat

      Thermodynamic constants

Additional data where available were included in addition to those listed above. References are listed in parentheses at the right, and in many cases units were converted to make the data directly comparable.

There are 8 chapters and 3 appendices as follows:

Chapter 1 - Borides

       2 - Carbides

       3 - Mixed Carbides

       4 - Elements

       5 - Nitrides

       6 - Oxides

       7 - Mixed Oxides

       8 - Silicides

Appendix A - Bibliography, Numerical

B - Bibliography, Alphabetical

C - Units and Conversion Factors

Within each chapter, compounds are arranged according to the procedure used in Chemical Abstracts, neglecting those atoms common in a chapter (boron in borides, etc.) References are alphabetized by author, each author being listed separately, together with a somewhat abbreviated reference and the reference number. The numerically arranged references are as complete as the available information permits.

Each compound was placed on a separate page to permit the reader to add data as it became available, and also to facilitate typing and assembly.

DATA TABLES

I BORIDES

$$AlB_2$$

ALUMINUM DIBORIDE

Formula weight:	48.61 g/mole	
Formula volume:	15.2 cc/mole	
X-ray density:	3.19 g/cc	(127)

CHEMICAL

Theoretical analysis: 44.6% boron
55.4% aluminum

STRUCTURE

Hexagonal   (127)

MECHANICAL

Hardness: Knoop 100 g : 2410 kg/mm^2 (2040-2910)   (38)

AlB_{12}

ALUMINUM DODECABORIDE

Formula weight: 156.83 g/mole

Formula volume: 60.8 cc/mole

X-ray density: 2.58 g/cc   (127)

CHEMICAL

Theoretical analysis: 82.8% boron
                         17.2% aluminum

MECHANICAL

Hardness: Knoop 100 g :   2370 kg/mm^2 (red form)   (127)
                              2620 kg/mm^2 (yellow form)   (127)
                              2650 kg/mm^2 (38)

OPTICAL

Color: Red or yellow, two forms known

BaB_6

<u>BARIUM HEXABORIDE</u>

Formula weight: 202.28 g/mole

Melting point: $>2100^{\circ}C$ (227, 261, 298)

<u>CHEMICAL</u>

Theoretical analysis: 32.1% boron
68.9% barium

<u>THERMAL</u>

Expansion: 5.89×10^{-6} per $^{\circ}C$; 25-500$^{\circ}C$ (329)

6.47×10^{-6} per $^{\circ}C$; 25-1000$^{\circ}C$ (329)

CALCIUM HEXABORIDE

Formula weight:	105.00 g/mole	
Formula volume:	42.8 cc/mole	(127)
Melting point:	>2100°C	(227, 261, 298)
Pycnometric density:	2.45 g/cc	(127)

CHEMICAL

Theoretical analysis: 61.9% boron
39.1% calcium

MECHANICAL

Strength (MOR): 27,000 psi   (127)

Hardness: Knoop 100 g :   1640 kg/mm^2 (127)

STRUCTURE

Cubic   (127)

THERMAL

Expansion: 5.05×10^{-6} per °C; 25-500°C   (329)

5.85×10^{-6} per °C; 25-1000°C   (329)

CeB_4

CERIUM TETRABORIDE

Formula weight::   183.41 g/mole

Formula volume: 33.6 cc/mole   (127)

Melting point: >2100°C   (227, 261, 298)

Pycnometric density: 5.74 g/cc   (127)

CHEMICAL

Theoretical analysis: 23.6% boron
                                     76.4% cerium

STRUCTURE

Tetragonal   (127)

CeB_6

CERIUM HEXABORIDE

Formula weight: 205.05 g/mole

CHEMICAL

Theoretical analysis: 31.7% boron
68.3% cerium

THERMAL

Expansion: 6.31×10^{-6} per $^{\circ}C$; $25\text{-}500^{\circ}C$ (329)

6.67×10^{-6} per $^{\circ}C$; $25\text{-}1000^{\circ}C$ (329)

CoB_x

<u>COBALT BORIDE</u>

Activation energy of boride formation by diffusion of B into Co:

33.40 kcal/mole   (474)

CHROMIUM MONOBORIDE

Formula weight:	62.83 g/mole	
Formula volume:	10.2 cc/mole	
Melting point:	$1550 \pm 50^{\circ}C$	(65, 71)
	$1515^{\circ}C$	(127)
X-ray density:	6.2 g/cc	(344)
	6.15 - 6.20 g/cc	(66)
	6.17 g/cc	(127)
Pycnometric density:	6.05 g/cc	(19, 65)

CHEMICAL

Theoretical analysis: 17.2% boron
82.8% chromium

Synthesis:
a. $Cr + B$ Fusion   (345, 346)

b. $Cr + B$ Sintering   (347)

c. $Cr_2O_3 + B_2O_3 + Al$, Aluminothermic (66, 348, 350)

d. $Cr_2O_3 + B$ (349)

e. $Cr + BCl_3 + H_2$ (349)

f. Fused salt electrolysis   (66)

g. $Cr + B$ vac. fusion, $1600^{\circ}C$ (343)

h. $Cr + B$ vac. sintering, $1150^{\circ}C$ and many others   (343)

Reacts with dry NH_3 at $1180^{\circ}C \longrightarrow Cr_2N + BN$ (351)

ELECTRICAL

Resistivity: 64×10^{-6} ohm-cm   (202)

MECHANICAL

Hardness: Mohs   : 8   (67)
                      8.5   (66)

           Knoop 30 g : 2135 kg/mm^2 (8)

OPTICAL

Color: Gray, metallic luster   (19)

STRUCTURE

Orthorhombic   (343)

Forms plates or needles of square cross section   (343)

a = 2.96 A,   b = 7.81 A,   c = 2.94 A   (344)
    2.969   7.858   2.932   (343)
    2.95   7.80   2.93   (66)

Isomorphous with monoborides of metals of Group V,
but not Group IV.

THERMAL

Activation energy: Cr + B: 20.52 kcal/mole   (474)

- 15 -   CrB$_2$
<u>CHROMIUM DIBORIDE</u>

Formula weight: 73.65 g/mole

Formula volume: 14.3 cc/mole

Melting point: 1960°C   (38)
                     1850 $\pm$ 50°C   (65, 68, 71)
                     1900°C   (72)

X-ray density: 5.6 g/cc   (343)

<u>CHEMICAL</u>

Theoretical analysis: 29.3% boron
                      70.7% chromium

Synthesis: See CrB

Reacts with dry NH$_3$ at 1180°C $\rightarrow$ Cr$_2$N + BN   (351)

<u>ELECTRICAL</u>

Resistivity: 21 x 10^{-6} ohm-cm   (202)

<u>MECHANICAL</u>

Hardness: Micro 100 g : 1800   (71)
             Vickers 50 g : 1800   (8)
             Knoop 100 g : 1700   (65, 68, 71)

Strength: Tensile: 106,000 psi

Young's modulus: 30.6 x 10^6 psi   (166)

<u>STRUCTURE</u>

Hexagonal, AlB$_2$ structure (C-32 type)   (343)

Isomorphous with other transition metal diborides   (343)
    a = 2.969A;   c = 3.066A;   c/a = 1.03   (343)

THERMAL

Conductivity: 0.049 cgs   (45)

Expansion: 4.6×10^{-6} per $^{\circ}$C   (45)

                         5.4×10^{-6} per $^{\circ}$C; 25-1000°C   (43)

                         11.1×10^{-6} per $^{\circ}$C   (166)

Heat content:

$$H_T - H_{293} = 9.61T + 5.36 \times 10^{-3} \, T^2 - 3.342$$
$$\text{cal/mole}; \quad 494\text{-}1010^{\circ}\text{K} \tag{128}$$

Heat capacity:

$$Cp = 9.61 + 10.72 \times 10^{-3}T \; \text{cal/mole/}^{\circ}\text{K}$$
$$494\text{-}1010^{\circ}\text{K} \tag{128}$$

Heat of formation: $\Delta H^{\circ}_{298} = -17$ kcal/mole   (444)

Activation energy: Cr + B : 20.52 kcal/mole   (474)

DICHROMIUM MONOBORIDE

Formula weight: 114.84 g/mole

CHEMICAL

Theoretical analysis: 9.4% boron
                        90.6% chromium

Synthesis: a. $Cr_2O_3 + B$ (349)

            b. $Cr + BCl_3 + H_2$ (349)

Reacts with NH_3 (dry) at $1180^oC \rightarrow Cr_2N + BN$ (351)

STRUCTURE

Orthorhombic   (343)
    a = 14.7A;   b = 7.34A;   c = 4.29A   (341)

Not isomorphous with M_2B phases of other transition
metals   (341)

THERMAL

Activation energy: See CrB_2

Cr_3B_2

<u>TRICHROMIUM DIBORIDE</u>

Formula weight:	177.67 g/mole	
Melting point:	1960°C	(71)
Pycnometric density:	6.7 g/cc	(349)
	6.13 g/cc	(67)

CHEMICAL

Theoretical analysis: 12.18% boron
87.82% chromium

Synthesis: a. $BCl_3 + Cr + H_2$ (349)

          b. $Cr_2O_3 + B$ (349)

          c.   Fused salt electrolysis   (67, 70, 339)

V. stable toward: HF, HCl, HNO_3, H_2SO_4 (67)
         solutions of alkalis   (67)

Dissolved: Fused alkali hydroxides and carbonates   (67)
       perchloric acid   (67)

Reacts with dry NH_3 at 1180°C $\rightarrow Cr_2N + BN$ (351)

MECHANICAL

Hardness: Mohs : .9+   (67)

OPTICAL

Color: Gray, metallic luster   (19)

STRUCTURE

Tentatively indexed as tetragonal unit cell   (19)

THERMAL

Activation energy: See CrB_2

TRICHROMIUM TETRABORIDE

Formula weight: 199.31 g/mole

CHEMICAL

Theoretical analysis: 21.7% boron
78.3% chromium

Synthesis: a.   Cr + B, vacuum fusion, 1600^oC (343)

b.   Cr + B, vacuum sintering, 1150^oC (343)

Reacts with dry NH_3 at 1180 $\rightarrow Cr_2N$ + BN   (351)

STRUCTURE

Orthorhombic   (341)

Isomorphous with Ta_3B_4 and Nb_3B_4 (343)

a = 2.984A;   b = 13.02A;   c = 2.953A   (341)

THERMAL

Activation energy: See CrB_2

IRON MONOBORIDE

Formula weight: 66.67 g/mole

Melting point: ~1550°C   (227, 261, 298)

CHEMICAL

Theoretical analysis: 16.2% boron
83.8% iron

THERMAL

Activation energy: B + Υ Fe : 21.16 kcal/mole   (474)

Fe_2B

DIIRON MONOBORIDE

Formula weight: 122.52

Melting point: $\sim 1390^{\circ}C$ (227, 298)

CHEMICAL

Theoretical analysis: 8.84% boron
                          91.16% iron

THERMAL

Activation energy: B + γ Fe : 21.16 kcal/mole   (474)

GdB_6

GADOLINIUM HEXABORIDE

Formula weight: 221.82 g/mole

Formula volume: 42.0 cc/mole

X-ray density: 5.27 g/cc   (83)

CHEMICAL

Theoretical analysis: 29.2% boron
                         70.8% gadolinium

HAFNIUM MONOBORIDE

Formula weight: 189.42 g/mole

Formula volume: 14.8 cc/mole

Melting point: 3060°C (39, 72)

X-ray density: 12.8 g/cc (19)

CHEMICAL

Synthesis: Hot press Hf + B (19)

STRUCTURE

Cubic, NaCl type

  d = 4.62 ± 0.02 (44)

Radius ratio: 0.61 (47)

Formula weight: 200.24 g/mole

Formula volume: 17.9 cc/mole

Melting point: $3250 \pm 100^{\circ}C$ (44)
         $3060 - 3065^{\circ}C$ (2, 33, 42, 65)
         $3100^{\circ}C$ (52)

X-ray density: 11.2 g/cc   (19, 44)

Pycnometric density: 10.5 g/cc   (19)

CHEMICAL

Theoretical analysis: 10.8% boron
           89.2% hafnium

Synthesis: a. Hot wire: $HfCl_4 + BCl_3 + H_2$ at $1900\text{-}2700^{\circ}C$ (324)

     b. Hot press: Hf + B   (19)

     c. $HfO_2 + B_4C + C$, 1 hour at $2000^{\circ}C$ (38)

ELECTRICAL

Conductivity: 10×10^{-6} ohm-cm at rt   (42)

      12×10^{-6} ohm-cm at rt, 80% dense   (19)

      Not superconductive to $1.26^{\circ}K$ (333)

STRUCTURE

Simple hexagonal, AlB$_2$ type (C-32)

Isomorphous with TiB$_2$ and ZrB$_2$
  a = 3.14A;   c = 3.47A   (19)
    $3.141 \pm .002A$; $3.470 \pm .002A$; c/a = 1.105   (44)

Radius ratio: 0.61   (47)

THERMAL

Expansion: 5.5×10^{-6} per $^{\circ}C$; rt to $1000^{\circ}C$ (19)

5.26×10^{-6} per $^{\circ}C$; 25 - $500^{\circ}C$ (329)

5.54×10^{-6} per $^{\circ}C$; 25 - $1000^{\circ}C$ (329)

LaB$_6$

LANTHANUM HEXABORIDE

Formula weight:	203.84 g/mole	
Formula volume:	43.2 cc/mole	
Melting point:	>2100°C	(227, 261, 298)
X-ray density:	4.72 g/cc	(83)

CHEMICAL

Theoretical analysis: 31.9% boron
68.1% lanthanum

THERMAL

Expansion: 4.84×10^{-6} per °C; 25-500°C   (329)

5.75×10^{-6} per °C; 25-1000°C   (329)

MOLYBDENUM MONOBORIDE

Formula weight: 106 77 g/mole

Formula volume: 12 3 cc/mole

Melting point: 2180°C   (2, 86, 227, 261, 298, 360)
                      1930 ± 60°C   (68, 71)
                      >2000°C   (87)
                      2190°C

X-ray density: 8.8 g/cc   (19)
                      8.77 g/cc   (90)

CHEMICAL

    Theoretical analysis: 10 3% boron
                           89 7% molybdenum

    Synthesis: a.   Mo + B, fused or sintered in vacuum or
                     argon   (86, 87, 90, 202, 352, 353)

                b.   Fused salt electrolysis   (201, 355, 356)

                c.   Mo + BCl_3, hot wire and others   (324, 357)

    Poor oxidation resistance   (86)

ELECTRICAL

    Resistivity: (α) 45×10^{-6} ohm-cm at rt   (360)

                  (β) 25×10^{-6} ohm-cm at rt   (360)

    Becomes superconductive at 4.4°K   (340)

    No superconductivity observed to 1.8°K   (323)
    Thermionic work function   2 4 ev   (444)

MECHANICAL

    Hardness: Mohs   : 8   (201)
                Vickers 100 g : 1570 kg/mm^2 (68, 71)

OPTICAL

    Color : Dark gray, metallic luster   (201)

STRUCTURE

Tetragonal (α form)   (90)

Acicular crystals by fused salt electrolysis   (201)
     a = 3.110 A   c = 16.95 A   (90)

Not isomorphous with TaB, NbB, CrB, but structures
are closely related   (90)

Orthorhombic (β form)

Isomorphous with NbB, TaB, and CrB   (359, 360)

     a = 3.16A; b = 8.61A; c = 3.08A   (359, 360)

Radius ratio: 0.71   (47)

THERMAL

Activation energy: Formation of borides by diffusion
                      of B into Mo : 14.30 kcal/mole   (474)

MOLYBDENUM DIBORIDE

Formula weight: 117.59 g/mole

Formula volume: 16.5 cc/mole

Melting point: 2100^oC (19, 38, 72, 360)
                     2250 ± 50^oC (68, 71)

X-ray density: 7.8 g/cc   (19)
                     7.78 g/cc   (86)

CHEMICAL

Theoretical analysis: 18.4% boron
                        81.6% molybdenum

Synthesis: See MoB

Poor oxidation resistance   (86)

ELECTRICAL

Resistivity: 45×10^{-6} ohm-cm at room temperature   (360)
Thermionic work function: 2-4 ev   (444)

MECHANICAL

Hardness: Micro 100 g : 1280 kg/mm^2 (71)
             Knoop 100 g : 1200 kg/mm^2 (2)
             Vickers   : 1380 kg/mm^2 (8, 68, 71)

STRUCTURE

Hexagonal, AlB_2 structure   (357)

Isomorphous with transition metal diborides
     a = 3.05A;   c = 3.113A;   c/a = 1.02   (357)
        3.06   3.10   1.01   (86)

Radius ratio: 0.71   (47)

THERMAL

Heat of formation: ΔH^o_{298} = -23 kcal/mole   (444)

Activation energy: See MoB

Mo_2B

DIMOLYBDENUM MONOBORIDE

Formula weight: 202.72 g/mole

Formula volume: 21.9   cc/mole

Melting point: $2165^{\circ}C$ (12, 227)
               d. $2000^{\circ}C$ (19, 38, 360)

X-ray density: 9.3 g/cc   (19)
               9.31 g/cc   (90)

Pycnometric density: 9.26 g/cc   (127)

CHEMICAL

Theoretical analysis: 5.3% boron
               94.7% molybdenum

Synthesis: See MoB

No reaction: HCl   (201, 355)

Dissolves: Cold HNO_3, hot H_2SO_4, fused alkalis and
           oxidizing agents   (201, 355)

Poor oxidation resistance   (86)

ELECTRICAL

Resistivity: 40×10^{-6} ohm-cm at room temperature   (360)
Thermionic work function: 2-4 ev   (444)

MECHANICAL

Hardness: Mohs   : 8-9   (201)
            Vickers 100 g : 1660 kg/mm^2 (8, 71)

OPTICAL

Color: Light gray, metallic luster   (201)

STRUCTURE

Tetragonal, $CuAl_2$ structure, C-16 type   (90)

STRUCTURE (cont.)

Forms tabular crystals by fused salt electrolysis (201)

Isomorphous with Ta_2B (90)

 a = 5.543A; c = 4.735A; c/a = 0.854 (90)

Radius ratio: 0.71 (47)

THERMAL

Heat content:
$$H_T - H_{293} = 18.43T + 0.64 \times 10^{-4}T^2$$
$$-5,552 \text{ cal/mole}; \; 415-876^{\circ}K \quad (128)$$

Heat capacity:
$$Cp = 18.43 + 1.28 \times 10^{-4} \; T \text{ cal/mole}/^{\circ}K;$$
$$415-876^{\circ}K \quad (128)$$

Activation energy: See MoB

Mo_2B_5

DIMOLYBDENUM PENTABORIDE

Formula weight: 246.00 g/mole

X-ray density: 7.48 g/cc   (90)

Pycnometric density: 7.01 g/cc   (19)

CHEMICAL

Theoretical analysis: 22.0% boron
                        78.0% molybdenum

Synthesis: Electrolytic deposition of boron on Mo wires   (357)

Decomposes to MoB_2, 1600 - 1650°C   (86)

Poor oxidation resistance   (86)

ELECTRICAL

Resistivity: 25×10^{-6} ohm-cm at room temperature (360)

               Not superconductive to 1.32°K   (340)
               Superconductive below 1.9°K   (444)
               Thermionic work function 2-4 ev   (444)

STRUCTURE

Rhombohedral   (90)

Related to hexagonal MoB_2 structure
     r = 7.190A α = 24° 10'

Nearly analogous to hexagonal W_2B_5 structure-hexagonal
parameters would be:
     a = 3.011A;   c = 20.93A   (90)

Radius ratio: 0.71   (47)

THERMAL

Activation energy of formation: See MoB

Mo_3B_2

TRIMOLYBDENUM DIBORIDE

Formula weight: 309.49 g/mole

Melting point: 2075°C   (227, 298)

CHEMICAL

Theoretical analysis: 7.0% boron
                          93.0% molybdenum

Stability: 1910 - 2075°C   (2, 227)
             1850 - 2070°C   (19, 86)

Synthesis: Peretectic decomposition of Mo_2B

Poor oxidation resistance   (86)

Oxidation in air becomes severe between 1100-1400°C   (227)

STRUCTURE

Isomorphous with Cr_3B_2 (19)

Tentatively indexed on basis of tetragonal cell   (19)

Radius ratio: 0.71   (47)

THERMAL

Activation energy of formation: See MoB

NIOBIUM MONOBORIDE

Formula weight: 102.74 g/mole

Melting temperature: $>2000^{\circ}$C   (227, 261, 298)

CHEMICAL

This is the only Nb-B phase to melt undecomposed   (202)

ELECTRICAL

Resistivity: Superconductive at 6°K   (340)

Not superconductive to 1.8°K   (323)

64.5×10^{-6} ohm-cm at room temperature   (202)

STRUCTURE

Orthorhombic, isomorphous with CrB   (341)

a = 3.298A; b = 8.724A; c = 3.137A   (341)
    3.292   8.713   3.165   (87)

THERMAL

Activation energy of formation: See NbB_2

NIOBIUM DIBORIDE

Formula weight:	114.55 g/mole	
Formula volume:	16.43 cc/mole	
Melting point:	d. 2900°C	(68, 202)
X-ray density:	7.21 g/cc	(89)
Pycnometric density:	6.97 g/cc	(127, 129)
	6.6 g/cc	(68, 89)
	6.5 g/cc	(2)
	6.4 g/cc	(70, 339)

CHEMICAL

Theoretical analysis:	18.89% boron	
	81.11% niobium	
Synthesis: a.	Electrolysis at 1000°C	(70, 89)
b.	Nb + B + argon (vacuum)	(87)
c.	Hot press Nb + B	(202)
d.	$Nb_2O_5 + B_4C + C$, 1850°C, 1 hour	(38)
No reaction:	HCl, HNO_3, aqua regia	(19, 70)
Slow attack:	Hot H_2SO_4 and HF	(19, 70)
Oxidized in air at red heat		(19, 70)
Dissolves rapidly:	Alkali hydroxides, carbonates bisulfates, and Na_2O_2	(19, 70)

Only Nb-B phase stable in presence of carbon

ELECTRICAL

Resistivity:	65.5×10^{-6} ohm-cm	(127)
	32×10^{-6} ohm-cm	(202)
	Not superconductive at 1.27°K	(340)

MECHANICAL

Hardness:	Mohs	: 8+	(68, 70, 90, 339)
	Knoop 30 g	: 2594 kg/mm^2	(8)
	Micro	: 1700 kg/mm^2	(65)

OPTICAL

Color: Metallic gray

Spectral emissivity: 0.65μ 0.64 at 1263°C   (220)
0.67 at 1471°C
0.64 at 1640°C
0.62 at 2016°C

Total emissivity: 0.31 at 1327°C   (220)
0.35 at 1527°C
0.38 at 1727°C
0.40 at 1927°C
0.42 at 2127°C

0.29 at 1327°C   (221)
0.30 at 1527°C
0.33 at 1727°C
0.35 at 1927°C
0.37 at 2127°C

STRUCTURE

Simple hexagonal, AlB$_2$ type (C-32)   (89)

Isomorphous with TiB$_2$, etc.
    a = 3.086A;   c = 3.306A;   c/a = 1.071   (89)

Radius ratio: 0.68   (47)

THERMAL

Conductivity:	0.040 cgs at 20°C	(127)
	0.040 cgs at 24°C	(227, 2)
	0.047 - 0.062 cgs at 200°C	(227, 2)

Heat of formation: ΔH°_{298} = -36 kcal/mole   (444)

Activation energy of formation: Nb + B: 14.13 kcal/mole   (474)

Nb_2B

DINIOBIUM MONOBORIDE

Formula weight: 194.66 g/mole

STRUCTURE

Not isomorphous with Ta_2B (87)

THERMAL

Activation energy: See NbB_2

Nb_3B

<u>TRINIOBIUM MONOBORIDE</u>

Formula weight: 286.58 g/mole

<u>STRUCTURE</u>

Isomorphous with Ta_3B (87)

<u>THERMAL</u>

Activation energy: See NbB_2

TRINIOBIUM TETRABORIDE

Formula weight: 319.04 g/mole

ELECTRICAL

Resistivity: Not superconductive down to $1.27^{\circ}K$ (340)

STRUCTURE

Orthorhombic, isomorphous with Ta_3B_4 (341)

a = 3.305A;   b = 14.08A;   c = 3.137A   (341)
3.30   14.1   3.13   (87)

THERMAL

Activation energy: See NbB_2

PrB_6

PRASEODYMIUM HEXABORIDE

Formula weight: 205.84 g/mole

Formula volume: 42.5 cc/mole

X-ray density: 4.85 g/cc   (83)

CHEMICAL

Theoretical analysis: 22.0% boron
78.0% praseodymium

SiB_6

<u>SILICON HEXABORIDE</u>

Formula weight: 92.98 g/mole

<u>CHEMICAL</u>

    Theoretical analysis: 69.9% boron
                                     30.1% silicon

<u>THERMAL</u>

    Expansion: 5.26×10^{-6} per $^{\circ}$C; 25-500°C   (329)

                     5.95×10^{-6} per $^{\circ}$C; 25-1000°C   (329)

SAMARIUM HEXABORIDE

Formula weight: 215.35 g/mole

Formula volume: 42.5 cc/mole

Melting point: 2400 $\pm$ 100°C   (83)

X-ray density: 5.07 g/cc   (83)

Pycnometric density: 5.14 g/cc   (83)

CHEMICAL

Theoretical analysis: 30.1% boron
                        69.9% samarium

MECHANICAL

Hardness: Knoop   :   1391 $\pm$ 158 kg/mm^2 (83)

THERMAL

Expansion: 5.89 x 10^{-6} per $^{\circ}$C; 25-500°C   (329)

                6.47 x 10^{-6} per $^{\circ}$C; 25-1000°C   (329)

SrB_6

STRONTIUM HEXABORIDE

Formula weight: 152.55 g/mole

Melting point: $>$2100   (227, 261, 298)

CHEMICAL

Theoretical analysis: 42.6% boron
57.4% strontium

TANTALUM MONOBORIDE

Formula weight:	191.70 g/mole	
Formula volume:	13.4 cc/mole	
Melting point:	>2000°C	(227, 261, 298)
X-ray density:	14.29 g/cc	(342)
Pycnometric density:	14.0 g/cc	(19)

CHEMICAL

Theoretical analysis: 5.65% boron
94.35% tantalum

Synthesis: a. $Ta_2O_5 + B_2O_3 + C$, 2000°C (59)

  b. Ta + B, vacuum, sinter (342)

  c. Ta + B, argon, sinter (87)

  d. TaH + B, hot press (202)

Stable from room temperature to melting point (19)
Stable to 2050°C (227)

Oxidation in air becomes severe between 1100-1400°C (227)

ELECTRICAL

Resistivity: 100×10^{-6} ohm-cm (202)

  Not superconductive to 1.29°K (340)

STRUCTURE

Orthorhombic structure, isomorphous with NbB (342)
  a = 3.276A; b = 8.669A; c = 3.157A (342)

THERMAL

Activation energy: See TaB_2

TaB_2

TANTALUM DIBORIDE

Formula weight: 202.52 g/mole

Formula volume: 16.3 cc/mole

Melting point: >3000°C   (19)
                       3000°C   (32, 127, 129)
                       3150°C

X-ray density: 12.60 g/cc   (89)

Pycnometric density: 11.70 g/cc   (89)

CHEMICAL

Theoretical analysis: 10.68% boron
                       89.32% tantalum

Synthesis: a. Fused salt electrolysis   (70, 339)
             b. $Ta_2O_5 + B_2O_3 + C$, 2000°C   (59)
             c. Ta + B, vacuum   (342)
             d. Ta + B, argon   (87)
             e. TaH + B, hot press   (202)
             f. Direct vapor deposition from
                 $TaCl_5 + BBr_3 + H_2$, not successful (324, 52)

Homogeneity range: 61-72% boron   (342)
                     64-72% boron   (87)

No reaction: HCl, HNO_3, aqua regia   (87)
Slowly attacked: HF, H_2SO_4 mixtures   (87)
Rapidly dissolved: Fused alkali hydroxides, carbonates
                     bisulfates and peroxides   (87)

Oxidized in air at red heat   (19)

Only Ta-B phase stable in presence of carbon.
Stable from room temperature to melting point.   (19)

ELECTRICAL

Resistivity: 68 x 10^{-6} ohm-cm   (202)
                Not superconductive to 1.32°K   (340)
                86.5 x 10^{-6} ohm-cm   (127)
                Superconducts > 2.2°K   (444)

ELECTRICAL (cont.)

Thermionic emission: Measured by Goldwater   (337)
                      and Morgan   (336)

MECHANICAL

Hardness: Mohs   : 8+   (70)
             Knoop 30 g : 2537 kg/mm^2 (8)
                   100 g : 2615 $\pm$ 120 kg/mm^2 (2)
             Micro   : 1700 kg/mm^2 (65)

Young's modulus: 37 x 10^6 psi

OPTICAL

Color: Gray, metallic

STRUCTURE

Hexagonal, AlB$_2$ structure (C-32 type)

Isomorphous with other transition metal diborides
    a = 3.088A;   c = 3.241A;   c/a = 1.074   (89)
        3.078   3.265   1.06   (342)

Low boron composition; 64 atom % boron
    a = 3.099A;   c = 3.244   (342)
        3.097   3.277   (87)

High boron composition; 72 atom % boron
    a = 3.057A;   c = 3.291A   (342)
        3.060   3.290   (87)

Radius ratio: 0.68   (47)

THERMAL

Conductivity: 0.026 cgs at 20°C   (19, 37, 127)
                     0.026 cgs at 24°C   (227, 2)
                     0.033 cgs at 200°C   (227, 27)

THERMAL (cont.)

      Expansion: 5.1 x 10^{-6} per $^{\circ}$C at room temperature

      Specific heat: 0.04 cal/g/$^{\circ}$C

      Thermal diffusivity: 0.05 cm^2/sec

      Heat of formation: ΔH^o_{298} = -52 kcal/mole   (444)

      Activation energy of formation: Ta + B: 16.90 kcal/mole   (474)

DITANTALUM MONOBORIDE

Formula weight: 372.58 g/mole

CHEMICAL

Theoretical analysis: 2.91% boron
97.09% tantalum

Synthesis: a. Cannot be prepared pure. Presumed to be a high
temperature phase which always decomposes to
some extent to Ta + TaB   (342)

b. May be formed only on cooling, hence a low
temperature form   (87)

STRUCTURE

Tetragonal, $CuAl_2$ structure, C-16 type

a = 5.778A;   c = 4.864A   (342)
5.785   4.867   (87)

Not isomorphous with Nb_2B (87)

Isomorphous with W_2B, Mo_2B (87)

THERMAL

Activation energy: See TaB_2

TRITANTALUM MONOBORIDE

Formula weight: 553.46 g/mole

CHEMICAL

Theoretical analysis: 1.96% boron
98.04% tantalum

Synthesis: a. $Ta + Ta_2B$, 1950°C, 4×10^{-5}mm, quench   (87)

b.   Disproportionate to $Ta + Ta_2B$ on cooling
slowly   (87)

STRUCTURE

Isomorphous with Nb_3B (87)

THERMAL

Activation energy: See TaB_2

Ta_3B_4

TRITANTALUM TETRABORIDE

Formula weight:	585.92 g/mole	
Formula volume:	43.1 cc/mole	
X-ray density:	13.60 g/cc	(342)
Pycnometric density:	13.50 g/cc	(342)

CHEMICAL

Theoretical analysis: 7.4% boron
92.6% tantalum

Synthesis: TaH + B, hot press (202)

ELECTRICAL

Resistivity: Not superconductive to $1.30^{\circ}K$ (340)

STRUCTURE

Orthorhombic, isomorphous with Nb_3B_4

a = 3.29A; b = 14.0A; c = 3.13A (342)

THERMAL

Activation energy of formation: See TaB_2

THORIUM BORIDE

Formula weight: 242.94

Melting point: >2100°C   (227, 298)

THORIUM TETRABORIDE

Formula weight:	275.40 g/mole	
Formula volume:	32.6 cc/mole	(127)
Melting point:	>2500°C	(87, 227, 261, 298)
X-ray density:	8.45 g/cc	(366)
Pycnometric density:	7.5 g/cc	(2)

CHEMICAL

Theoretical analysis: 15.6% boron
                          84.4% thorium

Synthesis:	a.	Th + B, sinter	(87)
	b.	Can be produced only in the absence of carbon	(202)
	c.	It can be stabilized by presence of W or Mo	(19)
	d.	Narrow range of homogeneity	(87)

STRUCTURE

Tetragonal, D_{4h}^5 - P4/mbm	(366)
a = 7.256A; c = 4.113A; c/a = 0.567	(366)
Isomorphous with UB_4 and CeB_4	(366)
Radius ratio: 0.54	(47)

MECHANICAL

Hardness: Knoop 100 g : 1365 kg/mm^2 (1155-1690)	(38)

THORIUM HEXABORIDE

Formula weight: 297.04 g/mole

Formula volume: 48.8 cc/mole

Melting point: >2100°C   (227)
                         2100°C   (2)

X-ray density: 6.08 g/cc   (367)

Pycnometric density: 6.4 g/cc   (2)

CHEMICAL

Theoretical analysis: 21.9% boron
                         78.1% thorium

Synthesis: a. Fused salt electrolysis   (69, 367, 368)

            b. Th + B, hot pressed   (202)

ELECTRICAL

Thermionic emission: measured by Lafferty   (370)

STRUCTURE

Cubic, CaB$_6$ type (L2$_1$)
   a = 4.32A   (367)
       4.113A   (369)

Isomorphous with corresponding alkaline earth and rare
earth hexaborides   (19)

Radius ratio: 0.54   (47)

MECHANICAL

Hardness: Knoop 100 g : 1765 kg/mm^2 (1470-2000)   (38)

TITANIUM MONOBORIDE

Formula weight: 58.72 g/mole

CHEMICAL

Theoretical analysis: 18.4% boron
81.6% titanium

Stable only in the absence of carbon   (330)

ELECTRICAL

Resistivity: 40×10^{-6} ohm-cm at 85% density at rt   (202)

STRUCTURE

Face-centered cubic, NaCl type   (330)

d = 4.24A

THERMAL

Activation energy: See TiB_2

TiB_2

TITANIUM DIBORIDE

Formula weight:	69.54 g/mole	
Formula volume:	15.47 cc/mole	
Melting point:	2980°C	(19)
	2900 ± 80°C	(71)
X-ray density:	4.52 g/cc	(89)
Pycnometric density:	4.52 g/cc	(89)
	4.50 g/cc	(32)

CHEMICAL

Theoretical analysis: 31.12% boron (19)
68.88% titanium

Synthesis:
    a. Fused salt electrolysis of composition
        $1/2\ TiO_2 + 2B_2O_3 + CaO(MgO) + CaF_2(MgF_2)$ (67)

    b. $TiCl_4 + BBr_3 + H_2$ at 1130-1330°C (52)
        $TiCl_4 + BCl_3 + H_2$ at 1000-1300°C (334)

    c. $TiO_2 + B_4C + C$ at 2000°C, 1 hour (38)

Reactivity:

No reaction:	HCl, HF	(19, 67)
Reacts:	Hot H_2SO_4	(19, 67)
Easily soluble:	HNO_3-H_2O_2 or H_2SO_4-H.	(19, 67)
Decomposed:	Fused alkali, carbonates bisulfates	(19, 67)
Violent reaction:	Na_2O_2; brown lead oxide	(19, 67)

Air oxidation becomes severe 1100-1400°C (227)

Only boride of titanium stable in presence of carbon (330)

ELECTRICAL

Resistivity: 28.4×10^{-6} ohm-cm at room temperature
        85% density (89)

ELECTRICAL (cont.)

15.2×10^{-6} ohm-cm at room temperature   (42)

3.7×10^{-6} ohm-cm at liquid air temperature   (42)

Not superconductive at $1.26^{O}K$ (333)

Superconductive at $1.9^{O}K$ (444)

MECHANICAL

Modulus of rupture: 19,000 psi   (127)

Tensile strength: 18,400 psi at $1000^{O}C$

Compressive strength: 97,000 psi   (127)
                                         47,000 psi   (38)

Hardness: Mohs   : > 9   (67)
                   Vickers 50 g :   3400 kg/mm^2 (8, 65,71)
                   Knoop 30 g   :   3370 kg/mm^2 (8)
                            100 g :   2710 kg/mm^2 (127)
                                          2575 kg/mm^2 (2275-3025)   (38)
Young's modulus: 53.2×10^6 psi   (166)

OPTICAL

Color: Yellow, metallic luster, by fused salt electrolysis   (19)

Spectral emissivity: 0.26 at $725^{O}C$ to 0.24 at $2125^{O}C$ (203)

STRUCTURE

Simple hexagonal, AlB_2 type (C-32)

Isomorphous with ZrB_2

a = 3.028A;   c = 3.228A;   c/a = 1.064   (89)
     3.03   3.22   (197)
     3.030   3.227   (332, 87)

Radius ratio: 0.66   (47)

THERMAL

Conductivity: 0.0624 cgs at 200°C, 85% dense   (89)
    0.058 cgs at 24°C   (2, 227)
    0.063 cgs at 200°C   (2, 227)
    0.062 cgs   (45)
    0.060 cgs   (127)
    0.063 cgs at 23°C

Expansion: 6.39×10^{-6} per °C   (166)

    4.6×10^{-6} per °C   (45)

Heat content:
$$H_T - H_{293} = 10.39T + 3.54 \times 10^{-3}T^2 - 3,573 \text{ cal/mole } (420\text{-}1180°K)$$ (128)

Heat capacity:
$$Cp = 10.93 + 7.08 \times 10^{-3}T \text{ cal/mole/°K}$$
$$(420\text{-}1180°K)$$ (128)

Heat of vaporization:
$$\Delta H_{298} = 430 \text{ kcal/mole}$$ (199)
    428.4 - 431.4 kcal/mole   (205)

Heat of formation:
$$\Delta H°_{298} = -52 \text{ kcal/mole}$$ (199)
    -50.7 to 53.7 kcal/mole   (205)
    -72 kcal/mole   (444)

Activation energy of formation: Ti + B: 9.15 kcal/mole   (474)

Ti_2B

DITITANIUM MONOBORIDE

Formula weight: 106.62   g/mole

CHEMICAL

Theoretical analysis: 10,1% boron
                                   89,9% titanium

Stable only in the absence of carbon   (330)

STRUCTURE

Tetragonal   (330)

     $a = 6.11A$; $c = 4.56A$; $c/a = 0.75$ (330)

Hexagonal close packed   (331)

THERMAL

Activation energy: See TiB_2

$$Ti_2B_5$$

DITITANIUM PENTABORIDE

Formula weight: 149.90 g/mole

CHEMICAL

Theoretical analysis: 36.1% boron
63.9% titanium

Stable only in the absence of carbon   (330)

STRUCTURE

Hexagonal   (330)

a = 2.98A;   c = 13.98A;   c/a = 4.69   (330)

Isomorphous with W_2B_5

THERMAL

Activation energy: See TiB_2

Formula weight:	259.71 g/mole	
Formula volume:	20.4 cc/mole	
Melting point:	2365°C	(127)
X-ray density	12.71 g/cc	(202, 373)

CHEMICAL

Theoretical analysis: 8.35% boron
91.65% uranium

Synthesis: a.   U + B, sinter, fuse   (253)

b.   U + B, Sinter   (202)

Narrow homogeneity range   (202)

MECHANICAL

Hardness: Knoop 100 g : 1390 kg/mm^2

OPTICAL

Color: Gray, metallic   (19)

STRUCTURE

Simple hexagonal, AlB$_2$ structure, C-32   (373, 202)

a = 3.136A;   c = 3.988A   (202)

Isomorphous with other transition metal diborides

UB_4

URANIUM TETRABORIDE

Formula weight:	281.35 g/mole	
Formula volume:	30.2 cc/mole	
Melting point:	$>1500^{\circ}C$	(195, 227)
X-ray density:	9.38 g/cc	(372, 366)
Pycnometric density:	9.32 g/cc	(372)

CHEMICAL

Theoretical analysis:	15.4% boron 84.6% uranium	
Synthesis: a. Fused salt electrolysis		(67, 339)
b. U + B, sintered		(202)
Narrow range of homogeneity		(202)

OPTICAL

Color: Gray, metallic	(19)

STRUCTURE

Tetragonal	(202)
a = 7.075A; c = 3.979A; c/a = 0.562	(372, 366)
Isomorphous with ThB_4	(372, 366)

UB_{12}

URANIUM DODECABORIDE

Formula weight:	367.91 g/mole	
Formula volume:	65.0 cc/mole	
X-ray density:	5.825 g/cc	(372)
Pycnometric density:	5.65 g/cc	(19)

CHEMICAL

Theoretical analysis: 35.3% boron
64.7% uranium

Synthesis: Fused salt electrolysis $\rightarrow UB_4 + UB_{12}$
chemically separated   (371)

Apparently decomposes at $\sim 1500^{\circ}$ losing boron   (19)

OPTICAL

Color: Gray, metallic   (19)

STRUCTURE

Face-centered cubic   (372)

Isomorphous with ZrB_{12} (372)

a = 7.473A

VANADIUM MONOBORIDE

Formula weight:	61.77 g/mole
Formula volume:	11.4 cc/mole
X-ray density:	5.28 g/cc	(198)
Pycnometric density:	5.44 g/cc	(198)

CHEMICAL

Theoretical analysis: 17.52% boron
82.48% vanadium

Synthesis: a. V + B, vacuum sintered; then fused (363)

b. $V_2O_5 + B_2O_3 + C$; in H_2, 1650°C (198)

ELECTRICAL

Resistivity: 65.5×10^{-6} ohm-cm at 65% density (19)

$35-40 \times 10^{-6}$ ohm-cm - extrapolated to 100% density (19)

STRUCTURE

Orthorhombic (198)

Isomorphous with CrB, NbB, TaB (198)

a = 3.14A; b = 3.24A; c = 3.00A (198)

Radius ratio: 0.75 (47)

Formula weight:	72.59 g/mole	
Formula volume:	14.2 cc/mole	
Melting point:	d. 2100°C	(19, 38)
	2100 $\pm$ 60°C	(71)
X-ray density:	5.10 g/cc	(89)
Pycnometric density:	4.61 g/cc	(89)
	5.28 g/cc	(67, 339)

CHEMICAL

Theoretical analysis:	29.8% boron	
	70.2% vanadium	
Synthesis: a.	Fused salt electrolysis; 910-1050°C	(67, 339)
b.	V + B, hot pressed	(89)
Insoluble: HCl, HF, H$_2$SO$_4$		(364)
Easily soluble: HNO$_3$		(364)
Decomposed: Fused alkali hydroxides, carbonates,		
nitrates, bisulfates		(364)
Reacts violently: Alkali peroxides, brown lead oxide		(364)
Partly decomposed to VB on fusion		(202)

ELECTRICAL

Resistivity:	16 x 10^{-6} ohm-cm at 20°C	(52, 202)
	3.5 x 10^{-6} ohm-cm at liquid air temperature	(52)
	Superconductive below 1.9°K	(444)

MECHANICAL

Hardness:	Mohs	: 8-9	(19)
	Knoop 30 g	: 2077 kg/mm^2	(8)

STRUCTURE

Hexagonal, AlB$_2$ structure, C-32 type (89)

a = 2.998A;   c = 3.057A;   c/a = 1.020 (89)

Radius ratio: 0.75 (47)

TUNGSTEN MONOBORIDE

Formula weight:	194.74 g/mole	
Formula volume:	12.4 cc/mole	
Melting point:	2922 ± 50°C	(39, 73)
	2850 ± 80°C	(23, 71)
	2920 ± 50°C	(15, 141, 200, 227, 261)
	2860 ± 80°C	(65)
X-ray density:	16.0 g/cc	(90)
	15.73 g/cc	(127)
Pycnometric density:	15.41 g/cc	(65)
	15.3 g/cc	(23, 85)

CHEMICAL

Theoretical analysis: 5.55% boron
                         94.45% tungsten

Synthesis: a. W + B, vacuum   (90)

b. W + B, argon   (87)

c. W + B, hot press   (202)

d. Fused salt electrolysis, 960°C   (355, 67, 356, 201)

Homogeneity range: 48-51 atom percent boron   (90)

Stable to 2050°C   (2, 227)

No reaction with carbon at 1800°C   (44)

Oxidation in air becomes severe 800-1400°C   (227)

No reaction: HCl   (355)

Dissolved: Hot H_2SO_4, HNO_3 (355)

Rapidly dissolved: Aqua regia, HF-HNO_3 (355)
                   fused alkali hydroxides, and nitrides (355)

Reacts with dry NH_3 at 1100→W + BN   (351)

ELECTRICAL

Resistivity: Not superconductive to $1.8^{\circ}K$ (323)
Thermionic work function: 2-4 ev   (444)

MECHANICAL

Hardness: Mohs: 8-9   (355)

OPTICAL

Color: Gray, metallic luster

STRUCTURE

Tetragonal   (19)

a = 3.115A;   c = 16.92A   (19)

Isomorphous with α MoB   (19)

β WB- Orthorhombic high temperature from isomorphous
with β MoB, NbB, TaB, and CrB   (330)

a = 3.07A;   3.19A;   c = 8.46A   (330)

Radius ratio: 0.71   (47)

THERMAL

Activation energy: See WB_2

Formula weight:	205.56 g/mole	
Formula volume:	16.1 cc/mole	
Melting point:	2900°C	(127)

CHEMICAL

Theoretical analysis: 10.5% boron
89..5% tungsten

Synthesis: W + B, hot press   (202)

Reacts with dry NH$_3$ at 1100$\longrightarrow$W + BN   (351)

ELECTRICAL

Resistivity: Not superconductive to 1.8°K   (323)
Thermionic work function: 2-4 ev   (444)

MECHANICAL

Hardness: Knoop 30 g   :   2663 kg/mm^2 (8)

STRUCTURE

Tetragonal   (127)

Radius ratio: 0.71   (47)

THERMAL

Activation energy of formation: W + B: 20.40 kcal/mole   (474)

WB_6

<u>TUNGSTEN HEXABORIDE</u>

Formula weight: 248.84 g/mole

Melting point: 2920°C   (33)

CHEMICAL

Theoretical analysis: 26.1% boron
73.9% tungsten

All W-B phases react with dry NH_3 at 1100⟶W + BN   (351)

ELECTRICAL

Resistivity: Not superconductive to 1.8°K   (323)

Thermionic work function: 2-4 ev   (444)

STRUCTURE

Radius ratio: 0.71   (47)

THERMAL

Activation energy: See WB_2

W_2B

DITUNGSTEN MONOBORIDE

Formula weight:	378.66 g/mole	
Formula volume:	22.1 cc/mole	(127)
Melting point:	2770 $\pm$ 80°C	(23, 71)
X-ray density:	16.72 g/cc	(90)
Pycnometric density:	17.17 g/cc	(127)
	16.0 g/cc	(23, 85, 90)
	15.98 g/cc	(65)

CHEMICAL

Theoretical analysis: 2.9% boron
                       97.1% tungsten

Synthesis: a. W + B, sinter, vacuum   (90)

            b. W + B, sinter, argon   (87)

            c. W + B, hot press   (202)

Homogeneous at 33.3 atom percent boron   (90)

Reacts with dry NH_3 at 1100°C$\rightarrow$W + BN   (351)

ELECTRICAL

Resistivity: Not superconductive to 1.8°K   (323)
Thermionic work function: 2-4 ev   (444)

STRUCTURE

Tetragonal, $CuAl_2$ structure, (C-16 type)   (90)

Isomorphous with Mo_2B (90)
    a = 5.564A;   c = 4.740A   (90)

Radius ratio: 0.71   (47)

THERMAL

Activation energy: See WB_2

DITUNGSTEN PENTABORIDE

Formula weight:	421.94 g/mole	
Formula volume:	38.3 cc/mole	
Melting point:	2980°C	(202)
	2200°C	(72)
X-ray density:	13.1 g/cc	(19)
Pycnometric density:	11.0 g/cc	(23, 85)
	10.77 g/cc	(19)

CHEMICAL

Theoretical analysis: 12.9% boron
87.1% tungsten

Synthesis:	a. W + B, vacuum	(90)
	b. W + B, argon	(87)
	c. W + B, hot press	(202)

Homogeneity range 66.6 - 68 atom percent boron (90)

Reacts with dry NH_3 at 1100°C   W + BN (351)

Stable with B_4C to 2000°C (202)

ELECTRICAL

Resistivity: 21×10^{-6} ohm-cm	(202)
Not superconductive to 1.8°K	(323)
Superconductive below 1.4°K	(444)
Thermionic work function: 2-4 ev	(444)

STRUCTURE

Hexagonal	(19)
a = 2.982 A, c = 13.87 A	(19)
Closely related to Mo_2B_5, but not isomorphous	(19)
Isomorphous with Ti_2B_5	(19)

THERMAL

Heat of formation: ΔH_{298}^O = -29.5 kcal/mole (444)

Activation energy: See WB_2

YbB$_6$

YTTERBIUM HEXABORIDE

Formula weight: 237.96 g/mole

Formula volume: 42.8 cc/mole

Melting point: 1538 $\pm$ 33°C   (83)

X-ray density: 5.56 g/cc   (83)

Pycnometric density: 5.45 g/cc   (83)

CHEMICAL

Theoretical analysis: 27.1% boron
                        72.9% ytterbium

ZIRCONIUM MONOBORIDE

Formula weight:	102.04 g/mole	
Formula volume:	15.2 cc/mole	
Melting point:	2992 ± 50°C	(73, 185)
	2922°C	(39)
X-ray density:	6.7 g/cc	(338)
Pycnometric density:	5.7 g/cc	(338)

CHEMICAL

Theoretical analysis: 10.6% boron
                          89.4% zirconium

Synthesis: a.   Zr + B in tungsten furnace at 1800-2000°C   (361)

           b.   ZrH + B, hot pressed   (202)

Stability range: 800 - 1250°C   (19)

ELECTRICAL

Resistivity: 30-35 x 10^{-6} ohm-cm at rt   (338)

           Superconductive at 2.82-3.2°K   (333)

MECHANICAL

Hardness: Rockwell A: 69-72   (338)

STRUCTURE

Cubic, NaCl type   (330)

    d = 4.65A   (338)

ZIRCONIUM DIBORIDE

Formula weight: 112.86 g/mole

Formula volume: 18.53 cc/mole   (127)

Melting point: 3040 $\pm$ 50°C   (19)
                       2990 $\pm$ 50°C   (5, 6)
                       2980°C   (43)

Vapor pressure: P_{Zr} = 2.38 x 10^{-10} over ZrB_2
                       at 1727°C   (207)

                       $ZrB_{1.906}$ (s) = Zr (g) + 1.906 B (g)   (502)

X-ray density: 6.09 g/cc   (89)
                       6.102 g/cc   (126)

CHEMICAL

Theoretical analysis: 19.2% boron   (19)
                       80.8% zirconium

Synthesis: a. Fused salt electrolysis of ZrO_2, B_2O_3,
                CaO, CaF_2 (59)

            b. Hot press ZrH + B   (202)

            c. ZrO_2 + B_4C + C; 1 hour at 2000°C   (38)

Reacts with tungsten and graphite at 2100°C   (192)

Air oxidation becomes severe between 1100-1400°C   (227)

Little attack: cold HCl   (19, 67)

Reacts: HNO_3, aqua regia, hot H_2SO_4, fused alkali
        hydroxides, carbonates, and bisulfates   (19, 67)

Violent reaction: Na_2O_2, brown lead oxide   (19, 67)

ELECTRICAL

Resistivity: 9.2 x 10^{-6} ohm-cm at 20°C   (42)
               1.8 x 10^{-6} ohm-cm at liquid air temperature   (42)
               38.8 x 10^{-6} ohm-cm at 85% density   (62)
               Superconductive below 1.61°K   (461)
                                       1.8°K   (323)
                                       1.9°K   (444)

ELECTRICAL (cont.)

Becomes superconductive 2.8 - 3.2^{O}K   (333)

Does not become superconductive at 1.61^{O}K   (19)

Does not become superconductive at 1.8^{O}K   (323)

Thermionic emission: Goldwater   (337)

MECHANICAL

Strength:	MOR:	29,000 psi	(127)
		8,000 - 25,100 psi	(19)
	Tensile :	28,700 psi	(9)
Speed of sound:		29,130 fps	(126)
Hardness:	Mohs	: 8	(67, 69)
	Rockwell A	: 87-89	(19)
	Vickers 50 g	: 2200 kg/mm^2	(8, 65, 71)
	Knoop 100 g	: 1560 kg/mm^2	(1, 3)
Young's modulus:		63.8 x 10^6 psi	(126)
		49.8 x 10^6 psi	(166)
		3.305 x 10^6 psi, 22.4% dense (foam)	(38)
Bulk modulus:		30.12 x 10^6 psi	(126)
Shear modulus:		27.88 x 10^6 psi	(126)
Poisson's ratio:		0.144	(126)

OPTICAL

Color: Metallic, gray

Emission: Measured by Morgan   (336)

STRUCTURE

Simple Hexagonal, AlB$_2$ type (C-32)

STRUCTURE (cont.)

Isomorphous with TiB$_2$

a = 3.170 $\pm$ .002 A;	c = 3.533 $\pm$ .002 A; c/a = 1.114	(89)
3.169	3.530	(335)
3.169	3.528	(87)
3.1666	3.5365	(126)
3.1694 $\pm$ .0003	3.5303 $\pm$ .0004	(207)

Radius ratio: 0.62   (47)

THERMAL

Conductivity: 0.0550 CGS at 200°C   (89)
                  0.058 CGS at 24°C   (227, 2)
                  0.055-.060 CGS at 200°C   (227, 2)

Diffusivity: 0.08 cm^2/sec

Expansion: 5.5 x 10^{-6} per °C; rt - 1000°C   (19)

                  4.5 x 10^{-6} per °C (0-1000°C)   (45)

                  6.73 x 10^{-6} per °C at rt

                  5.69 x 10^{-6} per °C; 25-500°C   (329)

                  6.57 x 10^{-6} per °C; 25-1000°C   (329)

                  6.98 x 10^{-6} per °C; 25-1500°C   (329)

Heat content: $H_T - H_{293}$ = 15.81 T + 2.10 x 10^{-3}T^2 + 3.52 x
                  10^{5}T^{-1} - 6.081 cal/mole
                  (429 - 1171°K)   (128)

Heat capacity: Cp = 15.81 T + 4.20 x 10^{-3}T - 3.52 x
                  10^{5}T^{-2} cal/mole/°K (429 - 1171°K)   (128)

Heat of formation: -45 kcal/mole   (206)
                       at 25°C, -55 kcal/mole   (192)
                       ΔH^o_{298} = -62.3 kcal/mole   (444)

Heat of vaporization: at 25°C, 460 kcal/mole   (192)
                       ΔH_o 458.3 $\pm$ 6.5 kcal/mole   (207, 502)

Entropy: $\Delta S^o_{vap.}$ = 98.4 $\pm$ 3.0 eu   (502)

ZIRCONIUM DODECABORIDE

Formula weight: 318.44 g/mole

Formula volume: 87.7   cc/mole

Melting point: 2680°C   (19)

X-ray density: 3.63 g/cc   (338)

Pycnometric density: 3.7 g/cc   (338)

CHEMICAL

Theoretical analysis: 71.4% boron
                        28.6% zirconium

ELECTRICAL

Resistivity: 60-80 x 10^{-6} ohm-cm at room temperature   (19)

STRUCTURE

Face-centered cubic   (338)

    a = 7.408   (338)

Isomorphous with UB_{12} (338)

THERMAL

Conductivity: 0.029 CGS   (338)

DATA TABLES

II CARBIDES

Al_4C_3

ALUMINUM CARBIDE

Formula weight:	143.91 g/mole	(3)
Formula volume:	48.1 cc/mole	(2)
Melting point:	2800°C	(2, 227)
	d > 2200°C	(3)
	2200°C	(1, 377)
Vapor pressure:	Sublime in vacuo at 1800°C	(1, 377)
Pycnometric density:	2.99 g/cc	(2)
	2.95 g/cc	(3)

CHEMICAL

Theoretical analysis: 25.1% carbon
74.9% aluminum

Synthesis:	Al + C; slowly at 1000°C	(1, 374)
	20 min. at 2000°C	(1, 374)

Reacts with NH_3 > 1200°C	(227)
Reacts with H_2O to give methane	(73)
Dissolves in acids	(3)
Insoluble in acetone	(3)

OPTICAL

Color: Pale yellow crystals by vacuum sublimation	(1, 377)
Refractive index: 2.70	(3)

STRUCTURE

Rhombic	(2)
Al-C distance 1.90 - 2.22 A	(73)

STRUCTURE (cont.)

C-C distance 3.16 A minimum (1, 376)

Hexagonal (3)

THERMAL

Heat of formation: 50 kcals/mole (1, 375)

BORON CARBIDE

Formula weight: 55.29 g/mole

Formula volume: 22.30 cc/mole

Melting point: 2470 ± 20°C   (326)
                        2450°C   (3, 127, 129, 227, 261, 298)
                        2350 - 2500°C   (9)
                        2350°C   (38, 39)

Boiling point: >3500°C   (3)

X-ray density: 2.51 g/cc

Pycnometric density: 2.50 g/cc   (127, 129)
                        2.54 g/cc   (3)

CHEMICAL

Theoretical analysis: 21.7% carbon
                        78.3% boron

Synthesis: C + B$_2$O$_3$, electric furnace   (73)

Stable to 2250°C in helium   (227)

Stable to 540°C in oxygen   (227)

Air oxidation becomes severe 1100-1400°C   (227)

32 percent wt. change   3 hrs. in air at 1400°C   (227)

Maximum use temperature in air, 1000°C   (227)

ELECTRICAL

Resistivity: 0.3 - 0.8 ohm-cm   (127)

MECHANICAL

Strength (MOR):   44,000 psi at rt   (45, 127)
                  50,000 psi at rt   (38)
                  42,000 psi at 650°C   (38)
                  35,000 psi at 1100°C   (38)

MECHANICAL (cont.)

Tensile:	22,500 psi at 980°C	(9, 227, 297)
Compression:	414,000 psi at rt	(2, 127)

Hardness:	Mohs	: 9.32	(73)
		: 9.3	(2, 88, 227)
	Vickers	: 2400 kg/mm^2	(8)
		: 3700 kg/mm^2	(8)
	Knoop 100 g	: 2800 kg/mm^2	(8)
	Knoop 1000 g	: 2230 kg/mm^2	(8)
	100 g	: 3060 kg/mm^2 (2580-3940)	(38)

Young's modulus:	65×10^6 psi at 20°C	(5)
	64.79×10^6 psi at rt	(126)
	65.2×10^6 psi	(127)
	42×10^6 psi	(227, 297)
Shear modulus:	26.83×10^6 psi at rt	(126)
Bulk modulus:	36.80×10^6 psi at rt	(126)
Poisson's ratio:	0.207	(126)
Speed of sound:	48,230 fps	(126)

OPTICAL

Color: Black

Spectral emissivity: 0.65μ :	0.76 at 880°C	(158)
	0.56 at 1880°C	(158)

STRUCTURE

Rhombic	(127)
C_3 chains and B_{12} icosahedra in a NaCl type structure, extended along a body diagonal	(73, 378, 379)

STRUCTURE (cont.)

Continuous network of boron atoms accounts for its low
conductivity and great hardness   (1)

THERMAL

Conductivity:	0.065 cgs at 20°C	(227, 287)
		(5, 127, 129)
	0.069 cgs at 20°C	(227, 2)
	0.198 cgs at 425°C	(227, 2)
Expansion:	4.5×10^{-6} per °C; rt - 800°C	(2, 5, 127, 129, 227)
	4.78×10^{-6} per °C; 25-500°C	(329)
	5.54×10^{-6} per °C; 25-1000°C	(329)
	6.02×10^{-6} per °C; 25-1500°C	(329)
	6.53×10^{-6} per °C; 25-2000°C	(329)
	7.08×10^{-6} per °C; 25-2500°C	(329)

$$BaC_2$$

BARIUM CARBIDE

Formula weight:	161.38 g/mole	
Formula volume:	43.1 cc/mole	
Melting point:	1780°C	(227, 261)
Pycnometric density:	3.75 g/cc	

CHEMICAL

Theoretical analysis:	14.8% carbon	
	85.2% barium	
Decomposed by cold water or acids		(3)
Hydrolysis gives acetylene		(73)
Absorbs atmospheric nitrogen $\longrightarrow$ Ba(CN)$_2$		(73)
Reacts readily with O_2		(227)
Hydrolyzes in moist air		(227)

Be_2C

BERYLLIUM CARBIDE

Formula weight: 30.05 g/mole

Formula volume: 15.8 cc/mole

Melting point: d > 2100^oC (3)
                       d 2100^oC (38, 227)
                       d 2050^oC (227, 298)
                       d 2150^oC (227, 261)

Pycnometric density: 1.9 g/cc at 20^oC (3)

CHEMICAL

Theoretical analysis: 40.0% carbon
                       60.0% beryllium

Synthesis: a. Be + C or an organic material at
                1300^oC (380, 381, 382, 383, 384)

            b. BeO + C, at 1930^oC (380, 381, 382, 383, 384)

Stable in dry air at rt   (227)

Slowly decomposed by water   (3)

Slowly decomposed by acids   (3)

Rapidly decomposed by alkalis   (73)

Yields methane on hydrolysis   (73)

Be_2C at 2100^oC decomposes yielding
graphite   (380, 381, 382, 383, 384)

Dissociates in argon at 2150^oC (227)

Stable to dry oxygen to 1000^oC (227)

Stable in dry hydrogen to 1000^oC

Absorbs N_2 above 1200^oC, to give Be_3N_2 which decomposes
at 1100^oC (227, 305)

Reacts with NH_3 above 1200^oC (227)

Reacts with CO_2 above 1375^oC (227)

MECHANICAL

Strength: Compression: 105,000 psi at rt   (88, 227, 319)

Hardness: Mohs: 9+   (2, 88, 227)

Young's modulus: 45 x 10^6 psi at 20°C, 6-8% porous

45.6 x 10^6 psi at 0°C   (88, 227, 319)

Poisson's ratio: 0.01   (88, 227, 314)

OPTICAL

Color: Brick red   (73)

STRUCTURE

Forms regular octahedral crystals   (73)

Antifluorite structure, each C surrounded cubically by
8 Be atoms   (73)

THERMAL

Conductivity: 0.056 cgs at 20°C
0.054 cgs at 100°C   (227, 313)
0.050 cgs at 20-425°C   (227, 2, 289, 88)

Expansion: 10.8 x 10^{-6} per °C; 38-982°C

10.1 x 10^{-6} per °C; 25-500°C   (329)

10.4 x 10^{-6} per °C; 25-800°C   (227, 2)

CaC_2

CALCIUM DICARBIDE

Formula weight:	64.10 g/mole	
Formula volume:	28.8 cc/mole	
Melting point:	2300°C	(3, 227, 261)
Pycnometric density:	2.22 g/cc	(3)

CHEMICAL

Theoretical analysis: 37.4% carbon
62.6% calcium

Synthesis: a.   CaO + C   (385)

b. $CaCO_3$ + C, >2000°C   (1)

c.   Ca + C, >2000°C   (1)

Reacts with water, releasing acetylene

ELECTRICAL

Resistivity: An insulator   (248)

OPTICAL

Color: Gray   (3)

Refractive index: 1.75   (3)

STRUCTURE

Rhombohedral   (3)

Tetragonal - may be described as a NaCl type structure,
Na replaced by Ca, Cl by C_2 groups, all C_2 groups are
arranged with their axes parallel, hence it may be thought
of as a cubic lattice, extended along one axis.   (248)

CeC_2

CERIUM DICARBIDE

Formula weight:	164.15 g/mole	
Melting point:	>2500°C	(227)
Pycnometric density:	5.23 g/cc	(3)

CHEMICAL

Theoretical analysis: 14.6% carbon
85.4% cerium

Synthesis:	CeO_2 + C, electric furnace	(173)
Reactivity:	Stable to 2500°C in vacuum	(227)
	Decomposed by water and acids	(3)
	Liberates acetylene, ethylene, and methane by hydrolysis	(73)
	Can be used to 2000°C in H_2	(227)

OPTICAL

Color: Red   (3)

STRUCTURE

Hexagonal   (3)

Tetragonal, derived from a NaCl structure   (173)

$$Ce_2C_6$$

DICERIUM HEXACARBIDE

Formula weight:　　　352.32 g/mole

CHEMICAL

Theoretical analysis:　　20.4% carbon
79.6% cerium

Reactivity:　On hydrolysis, yields only acetylene　　　　(173)

STRUCTURE

Tetragonal, derived from a NaCl structure　　　　(173)

COBALT CARBIDE

Activation energy of formation from elements: 39.85 kcal/mole   (474)

CHROMIUM MONOCARBIDE

Formula weight: 64.02 g/mole

Melting point: 1550°C   (227, 298)

CHEMICAL

Theoretical analysis: 18.8% carbon
81.2% chromium

Reactivity: Oxidation in air becomes severe, 1100-1400°C   (227)

THERMAL

Activation energy of formation from elements:
26.00 kcal/mole   (474)

Cr_3C_2

TRICHROMIUM DICARBIDE

Formula weight: 180.05 g/mole

Formula volume: 26.9 cc/mole

Melting point: 1830-1890°C   (227, 261, 298)
                       1890°C   (3)

Boiling point: 3800°C   (3)

X-ray density: 6.70 g/cc   (9)

Pycnometric density: 6.68 g/cc   (3)

CHEMICAL

Theoretical analysis: 13.4% carbon
                        86.6% chromium

Reactivity: Oxidation in air becomes severe, 1100-1400°C   (227)
             Soluble in dilute HCl   (3)

MECHANICAL

Strength: Compression: 600,000 psi   (227)

OPTICAL

Color: Gray, metallic   (3)

STRUCTURE

Orthorhombic D5$_{10}$ type   (19)
a = 2.82 A,   b = 5.53 A,   c = 11.47 A   (19)

THERMAL

Expansion: 8.00×10^{-6} per °C; 25-500°C   (329)
              9.95×10^{-6} per °C; 25-1000°C   (329)
              8.8×10^{-6} per °C; 25-120°C   (227, 314)
              10.9×10^{-6} per °C; 150-980°C   (227, 314)

Heat of formation: ΔH^{o}_{298} = -21 kcal/mole   (444)

Activation energy of formation from elements:

                       26.00 kcal/mole   (474)

Fe_3C

TRIIRON MONOCARBIDE

Formula weight: 179.56 g/mole (Cementite)

Melting point: 1837 ^{o}C   (3)
                       1650°C   (2, 227, 261)

Pycnometric density: 7.4 g/cc   (3)

CHEMICAL

Theoretical analysis: 6.7% carbon
                         93.3% iron

Reactivity: Soluble in dilute acids   (3)

STRUCTURE

Pseudo hexagonal   (3)

THERMAL

Activation energy of formation from elements:
                    (γ Fe) 32.00 kcal/mole   (475)

HAFNIUM MONOCARBIDE

Formula weight: 190.61 g/mole

Formula volume: 15.0 cc/mole

Melting point: $3887 \pm 150^{o}C$ (54)
                         $3885 \pm 150^{o}C$ (227, 261, 298)
                         $3887^{o}C$ (39, 74, 78)
                         $3890^{o}C$ (38, 167)

X-ray density: 12.70 g/cc   (44)
                         12.52 g/cc   (76)

Pycnometric density: 12.20 g/cc   (75)

CHEMICAL

Theoretical analysis: 6.30% carbon
                         93.70% hafnium

Synthesis: a. $HfO_2 + C$; 1 hr at $2400^{o}C$ (38)

            b. $HfCl_4 + H_2 + CH_4$ etc, $2100 + ^{o}C$ (324)

Solubility of C in HfC at high temperature greatly reduces MP   (19)

Probably stable to H_2O at red heat   (227)

Rapid oxidation in O_2 at $1370^{o}C$ (227)

Oxidation in air becomes severe, $1100-1400^{o}C$ (227)

Stable in helium above $2230^{o}C$ (227)

An irreversible change has been observed on heating
between $1925-2025^{o}C$ (204)

Probably stable to $2000^{o}C$ in H_2 (227)

Likely to react with $N_2 \rightarrow HfN$ (227)

ELECTRICAL

Resistivity: 109×10^{-6} ohm-cm   (2, 42)

ELECTRICAL (cont.)

Resistivity:
(cont.)
41×10^{-6} ohm-cm at 4.2°K   (206)

41×10^{-6} ohm-cm at 80°K   (206)

45×10^{-6} ohm-cm at 160°K   (206)

49×10^{-6} ohm-cm at 240°K   (206)

60×10^{-6} ohm-cm at 300°K   (206)

$\rho_{(T)} = (30 + 0.0628T) \times 10^{-6}$ ohm-cm   (204)
from $300\text{-}2000^\circ$K

Does not become superconducting at 1.23°K   (333)

Hall coefficient: -21×10^{10} m^3/coul at 4.2°K; 80% dense   (206)

20×10^{10} m^3/coul at 80°K; 80% dense   (206)

Thermoelectric power: $10\text{-}11 \times 10^{-6}$ volt/$^\circ$C at rt   (204)

MECHANICAL

Strength: Bending (MOR):   34,670 psi at rt; $\rho = 11.9$ g/cc   (170)
12,640 psi at 2000°C; $\rho = 11.9$ g/cc (170)
4,780 psi at 2200°C; $\rho = 11.9$ g/cc (170)

Creep: Slight at 2200°C   (170)

Hardness: Vickers 50 g :   2533-3202   (76)
2913   (8)
Knoop   :   1790   (362)
1870 kg/mm^2 (38)

Young's modulus: 61.55×10^6 psi at rt; $\rho = 11.94$ (170)

Torsion modulus: 26.39×10^6 psi at rt; $\rho = 11.94$ (170)

Poisson's ratio: 0.166   (170)

OPTICAL

Color: Metallic gray

STRUCTURE

Face-centered cubic (B1)   (44)

NaCl type, isomorphous with HfB, HfN   (77)

a = 4.64A   (44)
4.46A   (77, 78)
4.619A at 20°C   (169)
4.632A at 480°C   (169)
4.638A at 680°C   (169)
4.643A at 890°C   (169)

Radius ratio: 0.49   (47)

THERMAL

Conductivity: 0.053 cgs
$0.15 + 1.20 \times 10^{-4}T$ watts/cm/^{o}C   (204)
($1000-2000^{o}$K)

Expansion: 7.75×10^{-6} per ^{o}C

$6.59 \pm 0.04 \times 10^{-6}$ per ^{o}C; $25-612^{o}$C   (168)

6.27×10^{-6} per ^{o}C; to 650°C   (68)

5.08×10^{-6} per ^{o}C; to 900°C (x-ray)   (169)

6.31×10^{-6} per ^{o}C; $25-500^{o}$C   (329)

6.25×10^{-6} per ^{o}C; $25-1000^{o}$C   (329)

Heat capacity: C_p = 15 cal/mole/^{o}C $\pm$ 10% at 925°C   (204)
16 cal/mole/^{o}C $\pm$ 10% at 1525°C   (204)
0.05 cal/gm/^{o}C

Thermal diffusivity: 0.04 cm^{2}/sec

Mn_3C

TRIMANGANESE MONOCARBIDE

Formula weight:	176.80 g/mole	
Melting point:	1520°C	(227, 261)
Pycnometric density:	6.89 g/cc	(3)

CHEMICAL

Theoretical analysis:　6.8% carbon
93.2% manganese

Reactivity:　Decomposed by water and acids　(3)

STRUCTURE

Tetragonal　(3)

MOLYBDENUM MONOCARBIDE

Formula weight: 107.96 g/mole

Formula volume: 12.7 cc/mole   (88)

Melting point: 2692 ± 50°C   (54)
                       2700°C   (38)
                       2690-2695°C   (2, 88, 127, 129)
                       2570°C   (3, 39)

Pycnometric density: 8.5 g/cc   (88)
                       8.78 g/cc   (127, 129)
                       8.48 g/cc   (3)

CHEMICAL

Theoretical analysis: 11.1% carbon
                       88.9% molybdenum

Reactivity: Soluble: HNO_3, HF, hot H_2SO_4
               Not attacked by H_2O at 600°C   (227, 88)
               Decomposes to Mo_2C at 900°C   (227)
               Oxidation in air becomes severe 500-800°C   (227)
               Useful to 2000°C in H_2 (227, 261)
               Decarburizes rapidly in H_2 above 1500°C (227, 265)
               Stable to 1500-1600°C in N_2 (227, 312)

ELECTRICAL

Superconductive at 8°K   (466)

Superconductive at 7.6-8.3°K   (467)

Thermionic work function: 2-4 ev   (444)

MECHANICAL

Strength: Bending (MOR): 16,400 psi at rt; ρ = 7.70   (170)

Creep at 2000°C   (170)

Hardness: Mohs: 7-8   (88, 227)

Young's modulus: 28.59×10^6 psi at rt; ρ = 7.60   (170)

MECHANICAL (cont.)

      Torsion modulus: 11.88×10^6 psi at rt; ρ = 7.60   (170)

      Poisson's ratio: 0.2038 at rt; ρ 7.60   (170)

OPTICAL

      Color: Metallic gray   (3)

      Form: Prismatic   (3)

STRUCTURE

      Hexagonal   (2, 88)

      a = 2.901 A,   c = 2.768 A   (88)

      Radius ratio: 0.57   (47)

THERMAL

      Expansion: 5.95×10^{-6} per $^{\circ}$C; $0\text{-}800^{\circ}$C   (7)

                       5.67×10^{-6} per $^{\circ}$C;   rt-1000°C   (170)

                       8.23×10^{-6} per $^{\circ}$C;   rt-1500°C   (170)

                       11.50×10^{-6} per $^{\circ}$C;   rt-2000°C   (170)

      Activation energy of formation from Mo + C :

                    34.40 kcal/mole   (474)

DIMOLYBDENUM MONOCARBIDE

Formula weight: 203.91 g/mole

Formula volume: 22.9   cc/mole

Melting point: 2687 $\pm$ 50°C   (54)
                        2685 - 90°C   (127, 88)
                        2687°C   (2)
                        2565°C   (227)
                        2500°C   (35)
                        2380°C   (39, 127)

X-ray density: 8.9 g/cc   (88)

Pycnometric density: 8.82 g/cc   (35)
                        9.18 g/cc   (127)

CHEMICAL

Theoretical analysis: 5.90% carbon
                        94.10% molybdenum

Reactivity: Oxidation in air becomes severe, 500-800°C   (227)

                Not attacked by H_2O at 700°C   (227, 88)

                Useful to 2000°C in H_2 (227, 261)

                Decarburizes rapidly in H_2 above 1500°C (227, 265)

                Stable to 1500-1600°C in N_2 (227, 312)

ELECTRICAL

Superconductive at 2.9°K   (460)
Superconductive at 2.4-3.2°K   (465)
Thermionic work function: 2-4 ev   (444)

MECHANICAL

Hardness: Mohs   : 7-9   (88, 227)
          Vickers   : 2000 kg/mm^2 (8)
          Knoop   : 1800 kg/mm^2 (19, 88, 227)
          Rockwell A : 88   (35)

MECHANICAL (cont.)

Young's modulus:	77.4 x 10^6 psi	(166)
	32.7 x 10^6 psi	(88, 227, 297)
	32.3 x 10^6 psi	(35, 185)

STRUCTURE

Hexagonal   (127)

Hexagonal close packed   (35)

a = 2.994 A,   c = 4.722 A   (88)
      3.012   4.35   (35)

Radius ratio: 0.57   (47)

THERMAL

Expansion: 4.40 x 10^{-6} per °C   (166)

(β form)   5.48 x 10^{-6} per °C; 25-500°C   (329)

(β form)   6.15 x 10^{-6} per °C; 25-1000°C   (329)

Heat of formation: ΔH^o_{298} = +4.2 kcal/mole   (444)

Entropy: 4.8 kcal/mole

Free energy: 2.8 kcal/mole

Activation energy of formation from elements:
                34.40 kcal/mole   (474)

NIOBIUM MONOCARBIDE

Formula weight:	103.93 g/mole	
Formula volume:	13.3 cc/mole	
Melting point:	$3497 \pm 125^{\circ}C$	(54)
	$3500^{\circ}C$	(2, 32, 33, 38, 78, 88, 127, 129)
	$3760^{\circ}C$	(227)
	$3800^{\circ}C$	(35)
X-ray density:	7.85 g/cc	(60)
Pycnometric density:	7.82 g/cc	(2, 32, 59, 88, 127, 129)
	7.56 g/cc	(8)

CHEMICAL

Theoretical analysis: 11.45% carbon
88.55% niobium

Synthesis: a. $Nb + C$ in H_2 at $1700^{\circ}C$ (52)

b. Nb_2O_5, Nb_2O_3, $Nb + C$ at $1300-1400^{\circ}C$ (19)

Reactivity: Oxidation in air becomes severe, $1100-1400^{\circ}C$ (227)

Useful to $3430^{\circ}C$ in helium   (227)

Stable to MP in helium   (227, 181)

Useful to $2000^{\circ}C$ in helium   (227, 261)

Decarburizes slightly in air   (227, 2)

Probably stable to $2200^{\circ}C$ in H_2 (227)

Useful to $2000^{\circ}C$ in H_2 (227, 261)

Attacked by H_2 above $1500^{\circ}C$ (227, 265)

Stable in N_2 to $2500^{\circ}C$ (88, 227, 261)

Stable in N_2 to $2400^{\circ}C$ (227, 2)

Dissolves C to form eutectic at $3250^{\circ}C$ (400)

Only NbC exists as true compound   (51)

ELECTRICAL

Resistivity: 147×10^{-6} ohm-cm at rt   (19, 53)
254×10^{-6} ohm-cm at mp   (19, 53)
74×10^{-6} ohm-cm   (19, 46)
21×10^{-6} ohm-cm at 4.2°K; 86% dense   (206)
22×10^{-6} ohm-cm at 80°K; 86% dense   (206)
30×10^{-6} ohm-cm at 160°K; 86% dense   (206)
36×10^{-6} ohm-cm at 240°K; 86% dense   (206)
40×10^{-6} ohm-cm at 300°K; 86% dense   (206)

Superconductive below 10°K   (283)

Becomes superconductive at 10.1   (469)
10.5   (463)
6   (464)
<4.2 (444)

Hall coefficient: 0.8×10^{-10} m^3/coul at 4.2°K; 86% dense   (206)

MECHANICAL

Strength: Bending (MOR): 35,600 psi at rt; $\rho = 7.50$ (170)
2,800 psi at 2000°C; $\rho = 7.49$ (170)

Creep: Severe at 2190°C   (170)

Hardness: Mohs   : 9-10   (2, 88, 227)
Vickers 50 g : 2400 kg/mm^2 (2)
Knoop 100 g : ~2000 kg/mm^2 (127)
Rockwell A : 91   (35)
Micro   : 2470 kg/mm^2 (173)
2400 kg/mm^2 (172)

Young's modulus: 49.1×10^6 psi   (166)
49×10^6 psi
49.4×10^6 psi   (19, 88, 227)

OPTICAL

Color: Light brownish powder to gray brown with purple cast (19)

Spectral emissivity: 0.28 at 725°C   (203)
0.22 at 2125°C   (203)

STRUCTURE

Face-centered cubic   (2)

NaCl type   (B1)   (60)

    d = 4.461A   (60)
       4.40A   (78)
       4.48A   (35)
       4.4584-4.462A   (88)

Radius ratio: 0.54   (47)

THERMAL

Conductivity: 0.034 CGS   (2, 19, 88, 127, 129, 174, 227)
             0.0405 CGS

Expansion: 7.02×10^{-6} per $^{\circ}$C (0-1000°C)   (7)

7.17×10^{-6} per $^{\circ}$C (0-1200°C)   (7)

7.29×10^{-6} per $^{\circ}$C (0-1400°C)   (7)

6.91×10^{-6} per $^{\circ}$C; at rt

6.52×10^{-6} per $^{\circ}$C; 25-500°C   (329)

7.07×10^{-6} per $^{\circ}$C; 25-1000°C   (329)

7.46×10^{-6} per $^{\circ}$C; 25-1500°C   (329)

Thermal diffusivity: 0.13 cm^2/sec

Heat of formation: ΔH^{o}_{298} = -33.6 kcal/mole   (444)

Activation energy of formation from elements:

             18.90 kcal/mole   (474)

NiC_x

NICKEL CARBIDE

Activation energy of formation from element: 32.20 kcal/mole (475)

PLUTONIUM MONOCARBIDE

Formula weight: 254 g/mole

THERMAL

Expansion: 10.8×10^{-6} per $^{\circ}C$ (432)

DIPLUTONIUM TRICARBIDE

Formula weight: 520 g/mole

THERMAL

   Expansion: 14.8×10^{-6} per ^{o}C (432)

SILICON CARBIDE

Note: A complete review is given in Reference 503.

Formula weight:	40. 07 g/mole	
Formula volume:	12. 46 cc/mole	(127)
Melting point:	Decomposes	
	d 2500°C	(5, 38)
	d 2600°C	(127)
	d 2537°C	(39)
	2700°C	(161, 227, 298)
	d 2200°C	(227, 298)
Subl.	2350°C	(227)
X-ray density: (6H Type)	3.217 g/cc	(503)
(Cubic)	3.210 g/cc	(503)

CHEMICAL

Theoretical analysis: 30. 0% carbon
70. 0% silicon

No evidence for non-stoichiometric compositions.

Reactivity:		
	Oxidation in air becomes severe, 1400-1700°C	(227)
	2% weight change; 3 hrs at 1400°C in air	(227)
	Maximum use temperature in air: 1100-1600°C (depending on binder)	(227)
	Sublines in helium at 2350°C	(227)
	Attacked by H_2, probable failure at 1350°C	(227)
	Stable in N_2	(227)
	Reacts with CO_2	(227)

ELECTRICAL

Resistivity: 10^3 - 10^5 ohm-cm   (503)

MECHANICAL

Strength: Bending: 50,000 psi at 20°C   (45)
                       64,000 psi at 1400°C   (45)
                       30-35,000 psi (practical limit) (38)
                       50,000 psi (absolute limit)   (38)
                       27,000 psi at rt (typical data) (38)
                       25,000 psi at 1300°C   (38)
                       11,000 psi at 1400°C   (38)
                       15,000 psi at 1800°C   (38)

          Compression: 82,000 psi   (2, 127, 227)
                        120,000 psi   (38)

Hardness: Mohs   : 9.2   (2, 227)
          Vickers 25 g : 3000-3500 kg/mm^2 (8)
          Knoop 100 g : 2500-2550 kg/mm^2 (8)
                     : 2500 kg/mm^2 (156)
                     : 2960 kg/mm^2 (black)   (503)
                     : 2745 kg/mm^2 (green)   (503)

Young's modulus: 58.2 x 10^6 psi at rt; ρ = 3.128   (167)

                    59.52 x 10^6 psi at rt; ρ = 3.120   (38)

                    61.5 x 10^6 psi at 1200°C   (227)

                    49.4 x 10^6 psi at 1400°C   (227)

                    56 x 10^6 psi at 20°C   (156)

                    55 x 10^6 psi at 400°C

                    53 x 10^6 psi at 800°C

                    51 x 10^6 psi at 1200°C

Shear modulus: 24.41 x 10^6 psi at rt; ρ = 3.128   (167)

Bulk modulus: 14.01 x 10^6 psi at rt; ρ = 3.128   (167)

MECHANICAL (cont.)

Poisson's ratio: 0.192 at rt; ρ = 3.128   (167)

                        0.183 at rt; ρ = 3.128   (156)

OPTICAL

Color: colorless when pure to black

Spectral emissivity: 0.65μ: 0.88; 1125-1525°C   (158)

                        0.79 at 800°C   (156)

                        0.84 at 1000°C   (156)

                        0.86 at 1150°C   (156)

                        0.84 at 1300°C   (156)

                        0.82 at 1400°C   (156)

                        0.77 at 1550°C   (156)

STRUCTURE

Low temperature form (β) cubic

High temperature form (α) hexagonal

(β form) - T_d^2 - F 43 m Space Group

   a = 4.349 A
   4 moles/unit cell   (503)

(α form) - Several poly types known
   6 H most common

Hexagonal - C_{6V}^4 - C6MC - Space Group

       a = 3.073 A
       c = 15.07 A
     c/a = 4.899
       6 moles/unit cell   (503)

THERMAL

    Conductivity: 0.049 CGS at 600°C   (5)

                      0.098 CGS at 20°C   (227)

                      0.100 CGS at 20°C   (129)

  0.0245 CGS at 615°C   (156)

  0.0158 CGS at 850°C   (156)

  0.0093 CGS at 1110°C   (156)

  0.0085 CGS at 1140°C   (156)

  0.0059 CGS at 1250°C   (156)

  0.0032 CGS at 1530°C   (152)

  0.080 CGS at 600°C   (227)

  0.061 CGS at 800°C   (227)

  0.051 CGS at 1000°C   (227)

  0.037-0.038 from 600-1500°C

    Expansion: 4.63×10^{-6} per $^{\circ}$C; 25-500°C (329)

  5.12×10^{-6} per $^{\circ}$C; 25-1000°C (329)

  5.48×10^{-6} per $^{\circ}$C; 25-1500°C (329)

  5.77×10^{-6} per $^{\circ}$C; 25-2000°C (329)

  5.94×10^{-6} per $^{\circ}$C; 25-2500°C (329)

  4.7×10^{-6} per $^{\circ}$C; 20-1500°C (5, 129)

  5.22×10^{-6} per $^{\circ}$C; to 275°C (156)

THERMAL (cont.)

Expansion: (cont.)

5.40×10^{-6} per $^{\circ}$C; to 1550°C   (156)

4.70×10^{-6} per $^{\circ}$C;   0-1700°C   (227, 297)

4.68×10^{-6} per $^{\circ}$C: 27-800°C   (2, 227)

4.51×10^{-6} per $^{\circ}$C; 20-1090°C   (227)

Specific heat: 0.26 cal/gm/$^{\circ}$C at   540°C   (156)

0.27 cal/gm/$^{\circ}$C at   700°C   (156)

0.30 cal/gm/$^{\circ}$C at 1000°C   (156)

0.32 cal/gm/$^{\circ}$C at 1200°C   (156)

0.33 cal/gm/$^{\circ}$C at 1350°C   (156)

0.35 cal/gm/$^{\circ}$C at 1550°C   (156)

SrC_2

STRONTIUM DICARBIDE

Formula weight: 111.65 g/mole

Formula volume: 35.0 cc/mole

Melting point: 1950°C   (38)
                     1930°C   (227, 261)

Pycnometric density: 3.19 g/cc   (3)
                     3.04 g/cc

CHEMICAL

Theoretical analysis: 10.74% carbon
                      89.26% strontium

Reactivity: Decomposed by water and acids   (3)

OPTICAL

Color: Black   (3)

STRUCTURE

Tetragonal

TANTALUM MONOCARBIDE

Formula weight:	192.89 g/mole	
Formula volume:	13.15 cc/mole	(127)
	13.3 cc/mole	(59)
	14.7 cc/mole	(436)
Melting point:	3877 ± 150°C	(54, 88)
	3877°C	(2, 39, 78)
	3880°C	(38, 54, 127, 129)
	3800°C	(35, 57)
	3740°C	(403)
Boiling point:	d. 4730-4830°C	(53)
Vapor pressure:	Evaporation rate: 10^{-6} g/cm^2/sec at 3030°C	(227, 290)
X-ray density:	14.48 g/cc	(59)
	14.49 g/cc	(88)
	14.53 g/cc	
	14.65 g/cc	(127)
Pycnometric density:	13.96 g/cc	

CHEMICAL

Theoretical analysis: 6.23% carbon
93.77% tantalum

Synthesis: a. Fusion of Ta_2O_5 + Na_2CO_3 + C at 1500°C (401, 402)

b. Ta_2O_5 + C, 1500°C or higher

c. Ta + C + Al melt, 2000°C

Reactivity: Oxidation in air becomes severe, 1100-1400°C (227)

Useful to 3760°C in helium (227)

Stable to MP in helium (227, 290)

Decomposes at 2230°C in helium (227, 230)

Useful to 2000°C in helium (227, 261)

Burns in air or H_2O at 800°C (227, 290)

Useful to 2000°C in H_2 (227, 261)

Attacked by H_2 above 1500°C (227, 265)

Not affected by N_2 at 3300°C if H_2 absent (227, 290)

ELECTRICAL

Resistivity: 20×10^{-6} ohm-cm   (2)
30×10^{-6} ohm-cm   (19, 46, 127)
100×10^{-6} ohm-cm
175×10^{-6} ohm-cm
8×10^{-6} ohm-cm at 4.2°K; 80% dense   (206)
10×10^{-6} ohm-cm at 80°K;   80% dense   (206)
15×10^{-6} ohm-cm at 160°K; 80% dense   (206)
20×10^{-6} ohm-cm at 240°K; 80% dense   (206)
25×10^{-6} ohm-cm at 300°K; 80% dense   (206)

Becomes superconductive between
$9.5-7.3^{\circ}$K   (283, 333)
at 9.3   (460)
9.3-9.5   (465)
9.5   (463)
9.2   (469)
Not superconductive to 1.8°K   (323)

Hall coefficient: 0.5×10^{-10} m^3/coul at 4.2°K; 80% dense   (206)
0.9×10^{-10} m^3/coul at 80°K; 80% dense   (206)
1.2×10^{-10} m^3/coul at 160°K; 80% dense   (206)
1.3×10^{-10} m^3/coul at 240°K; 80% dense   (206)
1.3×10^{-10} m^3/coul at 300°K; 80% dense   (206)

Thermionic emission: (336, 337, 395)

MECHANICAL

Strength: Bending (MOR): 30,910 psi at rt; ρ = 12.58   (170)
17,600 psi at 2000°C; ρ = 12.58 (170)

Tensile: 2000-42000 psi   (88, 227)

Creep: Becomes severe at 2200°C   (170)

Hardness: Mohs   : 9-10   (178)
Vickers 50 g : 1800 kg/mm^2 (8)
Knoop 50 g : 1800 kg/mm^2 (35)
1952 kg/mm^2 (180)
Knoop 100 g :   825 kg/mm^2 (38)
Rockwell A : 89   (35)
Brinell   : 840   (180)

MECHANICAL (cont.)

Young's modulus: 55.41 x 10^6 psi at rt; ρ = 12.68   (170)
                  52.8 x 10^6 psi at rt; ρ = 12.45   (170)
                  91.3 x 10^6 psi   (127)
                  41.5 x 10^6 psi   (181, 185)
                  41.3 x 10^6 psi   (166)
                  57   x 10^6 psi   (35, 185)

Torsion modulus: 23.64 x 10^6 psi at rt; ρ = 12.68   (170)

Poisson's ratio: 0.1719 at rt; ρ = 12.68   (170)

OPTICAL

Color: Golden, metallic, needles from Na_2CO_3,
        Ta_2O_5 + C fusion   (401, 402)

Total emissivity: 0.36 at 1600°C   (227)
                   0.36 at 1800°C   (227)
                   0.37 at 2200°C   (227)
                   0.44 at 2700°C   (227)

Spectral emissivity: 0.655μ: 0.67   (19)

STRUCTURE

Face-centered cubic   (2, 88)
NaCl type   (B 1), Z = 4   (35, 436)

d = 4.42 A   (35)
    4.455   (58, 59)
    4.445   (88)
    4.45   (78)
    4.457   (38)
    4.456   (154)

    4.450 at 20°C   (169)
    4.455 at 390°C   (169)
    4.459 at 600°C   (169)
    4.464 at 815°C   (169)
    4.469 at 1030°C   (169)

STRUCTURE (cont.)

d = 4.445 at 23°   (176)
     4.458 at 498°C   (176)
     4.465 at 705°C   (176)
     4.474 at 994°C   (176)

Radius ratio: 0.54   (47)

THERMAL

Conductivity: 0.053 CGS   (2, 5, 127)

Expansion: 8.2×10^{-6} per °C; 25-800°C   (2, 88, 127, 129, 227)
              $8.2 \pm 0.8 \times 10^{-6}$ per °C; 20-2380°C   (5)
              6.29×10^{-6} per °C; to 500   (176)
              6.50×10^{-6} per °C; 0-1000   (7)
              6.64×10^{-6} per °C; 0-1200   (7)
              6.64×10^{-6} per °C; to 1000   (176)
              8.02×10^{-6} per °C; to 2625   (177)
              8.19×10^{-6} per °C; 25-2400°C   (227, 94)
              6.32×10^{-6} per °C; 25-500°C   (329)
              6.67×10^{-6} per °C; 25-1000°C   (329)
              7.12×10^{-6} per °C; 25-1500°C   (329)
              7.64×10^{-6} per °C; 25-2000°C   (329)
              8.40×10^{-6} per °C; 25-2500°C   (329)

Heat of formation: -63.8 kcal/mole

                      ΔH^{o}_{298} = -78.5 kcal/mole   (444)

Specific heat: 0.045 cal/gm/°C

Thermal diffusivity: 0.08 cm^2/sec

Thermodynamic data: (59, 404)

Activation energy of formation from elements:
                      19.30 kcal/mole   (474)

DITANTALUM MONOCARBIDE

Formula weight:	373.77 g/mole	
Melting point:	3400°C	(57, 2)
	3250°C	(227)
Pycnometric density:	15.22 g/cc	(58, 88)

CHEMICAL

Theoretical analysis: 3.21% carbon
                         96.79% tantalum

STRUCTURE

Hexagonal   (88)

  a = 3.091A;   c = 4.93A;   c/a = 1.59   (58, 88)

  (L'3 type)   (58)

  Two crystalline modifications have been proposed   (58)

Radius ratio: 0.54   (47)

THERMAL

Activation energy of formation from elements:
                19.30 kcal/mole   (474)

THORIUM MONOCARBIDE

Formula weight: 244.13 g/mole

Formula volume: 22.9 cc/mole

Melting point: 2625 ± 25°C   (227, 298)
                         2625°C   (2)

Pycnometric density: 10.65 g/cc

CHEMICAL

Theoretical analysis: 4.92% carbon
                         95.08% thorium

Reactivity: Oxidation in air becomes severe, 1100-1400°C   (227)

Decomposed by H_2O (227)

Burns in air when ignited   (227)

STRUCTURE

Face-centered cubic   (2)

(B1 type)

d = 5.34A

Radius ratio: 0.43   (47)

ThC$_2$

THORIUM DICARBIDE

Formula weight: 256.14 g/mole

Formula volume: 26.4 cc/mole

Melting point: 2655 ± 25°C   (227, 298)
                     2655°C   (2)
                     2650°C   (38)
                     2773°C   (39)
                     2777°C

Boiling point: 5000°C   (3)

X-ray density: 9.6 g/cc   (19)

Pycnometric density: 8.96 g/cc   (3)
                     9.7 g/cc

CHEMICAL

Theoretical analysis: 9.38% carbon
                         90.62% thorium

Reactivity: Air oxidation becomes severe, 500-800°C   (227)

              Burns vigorously in air or sulfur   (2)

              Very slightly soluble in concentrated acids   (3)

              Decomposed by water   (3, 227)

              Decomposed by dilute acids   (3)

OPTICAL

Color: Yellow   (3)

STRUCTURE

Monoclinic   (2)
    C2 /C type   (19)

    a = 6.53A;   b = 4.24A;   c = 6.56A   (19)
Radius ratio: 0.43   (47)

THERMAL

Heat of formation: ΔH (1/2 ThC_2) = -22.8 kcal/mole

Entropy: ΔS (1/2 ThC_2) = + 7.6 kcal/mole

Free energy: ΔF (1/2 ThC_2) = -25.1 kcal/mole

TITANIUM MONOCARBIDE

Formula weight:	59.91 g/mole	
Formula volume:	12.15 cc/mole	(158)
	11.8 cc/mole	(436)
Melting point:	3180°C	(3)
	3160°C	(19, 53, 178)
	3140°C	(19, 52, 361)
	3250°C	(19)
Boiling point:	4300°C est.	(3, 19, 388)
X-ray density:	4.938 g/cc	(197)
Pycnometric density:	4.25 g/cc	(3)
	4.93 g/cc	(5, 88, 129, 159)
	4.85 g/cc	(35)

CHEMICAL

Theoretical analysis: 20.05% carbon
79.95% titanium

Synthesis: a. TiO_2 + C; > 800°C

  b. Ti + C; high vacuum (197)

Single phase from $TiC_{0.54}$ - $TiC_{1.0}$ (135)

Soluble in HNO_3 (3)

Insoluble in HCl (3)

Probably stable in helium to 3000°C (227, 19)

Resistant to H_2O at least to 650°C (227)

Oxidation in air becomes severe at 1200°C (143)

Stable to 2500°C in N_2 (2)

Oxidation in air becomes severe, 1100-1400°C (227)

8.6 percent wt. change in 3 hrs at 1400°C in air (227)

Maximum use temperature; 3000°C in helium (227)

CHEMICAL (cont.)

Stable to 1200°C in H_2	(227, 88)
Decarburized in H_2 at high temperature	(227, 265)
Probably stable in H_2 to 2200°C	(227, 265)
Stable in N_2 to 1200°C	(227, 88)
to 2400°C	(227, 2)
to 2500°C	(227, 261)
Reacts with N_2 above 1500°C	(227)
Oxidizes in CO_2 or N_2O at 1200°C	(227, 88)
Hardly attacked by HCl or H_2SO_4	(19)
Easily soluble in mixtures of HNO_3 + HF	(19)
Dissolved by alkaline oxidizing melts	(19)
Reacts with N_2 above 1500	(19)
Reacts with Cl_2	(19)

ELECTRICAL

Resistivity:

173×10^{-6} ohm-cm at 23°C; single crystal	(192)
148×10^{-6} ohm-cm at -200°C; single crystal	(192)
90×10^{-6} ohm-cm at 23°C; polycrystalline	(192)
70×10^{-6} ohm-cm at 23°C; polycrystalline	(158)
71×10^{-6} ohm-cm at -200°C; polycrystalline	(192)
193×10^{-6} ohm-cm at -200°C; polycrystalline	(158)
105×10^{-6} ohm-cm	(2, 158)
65×10^{-6} ohm-cm at rt to 215×10^{-6} ohm-cm at 2100°C	(209)

ELECTRICAL (cont.)

Resistivity: 68×10^{-6} ohm-cm at 4.2°K; 86% dense   (206)
(cont.)

    68×10^{-6} ohm-cm at 80°K; 86% dense   (206)

    70×10^{-6} ohm-cm at 160°K; 86% dense   (206)

    76×10^{-6} ohm-cm at 240°K; 86% dense   (206)

    84×10^{-6} ohm-cm at 300°K; 86% dense   (206)

    $180-250 \times 10^{-6}$ ohm-cm at rt   (53)

    193×10^{-6} ohm-cm at rt   (42)

    239×10^{-6} ohm-cm at -60°C   (42)

    294×10^{-6} ohm-cm at liquid air temperature   (42)

    68.2 and 72.1×10^{-6} at rt, 100% dense   (95)

    72×10^{-6} at rt, 99.5% dense   (95)

    78.3×10^{-6} at rt, 98.6% dense   (95)

    Becomes superconducting below <1.9   (444)
                    1.15°K   (333)
    Becomes superconducting at 1.1°K   (283)

Hall coefficient: 13.9×10^{-10} m^3/coul at 23°C; single crystal (192)

    26.1×10^{-10} m^3/coul at -200°C; single crystal (192)

    9.0×10^{-10} m^3/coul at 23°C; single crystal (192)

    16.4×10^{-10} m^3/coul at -200°C; single crystal (192)

    17×10^{-10} m^3/coul at 4.2°K; 86% dense (206)

    16×10^{-10} m^3/coul at 80°K; 86% dense (206)

    14×10^{-10} m^3/coul at 160°K; 86% dense (206)

ELECTRICAL (cont.)

Hall coefficient: 12×10^{-10} m^3/coul at 240°K; 86% dense   (206)
    (cont.)

                     10×10^{-10} m^3/coul at 300°K; 86% dense   (206)

Magnetic susceptibility: 0.12×10^{-7} (19, 394)

MECHANICAL

      Strength: Review article - Ref. 88

               Bending (MOR): 32,670 psi at rt; ρ = 4.85   (170)
                               13,600 psi at 2000°C; ρ = 4.85   (170)
                               20,000 psi   (158)
                               68,000 psi   (45)
                               124,000 psi at 100% density,
                                 2-8μ original particle size   (95)
                               100,000 psi at 100% density,
                                 8-37μ original particle size   (95)
                             91,000 psi at 99.5% density,
                               37-44μ original particle size   (95)
                             73,500 psi at 98.6% density,
                               40-74μ original particle size   (95)

               Tensile: 17,200 psi at 1000°C   (9)
                        940 psi   (9)

               Compression: 109,000 psi   (2, 158)
                           190,000 psi   (2, 88, 227)

      Hardness:

Mohs	:	9-10	(19, 53, 88, 178, 227)
Vickers 50 g	:	2900-3200 kg/mm^2	(8)
Vickers 100 g	:	2850-3390 kg/mm^2	(8)
Knoop 100 g	:	2470 kg/mm^2	(8, 158)
Knoop 100 g	:	1905 kg/mm^2	(362)
Micro	:	3200 kg/mm^2	(35)
Micro 20 g	:	3200 kg/mm^2	(386)
Rockwell A	:	93.5	(35)
		88-89 at 98.6% density	(95)
		88-89 at 99.5% density	(95)
		91-92 at 100% density	(95)
		92.5-93.5 at 100% density	(95)

MECHANICAL (cont.)

Young's modulus:	63.715×10^6 psi at rt; $\rho = 4.85$	(170)
	55×10^6 psi at 1000°C	(9)
	45.5×10^6 psi	(166)
	45.8×10^6 psi	(35)
	51×10^6 psi	(35)
	45×10^6 psi	(19, 88, 175, 227, 297)

Poisson's ratio:	0.182 single crystal	(192)
	$0.189; \rho = 3.56$	(194)
	$0.187; \rho = 4.21$	(192)

Compressibility:	$3.29 \times 10^{-8} \ \dfrac{m^2}{lb}$	(387, 185)

Compressibility and elastic properties of single crystals
investigated   (192, 194)

OPTICAL

Color: Light gray, metallic

Spectral emissivity: 0.65μ : 0.96 at 1820°C
                       0.65μ : 0.40 at 725°C to 0.32 at 2125°C (203)
                       See Also   (336, 337, 395)

STRUCTURAL

Face-centered cubic,	(2, 158)
NaCl type (B-1), Z = 4	(436, 93)
d = 4.3316A	(135)
4.315A	(35)
4.32A	(88)

Isomorphous with TiO and TiN

Radius ratio: 0.53   (47)

THERMAL

Conductivity:	0.060 CGS	(129)
	0.041 CGS at 20°C	(2, 5, 46, 74, 158)
	0.049 CGS	
	0.074 CGS at 20°C	(227, 88)
	0.025 CGS at 20°C	(227, 88)
	0.0135 CGS at 1000°C	(227, 88)
Expansion:	6.52×10^{-6} per °C; 25-500°C	(192, 195)
	7.45×10^{-6} per °C; 25-750°C	(192, 195)
	8.82×10^{-6} per °C; 25-1000°C	(192, 195)
	9.32×10^{-6} per °C; 25-1250°C	(192, 195)
	9.45×10^{-6} per °C; 25-1500°C	(192, 195)
	9.86×10^{-6} per °C; 25-1700°C	(192, 195)
	6.82×10^{-6} per °C; 25-500°C	(7, 185)
	7.18×10^{-6} per °C; 25-750°C	(7, 185)
	7.80×10^{-6} per °C; 25-1000°C	(7, 185)
	8.65×10^{-6} per °C; 25-1250°C	(7, 185)
	9.50×10^{-6} per °C; 25-1500°C	(7, 185)
	7.4×10^{-6} per °C at rt	(7, 185)
	7.9×10^{-6} per °C at rt-1000°C	(170)
	8.94×10^{-6} per °C at rt-1500°C	(170)
	10.02×10^{-6} per °C at rt-2000°C	(170)
	7.4×10^{-6} per °C; 25-1000°C	(2, 4, 5, 9, 88, 129, 158)
	7.86×10^{-6} per °C; 0-1000°C	(7)
	8.10×10^{-6} per °C; 0-1200°C	(7)
	8.29×10^{-6} per °C; 0-1400°C	(7)
	7.42×10^{-6} per °C;	(166)
	7.38×10^{-6} per °C; 25-800°C	(2, 227)
	7.06×10^{-6} per °C; 0-750°C	(206)
	7.85×10^{-6} per °C; 0-1000°C	(206)
	8.02×10^{-6} per °C; 0-1275°C	(206)
	8.26×10^{-6} per °C; 0-1525°C	(206)
	8.40×10^{-6} per °C; 0-1775°C	(206)
	7.15×10^{-6} per °C; 25-500°C	(329)
	7.70×10^{-6} per °C; 25-1000°C	(329)
	8.15×10^{-6} per °C; 25-1500°C	(329)
	8.81×10^{-6} per °C; 25-2000°C	(329)
	7.90×10^{-6} per °C; 25-2500°C	(329)
Heat capacity:	0.150 cal/gm at 150°C	(146)
	0.170 cal/gm at 300°C	
	0.183 cal/gm at 450°C	

THERMAL (cont.)

Heat capacity: 0.192 cal/gm at 600°C   (146)
(cont.)   0.20 cal/gm at 750°C

              0.17 cal/gm at 150°C   (145)
              0.187 cal/gm at 300°C
              0.196 cal/gm at 450°C
              0.201 cal/gm at 600°C
              0.207 cal/gm at 750°C
              0.209 cal/gm at 900°C
              0.210 cal/gm at 1000°C
              0.211 cal/gm at 1100°C

Thermal diffusivity: 0.08 cm^2/sec at rt
              0.101 cm^2/sec at 1350°C (209)
              0.096 cm^2/sec at 1500°C (209)

Thermodynamic data: (145, 389-393)

Heat of formation: ΔH^o_{298} = -43.85 kcal/mole (444)

Activation energy of formation from elements:
     17.50 kcal/mole (474)

URANIUM MONOCARBIDE

Formula weight: 250.08 g/mole

Formula volume: 18.4 cc/mole

Melting point:

	$2450\text{-}2500^{o}C$	(63)
	$2350\text{-}2400^{o}C$	(227, 298)
	$2470^{o}C$	(38)
	$2277^{o}C$	(140, 227)
	2590 ± 50	(100)

X-ray density: 13.6 g/cc   (19)

Pycnometric density: 12.08 g/cc   (136)

CHEMICAL

Theoretical analysis: 4.8% carbon
95.2% uranium

Reactivity:	
Partial reaction with NH_3 at red heat	(227)
Reacts with N_2	(227, 277)
Thermodynamically stable in H_2	(227, 277)
Useful in helium to $2000^{o}C$	(227)
Disintegrates in water at $40^{o}C$	(136)
Oxidation in air becomes severe, $500\text{-}800^{o}C$	(227)

ELECTRICAL

Resistivity: 50.1×10^{-6} ohm-cm; $\rho = 12.15$ (136)

MECHANICAL

Strength: Bending (MOR)	23,000 psi at $25^{o}C$	(136)
	17,600 psi at $800^{o}C$	(136)
	10,000 psi at $1000^{o}C$	(136)
Hardness: Knoop 100 g :	660 kg/mm^2 (351-1136)	(38)
Creep: $\sim 1200^{o}C$		(136)

Young's modulus: 25.0×10^6 psi   (136)

MECHANICAL (cont.)

Shear modulus: 9.7×10^6 psi (136)

Poisson's ratio: 0.29 (136)

STRUCTURE

Cubic, B1 type (61)

a = 4.951A (61)
4.955A (62)
4.952A at 20^oC (169)
4.972 at 570^oC (169)
4.984 at 750^oC (169)
4.998 at 850^oC (169)
4.952A at 20^oC (137)
4.996 at 1000^oC (137)
5.005 at 1100^oC (137)
5.028 at 1400^oC (137)
5.035 at 1500^oC (137)

THERMAL

Conductivity: 0.07 CGS at 800^oC (136)
0.195 CGS at 20^oC (227, 245)

Expansion: 9.47×10^{-6} per ^oC; 25-200^oC (136)
10.30×10^{-6} per ^oC; 25-400^oC (136)
10.68×10^{-6} per ^oC; 25-600^oC (136)
11.01×10^{-6} per ^oC; 25-800^oC (136)
11.23×10^{-6} per ^oC; 25-1000^oC (136)
11.43×10^{-6} per ^oC; 25-1200^oC (136)
9.26×10^{-6} per ^oC; 25-500^oC (329)
10.67×10^{-6} per ^oC; 25-1000^oC (329)
11.79×10^{-6} per ^oC; 25-1500^oC (329)
13.05×10^{-6} per ^oC; 25-2000^oC (329)

Heat of formation: -40 kcal/mole

Free energy: -41 kcal/mole

Entropy: 2 kcal/mole

URANIUM DICARBIDE

Formula weight: 262.09 g/mole

Formula volume: 22.4 cc/mole

Melting point: 2427^oC
                      2400^oC (140, 227, 298)
                      2260^oC (39)

X-ray density: 11.7 g/cc   (19)

CHEMICAL

Theoretical analysis: 9.16% carbon
                             90.84% uranium

Reactivity: Thermodynamically stable to H_2 (227, 277)

                 Useful to 2000^oC in helium   (227)

                 Oxidation in air becomes severe; $500\text{-}800^oC$ (227)

                 Ignites in air at 400^oC (227)

                 Disintegrates in H_2O (227)

                 Reacts with N_2 at 1100^oC (227, 277)

                 Partial reaction with NH_3 at red heat   (227)

STRUCTURE

Body-centered tetragonal   (19)

$CaCl_2$ type   (19)

     a = 3.52A;   b = 5.99A   (19)

THERMAL

Conductivity: 0.082 cgs at 20^oC (5)
                 0.077 cgs at 45^oC (227, 277)
                 0.085 cgs at 150^oC (227)

THERMAL (cont.)

Expansion: 6.32 x 10^{-6} per $^{\circ}$C; 25-500°C   (329)

9.43 x 10^{-6} per $^{\circ}$C; 25-1000°C   (329)

12.60 x 10^{-6} per $^{\circ}$C; 25-1500°C   (329)

14.64 x 10^{-6} per $^{\circ}$C; 25-1800°C   (329)

12.4 x 10^{-6} per $^{\circ}$C; 20-236°C   (227, 277)

Heat of formation: Δ H(1/2 UC$_2$) = -18 kcal/mole

Entropy: Δ S(1/2 UC$_2$) = +2 kcal/mole

Free energy: Δ F(1/2 UC$_2$) = -19 kcal/mole

U_2C_3

<u>DIURANIUM TRICARBIDE</u>

Formula weight: 512.14 g/mole

Formula volume: 39.7 cc/mole

Melting point: 2427°C
                    2400°C   (3, 39, 227, 298)
                    2370°C   (140, 227)

X-ray density: 12.9 g/cc   (19)

Pycnometric density: 11.28 g/cc   (3)

CHEMICAL

Theoretical analysis: 7.05% carbon
                        92.95% uranium

Reactivity: Partial reaction with NH_3 at red heat   (227)

Oxidation in air becomes severe, 500-800°C   (227)

Maximum use temperature, 2000°C in inert atmosphere   (227)

Decomposed by water or dilute acids   (3)

Thermodynamically stable to H_2 (227, 277)

Reacts with N_2 at 1100°C   (227, 277)

STRUCTURE

Body-centered cubic   (19)

1$\bar{4}$3 d type   (19)

d = 8.09A   (19)

THERMAL

Expansion: 6.26×10^{-6} per °C; 25-500°C   (329)

9.13×10^{-6} per °C; 25-1000°C   (329)

THERMAL (cont.)

10.70 x 10^{-6} per $^{\circ}$C; 25-1500°C (329)

11.38 x 10^{-6} per $^{\circ}$C; 25-1800°C (329)

Heat of formation: $\Delta H(1/3\ U_2C_3)$ = -24 kcal/mole

Entropy: $\Delta S(1/3\ U_2C_3)$ = +3 kcal/mole

Free energy: $\Delta F(1/3\ U_2C_3)$ = -25 kcal/mole

VANADIUM MONOCARBIDE

Formula weight: 62.96 g/mole

Formula volume: 10.90 cc/mole   (127)

Melting point: 2810°C   (78, 127)
2830°C   (3, 19, 38, 53, 227)
2827°C   (39)
2850°C   (161, 227)
2865°C   (227)

Boiling point: 3900°C   (3)

X-ray density: 5.81 g/cc   (53)

Pycnometric density: 5.77 g/cc   (127, 129)
5.36 g/cc   (3, 35)

CHEMICAL

Theoretical analysis: 19.08% carbon
80.92% vanadium

Synthesis: a. V_2O_5 + C; 1000°C or higher   (19)

b. VCl_4 + H_2 + Hydrocarbons, 1500-2000°C   (324)

Only one carbide has been confirmed

Reactivity: Soluble: HNO_3, fused KNO_3 (3)

Insoluble: HCl, H_2SO_4 (3)

Stable to 2400°C in inert atmosphere   (227)

Burns in O_2 (227)

Decomposes in H_2O (227)

Stable in NH_3 to 2400°C   (227)

Stable to H_2O, H_2S, HCl to 800°C   (19)

Reacts with Cl_2 below 500°C   (19)

ELECTRICAL

Resistivity: 160×10^{-6} ohm-cm   (127)

320×10^{-6} ohm-cm at $2230^{\circ}C$ (53)

150×10^{-6} ohm-cm at rt   (53)

Not superconductive to $1.8^{\circ}K$ (323)

See also (283)

Not a superconductor   (463)

MECHANICAL

Strength: Bending: 10,000 psi   (127)
          Compression: 89,000 psi   (2, 127)

Hardness: Mohs   : 9+   (53)
            Vickers 50 g : 2800 kg/mm^2 (8)
               100 g   : 2084-2510 kg/mm^2 (8)
            Knoop 100 g : 2080 kg/mm^2 (8, 127)
            Rockwell A   : 92

Young's modulus: 39.3×10^6 psi   (166)
                    36.98×10^6 psi   (35, 185)
                    39×10^6 psi   (175)

OPTICAL

Color: Metallic gray

STRUCTURE

Face-centered cubic (B1)   (127)

NaCl type   (35)

d = 4.165A   (35)
    4.14   (78)
    4.16   (19)
Isomorphous with TiC, TiN, VN, VO, etc.
Radius ratio: 0.59   (47)

THERMAL

    Conductivity: For low temperature data see   (227, 324)

    Entropy: ΔS = -1.6 kcal/mole
        See also ref.   (399)

    Heat content: See ref. 399

    Heat of formation: H^{O}_{298} = -28 kcal/mole   (444)

TUNGSTEN MONOCARBIDE

Formula weight:	195.93 g/mole	
Formula volume:	12.51 cc/mole	(127)
Melting point:	$2867 \pm 50^{\circ}C$	(54)
	$2870^{\circ}C$	(127, 129, 178)
	$2900^{\circ}C$	(35)
	$2777^{\circ}C$	(3, 39)
	$2867^{\circ}C$	(2)
	$2627^{\circ}C$	
Boiling point:	$6000^{\circ}C$	(3)
X-ray density:	15.8 g/cc	
Pycnometric density:	15.63 g/cc	(127, 129)
	15.60 g/cc	(35)
	15.7 g/cc	(3)

CHEMICAL

Theoretical analysis: 6.23% carbon
                        93.77% tungsten

Reactivity: Oxidation in air becomes severe, 500-800°C	(227)
Soluble: Fluorine	(3)
Insoluble: Aqueous acids	(3)
Stable to 2850°C in argon	(227, 300)
Useful to 2000°C in helium	(227, 261)
Stable in air to 700°C	(227, 300)
More easily oxidized that W metal	(227, 316)
Decarburizes above 2500°C in H_2	(227, 2)
Decarburizes above 2500°C in hydrocarbons	(227, 300)
Decarburizes above 1500°C in H_2	(227, 300)
Unaffected by N_2 at high temperature	(227, 265, 300)

MECHANICAL

Strength: Bending (MOR):   84,000 psi   (127)

55,650 psi at rt: ρ = 15.35   (170)

Tensile: 50,000 psi   (227, 297)

Creep: at 2000°C   (170)

Hardness: Vickers 50 g : 2400   (8)

100 g : 1730   (8)

Knoop 100 g : 1870   (127)

: 1880   (8)

Rockwell A : 92   (35)

Young's modulus: 78.3×10^6 psi   (127)

87.2×10^6 psi   (166)

96.91×10^6 psi at rt; ρ = 15.35   (170)

75.27×10^6 psi   (35, 185)

102.5×10^6 psi at 20°C   (5, 227, 297)

OPTICAL

Color: Gray, metallic

STRUCTURE

Hexagonal   (35, 127)

Cubic   (2, 3)

a = 2.897A; c = 2.27A   (35)

2.90   2.83   (19)

Radius ratio: 0.57   (47)

THERMAL

Expansion: 4.9×10^{-6} per °C; rt -1000°C   (170)

5.8×10^{-6} per °C; rt - 1500°C   (170)

THERMAL (cont.)

Expansion: 7.4×10^{-6} per ^{o}C; rt - 2000°C   (170)
(cont.)
                  4.84×10^{-6} per ^{o}C; to 1000°C   (7)

                  4.94×10^{-6} per ^{o}C; to 1200°C   (7)

                  5.04×10^{-6} per ^{o}C; to 1400°C   (7)

                  6.2×10^{-6} per ^{o}C; to 800°C   (2, 127, 129)

                  5.2×10^{-6} per ^{o}C; at 20°C   (5)

                  7.3×10^{-6} per ^{o}C; at 1930°C   (5)

                  3.9×10^{-6} per ^{o}C   (166)

                  4.42×10^{-6} per ^{o}C; 25-500°C   (329)

                  4.92×10^{-6} per ^{o}C; 25-1000°C   (329)

                  5.35×10^{-6} per ^{o}C; 25-1500°C   (329)

                  5.82×10^{-6} per ^{o}C; 25-2000°C   (329)

                  6.14×10^{-6} per ^{o}C; 20-1930°C   (227)

                  $\begin{bmatrix} 20.5 \ (\text{Parallel 001}) \\ 2.16 \ (\text{Parallel 1001}) \end{bmatrix}$

Heat of formation: -3.9 kcal/mole
                      ΔH^{o}_{298} = + 8.4 kcal/mole   (444)
Entropy: -1.7 kcal/mole

Free energy: -3.4 kcal/mole

Activation energy of formation from elements:
                      39.50 kcal/mole   (474)

DITUNGSTEN MONOCARBIDE

Formula weight: 379.85 g/mole

Formula volume: 21.9 cc/mole

Melting point:	$2857 \pm 50^{\circ}C$	(54)
	$2857^{\circ}C$	(2, 39)
	$2877^{\circ}C$	(3)
	$2730^{\circ}C$	(38)
	$2860^{\circ}C$	(127)

Boiling point:	$6000^{\circ}C$	(3)

X-ray density: 17.3 g/cc

Pycnometric density:	17.15 g/cc	(127)
	16.05 g/cc	(3)

CHEMICAL

Theoretical analysis: 3.16% carbon
96.84% tungsten

Reactivity: Soluble: HNO_3 (hot)   (3)

Slightly soluble: HCl, H_2SO_4 (3)

Oxidation in air becomes severe, $500-800^{\circ}C$ (227)

Stable to $2850^{\circ}C$ in argon   (227, 300)

Useful to $2000^{\circ}C$ in helium   (227, 261)

Stable to air to $540^{\circ}C$ (227, 300)

More easily oxidized than W metal   (227, 316)

Decarburizes in H_2 above $2500^{\circ}C$ (227, 2)

Decarburizes in hydrocarbons above $2500^{\circ}C$ (227, 300)

Unaffected by N_2 at high temperatures   (227, 265, 300)

W_2C

(Cont.)

MECHANICAL

Hardness:	Mohs	: 9+	(19, 227)
	Vickers 50 g :	3000 kg/mm^2	(8)
	:	3200-3400 kg/mm^2	(8)
	Knoop 100 g :	2150 kg/mm^2	(127)

Young's modulus: 60.9 x 10^6 psi   (166)

OPTICAL

Color: Iron gray, metallic

Total emissivity: 0.256 at 1125°C   (227)
                       0.350 at 2325°C   (227)

STRUCTURE

Hexagonal   (127)

L'3 type   (19)

     a = 2.98A;   c = 4.71A   (19)

Radius ratio: 0.57   (47)

THERMAL

Expansion:	
6.0 x 10^{-6} per °C; to 800°C	(2, 127)
3.58 x 10^{-6} per °C; 25-500°C	(329)
3.80 x 10^{-6} per °C; 25-1000°C	(329)
4.00 x 10^{-6} per °C; 25-1500°C	(329)
4.25 x 10^{-6} per °C; 25-2000°C	(329)
4.72 x 10^{-6} per °C; 25-2500°C	(329)
5.95 x 10^{-6} per °C; 20-1930°C	(227)

$$\begin{bmatrix} 13.15 \text{ (Parallel 001)} \\ 9.38 \text{ (Parallel 1001)} \end{bmatrix}$$

Activation energy of formation from elements:
     39.50 kcal/mole   (474)

ZIRCONIUM MONOCARBIDE

Formula weight:	103.23 g/mole	
Formula volume:	15.38 cc/mole	(127)
	15.0 cc/mole	(436)
Melting point:	3532 $\pm$ 125°C	(54)
	3532°C	(39, 78)
	3530°C	(2)
	3250°C	(35)
	3540°C	(127, 129)
	3175°C	(19, 227)
Boiling point:	5100°C	(3)
X-ray density:	6.661 g/cc	(167)
	6.73 g/cc	(127, 129)
	6.70 g/cc	(35)
	6.44 g/cc	(131)
Pycnometric density:	6.9 g/cc	(19)

CHEMICAL

Theoretical analysis: 11.64% carbon
88.36% zirconium

Synthesis: a. ZrO_2 + C; 1900-2100°C   (396)

b. ZrC + C

c. $ZrCl_4$ + H_2 + CO, CH_4, etc.   (19)

ZrC dissolves excess carbon lowering MP from 3530 to
2430°C. This excess carbon exsolves on cooling to give
Zr C + graphite   (19, 52)

Stable to 2000°C   (227, 261)

Excess carbon lowers MP to 2430°C   (54)

Reacts with ZrO_2 at 2200°C   (40)

Oxidizes rapidly at 980°C   (130)

Nitrides in N_2 above 1500°C   (130)

CHEMICAL (cont.)

Powerful getter if deficient in carbon	(40)
Oxidation in air becomes severe, 1100-1400°C	(227)
40 percent weight change; 3 hrs, 1400°C in air	(227)
Maximum use temperature in helium 2230°C	(227)

Soluble: HNO_3 + HF, H_2SO_4
           dilute HF, HNO_3 (3)

Insoluble: HCl

Not attacked by H_2O at red heat	(227)
Stable in H_2 to 2000°C	(227, 261)
Probably stable in H_2 to 2200°C	(227)
Reacts with $N_2 \rightarrow$ ZrN at 1500°C	(227, 19)
Fine powder is pyrophoric	(19)

ELECTRICAL

Resistivity: 64×10^{-6} ohm-cm; rt   (327)
               97×10^{-6} ohm-cm; 500°C   (327)
               137×10^{-6} ohm-cm; 1000°C   (327)
               75×10^{-6} ohm-cm   (127)
               63.4×10^{-6} ohm-cm   (52)
               63×10^{-6} ohm-cm
               50×10^{-6} ohm-cm
               37.8×10^{-6} ohm-cm at liquid air temperature   (52)
               45×10^{-6} ohm-cm at 4.2°K; 98% density   (206)
               45×10^{-6} ohm-cm at 80°K; 98% density   (206)
               47×10^{-6} ohm-cm at 160°K; 98% density   (206)
               53×10^{-6} ohm-cm at 240°K; 98% density   (206)
               61×10^{-6} ohm-cm at 300°K; 98% density   (206)
               Becomes superconductive between 4.1°K
               and 2.1°K   (333)
               Does not become superconductive to 1.8°K   (323)

ELECTRICAL (cont.)

Hall coefficient: 14×10^{-10} m^3/coul at 4.2°K; 98% dense   (206)
                   14×10^{-10} m^3/coul at 80°K; 98% dense
                   13×10^{-10} m^3/coul at 160°K; 98% dense
                   12×10^{-10} m^3/coul at 240°K; 98% dense
                   11×10^{-10} m^3/coul at 300°K; 98% dense

Magnetic properties: See Reference 394

Thermionic emission: See Ref. 336, 337, 398

MECHANICAL

Strength:	Bending:	22,500 psi	(127)
		16,600 psi at rt	(170)
		8,300 psi at 1250°C	(170)
		5,140 psi at 1750°C	(170)
		2,500 psi at 2000°C	(170)
	Tensile:	16,000 psi at rt	
		11,700 psi at 980°C	(130)
		14,450 psi at 980°C	(130)
		12,950 psi at 1200°C	(130)
		15,850 psi at 1200°C	(130)
	Compression:	238,000 psi	(2, 127)

Hardness: Mohs   : 8-9   (178)
             Vickers 50 g : 2600 kg/mm^2 (8)
                    100 g : 2836-3480 kg/mm^2 (8)
             Knoop   : 2138 kg/mm^2 (362)
             Micro   : 2090 kg/mm^2 (127)
             Rockwell A   : 92.5   (35)

Young's modulus: 46.09×10^6 psi at rt; $\rho = 6.118$ (167)
                   28.3×10^6 psi   (35, 185)
                   69.6×10^6 psi   (127)
                   49×10^6 psi   (5)

Shear modulus: 17.98×10^6 psi at rt; $\rho = 6.118$ (167)

Bulk modulus: 31.07×10^6 psi at rt; $\rho = 6.118$ (167)

Poisson's ratio: 0.257; $\rho = 6.118$ (167)

MECHANICAL (cont.)

Elastic properties of single crystal ZrC (192)

OPTICAL

Color: Metallic gray

Spectral emissivity: 0.655μ : See Refs. 336, 337, 398

STRUCTURE

Face-centered cubic (B1)   (2, 127, 135, 167)
NaCl, Z = 4   (436)
  d = 4.689A   (56)
   4.69   (78)
   4.6865   (167)
   4.669   (35, 55)
   4.694   (154)

Isomorphous with ZrB, ZrN

Radius ratio: 0.49   (47)

THERMAL

Conductivity: 0.049 CGS at 20°C   (2, 5, 127, 129)
  0.098 CGS at 50°C   (132)
  0.069 CGS at 150°C   (132)
  0.073 CGS at 143°C   (133)
  0.065 CGS at 188°C   (133)
  0.061 CGS at 288°C   (133)
  0.080 CGS at 600°C   (146)
  0.083 CGS at 800°C   (146)
  0.086 CGS at 1000°C   (146)
  0.089 CGS at 1200°C   (146)
  0.092 CGS at 1400°C   (146)
  0.096 CGS at 1600°C   (146)
  0.099 CGS at 1800°C   (146)
  0.103 CGS at 2000°C   (209)
  0.105 CGS at 2200°C   (209)
  0.049 CGS at rt   (227, 287)
  0.075 CGS at 530°C to
  0.104 CGS at 2100°C   (326)

THERMAL (cont.)

Diffusivity:
 0.119 cm^2/sec at 1300°C (146)
 0.122 cm^2/sec at 1400°C (146)
 0.130 cm^2/sec at 1500°C (146)
 0.132 cm^2/sec at 1600°C (146)
 0.15 cm^2/sec at rt

Specific heat: 0.049

Expansion:
 6.32 x 10^{-6} per °C; 0-750°C (206)
 6.66 x 10^{-6} per °C; 0-1000°C (206)
 6.68 x 10^{-6} per °C; 0-1275°C (206)
 6.83 x 10^{-6} per °C; 0-1525°C (206)
 6.98 x 10^{-6} per °C; 0-1775°C (206)
 6.73 x 10^{-6} per °C; 24-500°C (5, 35)
 6.46 x 10^{-6} per °C; 0-1000°C (7)
 6.54 x 10^{-6} per °C; 0-1200°C (7)
 6.16 x 10^{-6} per °C; 0-1000°C
 6.60 x 10^{-6} per °C; 0-1400°C
 9.0 x 10^{-6} per °C; 1000-2000°C (182)
 7.27 x 10^{-6} per °C; to 1100°C
 6.47 x 10^{-6} per °C (166)
 7.27 x 10^{-6} per °C (2)
 6.10 x 10^{-6} per °C; 25-500°C (329)
 6.56 x 10^{-6} per °C; 25-1000°C (329)
 7.06 x 10^{-6} per °C; 25-1500°C (329)
 7.65 x 10^{-6} per °C; 25-2000°C (329)
 6.65 x 10^{-6} per °C; 25-800°C (2, 227)

Thermodynamic data: See Ref. (389, 390, 393, 397)

Heat of formation: ΔH^o_{298} = -44.1 kcal/mole (444)
                            -47.7 kcal/mole (525)

Activation energy of formation from elements:
                17.30 kcal/mole (474)

DATA TABLES

III MIXED CARBIDES

Formula weight: 383.50 g/mole

MECHANICAL

Hardness: Knoop 100 g : 1331 kg/mm^2 (1085-1770)   (38)

Formula weight:	769.28 g/mole
Melting point:	3930°C	(54)

Formula weight: 962.17 g/mole

Melting point: 3940 $\pm$ 150°C   (54)

Calculated density: 14.1 g/cc   (170)

Pycnometric density: 11.74 g/cc   (170)

CHEMICAL

Reactivity: Air oxidation becomes severe,
               1100-1400°C   (227)

               Useful   >2000°C in helium   (227)

MECHANICAL

Strength: Bending (MOR) :   18,400 psi at rt; ρ = 10.40   (170)
                            18,900 psi at 2000°C;
                                 ρ = 10.40   (170)
                            15,250 psi at 2200°C;
                                 ρ = 10.40   (170)

Creep: 2250°C   (170)

Young's modulus: 58.49 x 10^6 psi at rt; ρ = 11.74   (170)
                        32.12 x 10^6 psi at rt; ρ = 9.73   (170)

Torsion modulus: 23.82 x 10^6 psi at rt; ρ = 11.74   (170)
                        13.64 x 10^6 psi at rt; ρ = 9.73   (170)

Poisson's ratio: 0.219 at rt; ρ = 11.74   (170)

Hardness: Knoop 100 g   :   1150 kg/mm^2 (38)

HfC · 5 TaC

Formula weight: 1155.06 g/mole

Melting point: 3920°C   (54)

HfC · 8 TaC

Formula weight: 1733.73 g/mole

Melting point: 3910°C   (54)

NbC · TaC

Formula weight: 297.81 g/mole

Melting point: ~3627°C   (54)

Complete miscibility   (54)

$NbC \cdot 4\ TaC$

Formula weight: 876.84 g/mole

Melting point: $\sim$3757°C   (54)

Complete miscibility   (54)

NbC · 8 TaC

Formula weight: 1648.04 g/mole

Melting point: 3827°C   (54)

Complete miscibility   (54)

Maximum hardness	100 TiC - 0 NbC		(560)
	90 TiC	4. 324 A	(560)
	75 TiC	4. 348 A	
	60 TiC	4. 373 A	
	40 TiC	4. 398 A	
	25 TiC	4. 420 A	
	10 TiC	4. 438 A	
	0 TiC	4. 454 A	

Maximum hardness   60 VC - 40 NbC   (560)

90% VC   a = 4.187 A   (560)

75% VC   a = 4.242 A

60% VC   a = 4.286 A

40% VC   a = 4.337 A

25% VC   a = 4.367 A

10% VC   a = 4.427 A

Formula weight: 484.77 g/mole

Melting point: $\sim2960^{\circ}C$ (54)

$NbC \cdot 2 W_2C$

Formula weight: 864.62 g/mole

Melting point: ~2830°C   (54)

Formula weight: 589. 69 g/mole

Melting point: ~3130°C   (54)

Formula weight: 208.15 g/mole

Melting point: Essentially same as end members   (54)
                         ~3500°C

Complete solid solution

Formula weight: 313. 07 g/mole

Melting point: ~3500°C, essentially the same as
                         component carbides   (54)

Complete solid solution

Formula weight: 522.9.1 g/mole

Melting point: ~3500°C, essentially the same as
                         the end members   (54)

Complete solid solution

Maximum hardness   80 TiC - 20 TaC   (560)

Maximum hardness   75 VC - 25 TaC   (560)

$TaC \cdot W_2C$

Formula weight: 572.74 g/mole

Melting point: 3212°C   (54)

Two phase at room temperature   (54)

TaC$\cdot$2 W$_2$C

Formula weight: 952.69 g/mole

Melting point: 2947°C   (54)

Immiscibility at rt of compositions above 20% W$_2$C
suggested   (54)

TaC · 10 W$_2$C

Formula weight: 3991.39 g/mole

Melting point: 2777°C   (54)

Immiscibility at rt of compositions above 20% W$_2$C
suggested   (54)

2 TaC · W$_2$C

Formula weight: 765.63 g/mole

Melting point: 3407°C (54)

Immiscibility at rt of compositions above 20% W$_2$C
suggested (54)

3 TaC · 20 W$_2$C

Formula weight: 8175.67 g/mole

Melting point: 2877°C   (54)

Immiscibility at rt of compositions above 20% W$_2$C
suggested   (54)

$4\,TaC \cdot W_2C$

Formula weight: 1151.41 g/mole

Melting point: 3587°C   (54)

8 TaC $\cdot$ W$_2$C

Formula weight: 1922.97 g/mole

Melting point: 3702°C   (54)

Formula weight: 296.12 g/mole

Melting point: 3772°C   (54)

MECHANICAL

Hardness: Knoop 100 g : 1670 kg/mm^2 (1415-2010)   (38)

TaC · 2 ZrC

Formula weight: 399. 35 g/mole

MECHANICAL

Hardness: Knoop 100 g : 1710 kg/mm^2 (1650-1825)   (38)

Formula weight: 502.58 g/mole

MECHANICAL

Hardness: Knoop 100 g : 1385 kg/mm^2 (1260-1520)   (38)

Formula weight: 605.81 g/mole

MECHANICAL

Hardness: Knoop 100 g : 1895 kg/mm^2 (1700-2290)   (38)
Maximum hardness in this series   (560)

Formula weight: 915.50 g/mole

MECHANICAL

Hardness: Knoop 100 g : 1815 kg/mm^2 (1625-2020)   (38)

Formula weight: 489.01 g/mole

Melting point: 3807°C   (54)

MECHANICAL

Hardness: Knoop 100 g : 865 kg/mm^2 (740-960)   (38)

Young's modulus: 16.42×10^6 psi at rt; ρ= 7.075   (170)

Torsion modulus: 6.619×10^6 psi at rt; ρ= 7.075   (170)

Poisson's ratio: 0.180 at rt; ρ = 7.075   (170)

Formula weight: 681.90 g/mole

MECHANICAL

Hardness: Knoop 100 g : 1300 kg/mm^2 (1190-1590)   (38)

Formula weight: 874.79 g/mole

Formula volume: 69.35 cc/mole

Melting point: 3930 $\pm$ 150°C   (54)

Calculated density: 12.6 g/cc   (170)

Pycnometric density: 11.48 g/cc   (170)

CHEMICAL

Reactivity: Air oxidation becomes severe; 1100-1400°C (227)

Useful above 2000°C in helium   (227)

Stable in He, at least to 2000°C   (227)

Oxidizes rapidly in O_2 above 1375°C   (227)

Probably stable in H_2 to 2000°C   (227)

May react with N_2 to form ZrN   (227)

MECHANICAL

Strength: Bending (MOR): 40,000 psi at rt   (149)
>15,700 psi at 1900°C   (149)
11,380 psi at 2000°C   (149)
15,075 psi at 2000°C   (149)
10,650 psi at 2100°C   (149)
11,320 psi at 2190°C   (149)
41,000 psi at rt; ρ = 11.48   (170)
13,380 psi at 2000°C; ρ = 11.48   (170)
13,320 psi at 2190°C; ρ = 11.48   (170)
6,650 psi at 2300°C; ρ = 10.96   (170)

MECHANICAL

Hardness: Knoop 100 g : 1550 kg/mm^2 (1130-2175)   (38)

Young's modulus: 58.95 x 10^6 psi at rt; ρ = 11.48   (170)
                          33.51 x 10^6 psi at rt; ρ = 9.80   (170)

Torsion modulus: 25.81 x 10^6 psi at rt; ρ = 11.48   (170)
                          20.03 x 10^6 psi at rt; ρ = 10.73   (170)

Poisson's ratio: 0.212 at rt; ρ = 11.2   (170)

Formula weight: 1453.46 g/mole

MECHANICAL

Hardness: Knoop 100 g : 1290 kg/mm^2 (1160-1450)   (38)

Formula weight: 1646.35 g/mole

Formula volume: 12.99 cc/mole

Melting point: 3907°C   (54)

Calculated density: 12.68 g/cc

Pycnometric density: 11.70 g/cc   (170)

MECHANICAL

Strength: Bending (MOR): 42,900 psi at rt; ρ = 11.65   (170)
15,100 psi at 2000°C;
ρ = 11.65   (170)
18,300 psi at 2190°C;
ρ = 11.65   (170)

Hardness: Knoop 100 g :   1210 kg/mm^2 (1060-1290)   (38)

Young's modulus: 52.71 x 10^6 psi at rt; ρ = 11.70   (170)

Torsion modulus: 22.90 x 10^6 psi at rt; ρ = 11.70   (170)

Poisson's ratio: 0.1512 at rt; ρ = 11.70   (170)

THERMAL

Expansion: 6.70 x 10^{-6} per °C; rt-1000°C   (170)

7.53 x 10^{-6} per °C; rt-1500°C   (170)

8.10 x 10^{-6} per °C; rt-2000°C   (170)

Maximum hardness   80 TiC - 20 WC   (560)

Only 1:1 solid solution stable above 2100°C   (560)

Below 2100°C, 1:1, 3-7, 3:1 solid solutions
identified   (560)

Maximum hardness   3 TiC - ZrC   (560)

Maximum hardness   50 VC - 50 WC   (560)

$W_2C \cdot ZrC$

Formula weight: 471.07 g/mole

Melting point: 2857°C   (54)

Immiscible even at melting point   (54)

$W_2C \cdot 4\ ZrC$

Formula weight: 780.76 g/mole

Melting point: 2857°C   (54)

Immiscible even at melting point   (54)

DATA TABLES

IV ELEMENTS

ALUMINUM

Formula weight: 26.97 g/mole

ELECTRICAL

Resistivity:

T_c = 1.13 - 1.197°K   (557)

MAGNETIC

H_c = 106 Oersteds   (557)

STRUCTURE

Face-centered cubic   (557)

Formula weight: 197.20 g/mole

Formula volume: 10.22 cc/mole

Melting point: 1063°C   (3)

Boiling point: 2600°C   (3)

X-ray density: 19.3 g/cc   (3)

CHEMICAL

Reactivity: Soluble in aqua regia, KCN

Insoluble in most common reagents

OPTICAL

Color: Yellow

STRUCTURE

Cubic close packed   (248)

THERMAL

Expansion: 15.35×10^{-6} per °C; 25-500°C   (329)

16.95×10^{-6} per °C; 25-1000°C   (329)

17.3×10^{-6} per °C; 25-1063°C   (329)

BORON

Formula weight: 10.82 g/mole

Formula volume: 4.63 cc/mole

Melting point:	$2300^{\circ}C$	(3)
	$2050^{\circ}C$	(127)
	$2300 \pm 300^{\circ}C$	(227, 298)
	$2150^{\circ}C$	
Boiling point:	$2550^{\circ}C$	(3)
Vapor pressure:	4.94×10^{-6} mm of Hg at $1238^{\circ}C$	(227, 244)
X-ray density:	2.34 g/cc	(127)
Pycnometric density:	2.32 g/cc	(3)

CHEMICAL

Reactivity:	Insoluble:	H_2O, alcohol	(3)
	Soluble:	HNO_3	(3)

ELECTRICAL

Resistivity: 7000 ohm-cm   (127)

MECHANICAL

Strength: Bending:		76,000 psi	(127)
Hardness:	Mohs:	9.3	(248)
	Knoop 100g:	2500 kg/mm^2	(127)
		3340 kg/mm^2 (3140-3750)	(38)

STRUCTURE

Below $1100^{\circ}C$, essentially a close packed cubic array of B_{12} icosahedra.   (248)

STRUCTURE (cont.)

Above 1100°C, a number of complex structures have been
identified. (248)

THERMAL

Expansion: 8.29 x 10^{-6} per °C; 20-1000°C (227)

8.22 x 10^{-6} per °C; 25-500°C (329)

8.24 x 10^{-6} per °C; 25-1000°C (329)

Heat of vaporization: $\Delta H_{v_{298}}$ = 132 kcal/mole (206)

BERYLLIUM

Formula weight: 9.02 g/mole

Formula volume: 4.97 cc/mole

Melting point: $1330^{\circ}C$ (161, 227, 298)
                         $1284^{\circ}C$ (3)

Boiling point: $2767^{\circ}C$ (3)

Vapor pressure: 4.68×10^{-6} mm of Hg at $1130^{\circ}C$ (227, 244)
                         10^{-3} mm of Hg at $1093^{\circ}C$ (227, 315)

Pyconometric density: 1.816 g/cc

CHEMICAL

Reactivity: Slight reaction in hot H_2O (3)

              Soluble in dilute acids and alkalis   (3)

MECHANICAL

Complete review by White   (227, 319)

Data sheets available   (227, 292, 293)

STRUCTURE

Hexagonal close packed, mean Be-Be distance 2.25A   (248)

THERMAL

Conductivity: 0.394 CGS   (88, 185, 227)

Expansion: 11.5×10^{-6} per $^{\circ}C$; r.t.   (185, 227)

             13.4×10^{-6} per $^{\circ}C$; 20-255$^{\circ}C$ (227)

             15.9×10^{-6} per $^{\circ}C$; 20-492$^{\circ}C$ (227)

THERMAL

Expansion: (cont.)

18.75×10^{-6} per $^{\circ}$C; 20-1000°C   (227)

15.8×10^{-6} per $^{\circ}$C; 25-500°C   (329)

18.5×10^{-6} per $^{\circ}$C; 25-1000°C   (329)

Formula weight: 12.01 g/mole

Formula volume: 3.4168 $\pm$ 0.00001 cc/mole   (553)

Melting point: s. >3500°C   (3)
                     3700°C

Boiling point: 4200°C   (3)

Pycnometric density: 3.51 g/cc   (3)

MECHANICAL

Hardness: Mohs   : 10
          Vickers   : 10,600 kg/mm^2 (8)
          Knoop 100 g : 8000-8500 kg/mm^2 (8)
               1000 g : 5500-7000 kg/mm^2 (8)

OPTICAL

Color: colorless if pure

Form: cubes, octahedra

Refractive index: 2.4195   (3)

STRUCTURE

Cubic

C - C distance 1.54 A   (248)

a = 3.56688 $\pm$ 0.00009 A   at 298.16°K   (553)

THERMAL

Expansion: 3.48 x 10^{-6} per °C;   to 1000°C   (7)

                4.10 x 10^{-6} per °C;   to 1400°C   (7)

THERMAL (cont.)

Expansion: (cont.)

2.57×10^{-6} per $^{\circ}$C; 25-500°C (329)

3.53×10^{-6} per $^{\circ}$C; 25-1000°C (329)

4.38×10^{-6} per $^{\circ}$C; 25-1500°C (329)

$\propto$ linear $= 0.09613 \times 10^{-5} + 3.522 \times 10^{-9} T - 0.0888 T^{-2}$ (553)

$\propto$ volume $= 0.2884 \times 10^{-5} + 10.57 \times 10^{-9} T - 0.2665 T^{-2}$ (553)

298 - 973°K (553)

C

GRAPHITE

Vapor pressure: 3.86×10^{-6} mm of Hg at 2470°C (227, 244)

10^{-3} mm of Hg at 2315°C (227, 244)

CHEMICAL

Reactivity: Stable to 2000°C in He (227, 253)

Stable to 3000°C in neutral atm. (2, 227)

Reacts with O_2 at 450°C (227, 312)

Considerable reaction with H_2O above 800°C (227, 312)

Forms hydrocarbons in H_2 above 2000°C; above 900°C with CH_4 catalyst (227, 312)

Stable in N_2 to 3000°C (227, 2, 312)

Considerable reaction with CO_2 above 800°C (227, 312)

MECHANICAL

Reviewed by Currie (227, 242)

STRUCTURE

Hexagonal arrays in sheets

C - C distance in sheet 1.42 A

distance between sheets 3.35 A (248)

Hexagonal unit cell

a = 2.456 A; c = 6.696 A (248)

CADMIUM

Formula weight: 112.41 g/mole

ELECTRICAL

Resistivity:

$T_c = 0.54-0.65\,^{\circ}K$ (557)

MAGNETIC

$H_c = 27-33.8$ Oersteds   (557)

STRUCTURE

Close-packed hexagonal   (557)

<u>COBALT</u>

Formula weight:	58.94 g/mole	
Formula volume:	6.62 cc/mole	
Melting point:	1490 $\pm$ 1°C	(161, 227, 298)
	1480°C	(3)
Boiling point:	2900°C	(3)
Vapor pressure:	1.11 x 10^{-6} mm of Hg at 1360°C	(227, 244)
Pycnometric density:	8.9 g/cc	(3)

CHEMICAL

Reactivity: Not attacked by H_2O or O_2 below 300°C   (227)

Soluble in acids

ELECTRICAL

Resistivity: 9.79 x 10^{-6} ohm-cm   (444)

STRUCTURE

Below 500°C, a complex hexagonal close packed structure   (248)

Above 500°C, cubic close packed   (248)

THERMAL

Expansion: (β form)   13.25 x 10^{-6} per °C; 25-500°C   (329)

18.45 x 10^{-6} per °C; 25-1000°C   (329)

CHROMIUM

Formula weight:	52.01 g/mole	
Formula volume:	7.33 cc/mole	
Melting point:	1890 $\pm$ 10°C	(161, 227, 298)
	1615°C	(3)
Boiling point:	2200°C	
Vapor pressure:	10^{-3} mm of Hg at 1090°C	(227)
Evaporation rate:	1.22 x 10^{-7} g/cm^2/sec at 905°C	(227)
Pycnometric density:	7.1 g/cc	(3)

CHEMICAL

Reactivity:	Soluble HCl, dil H_2SO_4	(3)
	Insoluble HNO_3	(3)
	Maximum use temperature, 1090°C in vacuum	(227)
	Oxidized at red heat in H_2O vapor	(227)
	Oxidized severely above 1400°C in O_2	(227)
	High vapor pressure restricts use to 1100°C or lower	(227)

ELECTRICAL

Resistivity:	18.9 x 10^{-6} ohm-cm	(444)

OPTICAL

Emissivity:	Total: 0.08 at 100° C	(227)
	0.26 at 1000°C	(227)

STRUCTURE

Cubic	(3)
Body centered cubic	(248)

THERMAL

Conductivity: 0.069 CGS at rt   (227, 308)

Expansion: 6.49 x 10^{-6} per ^{o}C at rt   (227, 308)

6.61 x 10^{-6} per ^{o}C; 100-640°C   (1, 227)

8.85 x 10^{-6} per ^{o}C; 25-500°C   (329)

11.0 x 10^{-6} per ^{o}C; 25-1000°C   (329)

13.7 x 10^{-6} per ^{o}C; 25-1500°C   (329)

12.3 x 10^{-6} per ^{o}C; 25-1800°C   (329)

COPPER

Formula weight: 63.57 g/mole

Formula volume: 7.13 cc/mole

Melting point: 1083°C   (3)

Boiling point: 2300°C   (3)

Pycnometric density: 8.92 g/cc   (3)

CHEMICAL

Reactivity: Soluble HNO_3, HCl, hot H_2SO_7

STRUCTURE

Cubic   (3)

Close packed cubic   (248)

THERMAL

Expansion: 18.5×10^{-6} per °C;   25-500°C   (329)

                 21.5×10^{-6} per °C;   25-1000°C   (329)

                 22.2×10^{-6} per °C;   25-1083°C   (329)

Formula weight: 162.46 g/mole   (3)

STRUCTURE

Close packed hexagonal   (248)

THERMAL

Expansion: 10.3×10^{-6} per $^{\circ}$C; 25-500°C   (329)

12.3×10^{-6} per $^{\circ}$C; 25-1000°C   (329)

Formula weight:	167.2 g/mole	(3)
Formula volume:	35.1 cc/mole	
Pycnometric density:	4.77 g/cc	(3)

STRUCTURE

Close packed hexagonal   (248)

THERMAL

Expansion: 10.5×10^{-6} per $^{\circ}C$;   25-500°C   (329)

12.8×10^{-6} per $^{\circ}C$;   25-1000°C   (329)

Formula weight: 55.85 g/mole

Formula volume: 7.10 cc/mole

Melting point: 1535 $\pm$ 5°C (161, 227, 298)

Boiling point: 3000°C (3)

Evaporation rate: 1.29 x 10^{-7} g/cm^2/sec at 1090°C (227)

Pycnometric density: 7.86 g/cc (3)

CHEMICAL

Reactivity: Soluble in acids

Insoluble in alkalis

ELECTRICAL

Resistivity: 9.06 x 10^{-6} ohm-cm (444)

STRUCTURE

RT - 980°C and 1401 - mp body centered cubic (248)

980 - 1401°C - close packed cubic (248)

THERMAL

Conductivity: 0.168 CGS at rt (227, 308)

Expansion:
	12.1 x 10^{-6} per °C at rt	(227, 308)
	12.2 x 10^{-6} per °C; 100-800°C	(227, 1)
(α form)	14.3 x 10^{-6} per °C; 25-500°C	(329)
	15.1 x 10^{-6} per °C; 25-900°C	(329)
(γ form)	10.95 x 10^{-6} per °C; 25-900°C	(329)
	12.0 x 10^{-6} per °C; 25-1000°C	(329)
	15.3 x 10^{-6} per °C; 25-1390°C	(329)
(δ form)	16.6 x 10^{-6} per °C; 25-1390°C	(329)
	17.2 x 10^{-6} per °C; 25-1500°C	(329)

ELECTRICAL

  Resistivity: T_c = 1.06 - 1.103^{o}K   (557)

MAGNETIC

  H_c = 47 - 50.3 Oersteds   (557)

STRUCTURE

  Tetragonal   (557)

Formula weight: 156.9 g/mole

STRUCTURE

Close packed hexagonal   (248)

THERMAL

Expansion: 6.74×10^{-6} per $^{\circ}C$;   25-500°C   (329)

9.95×10^{-6} per $^{\circ}C$;   25-1000°C   (329)

<u>HAFNIUM</u>

Formula weight:	178.6 g/mole	
Formula volume:	14.75 cc/mole	
Melting point:	2205°C	(38)
	2020 ± 15°C	(227)
	2230°C	(227)
Boiling point:	5400°C	(38)
Vapor pressure:	0.053 x 10^{-3} mm Hg at 2205°C	(38)
	1000 x 10^{-3} mm Hg at 3500°C	(38)
Pycnometric density:	13.09 g/cc at 20°C	(38)

CHEMICAL

Reactivity: H_2O, steam: Excellent corrosion resistance (38)

O_2, pyrophoric if finely powdered

H_2, embrittled at elevated temperatures

N_2, NH_3; forms nitrides if heated

ELECTRICAL

Resistivity: 43 x 10^{-6} ohm-cm at 0°C	(38)
Temperature coefficient: 0.0044 per °C at 20°C	(38)
Superconductive: T_c = 0.37°K	(557)

MECHANICAL

Strength: Tensile   : 85,000 psi (38)
          Yield   : 38,000 psi (38)
          Elongation : 23% (38)
          Impact   : Izod, V notched: 2.38 ft lb at rt (38)
                                       8.83 ft lb at 316°C (38)

Hardness: Rockwell B: 78 (38)

Young's modulus: 20 x 10^6 psi (38)

NUCLEAR

Thermal neutron capture cross section: 115 ± 15 barns/atom (38)
$$4.71 \text{ cm}^2/\text{cm}^3 \qquad (38)$$

Scattering cross section (average Maxwellian distribution):
$$-8 \pm 2 \text{ barns} \qquad (38)$$

STRUCTURE

(α): rt - 1750°C: close-packed hexagonal (38)
$a_O = 3.1883$ A; $c_O = 5.0422$ A (38)
6 neighbors at 3.131 A (248)
6 neighbors at 3.198 A (248)

(β): Above 1750°C: body-centered cubic (38)

THERMAL

Conductivity: 0.056 CGS (38)

Expansion: 6.1×10^{-6} per °C; 0-100°C (38)

$3.26-3.46 \times 10^{-6}$ per °C; 20-205°C (185, 227, 258)

$5.90-6.22 \times 10^{-6}$ per °C; 20-315°C (185, 227, 258)

5.93×10^{-6} per °C; 20-425°C (185, 227, 258)

$5.88-6.03 \times 10^{-6}$ per °C; 20-540°C (185, 227, 258)

5.73×10^{-6} per °C; 20-650°C (185, 227, 258)

$5.70-6.00 \times 10^{-6}$ per °C; 20-760°C (185, 227, 258)

$5.69-5.91 \times 10^{-6}$ per °C; 20-870°C (185, 227, 258)

5.68×10^{-6} per °C; 20-980°C (185, 227, 258)

5.90×10^{-6} per °C; 25-500°C (329)

5.95×10^{-6} per °C; 25-1000°C (329)

Heat capacity: 0.034 cal/gm/°C (38)

ELECTRICAL

    Resistivity: T_c = 4.15 - 4.173^{O}K   (557)

MAGNETIC

    H_c = 400 - 419 Oersteds   (557)

STRUCTURE

    Rhombic   (557)

ELECTRICAL

Resistivity: T_c = 3.37 - 3.40^{o}K   (557)

MAGNETIC

H_c = 269 - 278 Oersteds   (557)

STRUCTURE

Face-centered tetragonal   (557)

Formula weight: 193.1 g/mole

Formula volume: 8.61 cc/mole

Melting point: $2455 \pm 15^\circ C$ (227, 298)
                       $2350^\circ C$ (3)
                       $2450^\circ C$
Boiling point: $>4800^\circ C$ (3)

Vapor pressure: 10^{-3} mm of Hg at $2340^\circ C$ (227)

Evaporation rate: 1.70×10^{-7} g/cm^2/sec at $1990^\circ C$ (227)

Pycnometric density: 22.4 g/cc

CHEMICAL

Reactivity: Slightly soluble in aqua regia and aqueous Cl_2 (3)

               Superficial oxidation but oxide dissociates
               at $1150^\circ C$ (227)

               Dissolves H_2 (227)

               N_2 insoluble in Ir   (227)

               Maximum use temperature, $2340^\circ C$ in
               argon or vacuum   (227)
ELECTRICAL
   Resistivity: $T_c = 0.14^\circ K$ (557)
MECHANICAL

Reviewed by Hampel, Ref. 258   (227)

STRUCTURE

Cubic   (3)

Close packed cubic   (248)

Face-centered cubic   (557)

THERMAL

    Conductivity: 0.133 CGS at rt   (227, 308)

    Expansion: 6.8×10^{-6} per $^{\circ}C$; 23-100°C   (227, 258, 308)

                  7.2×10^{-6} per $^{\circ}C$ at 500°C   (227, 308)

                  7.8×10^{-6} per $^{\circ}C$ at 1000°C   (227, 308)

                  7.17×10^{-6} per $^{\circ}C$; 25-500°C   (329)

                  7.90×10^{-6} per $^{\circ}C$; 25-1000°C   (329)

                  8.62×10^{-6} per $^{\circ}C$; 25-1500°C   (329)

                  9.12×10^{-6} per $^{\circ}C$; 25-2000°C   (329)

ELECTRICAL

Resistivity: T$_c$ (αLa) 4.2 - 6.0^{o}K (557)

(βLa) 4.2 - 6.0^{o}K (557)

STRUCTURE

(αform) Close packed hexagonal (557)

(βform) Face-centered cubic (557)

MANGANESE

Formula weight: 54.93 g/mole

Formula volume: 7.63 cc/mole

Melting point: 1260°C   (3)

Boiling point: 1900°C   (3)

Pycnometric density: 7.2 g/cc   (3)

CHEMICAL

Reactivity: Decomposed by water   (3)

                 Soluble in dilute acids   (3)

STRUCTURE

Several polymorphic forms, all with complex structures -
See Wells   (248)

THERMAL

Expansion:

(α form)	35.8×10^{-6} per °C;	25-500°C	(329)
	36.0×10^{-6} per °C;	25-727°C	(329)
(β form)	54.7×10^{-6} per °C;	25-727°C	(329)
	50.7×10^{-6} per °C;	25-1000°C	(329)
	49.4×10^{-6} per °C;	25-1101°C	(329)
(γ form)	52.6×10^{-6} per °C;	25-1101°C	(329)
	52.2×10^{-6} per °C;	25-1137°C	(329)
(δ form)	55.4×10^{-6} per °C;	25-1137°C	(329)
	54.2×10^{-6} per °C;	25-1244°C	(329)

MOLYBDENUM

Formula weight:	95.95 g/mole	
Formula volume:	9.41 cc/mole	
Melting point:	$2610^{o}C$	(29)
	$2630 \pm 50^{o}C$	(161, 227, 298)
	$2620 \pm 10^{o}C$	(3)
Boiling point:	$4830^{o}C$	(29)
	$3700^{o}C$	(3)
Vapor pressure:	10^{-3}mm of Hg at $2295^{o}C$	(227)
Evaporation rate:	1.29×10^{-7} g/cm^2/sec at $1925^{o}C$	(227)
X-ray density:	10.2 g/cc at $20^{o}C$	(29)

CHEMICAL

Etchant:	Alkaline $K_3Fe(CN)_6$ solution	(29)
Reactivity:	Useful to $2200^{o}C$	(227, 274)
	Appreciably volatile in He at high temperature	(227, 284)
	Oxidizes in O_2 above $500^{o}C$	(227, 258)
	Oxide volatile above $650^{o}C$	(227)
	Reacts with water vapor at $700^{o}C$	(227)
	H_2 slightly soluble	(227)
	Useful to $1650-2200^{o}C$ in H_2	(227, 284)
	Probable failure in H_2 at $1400^{o}C$, embrittled	(227)
	N_2 slightly soluble, forms nitride above $1500^{o}C$	(227, 258)
	Oxidize in CO_2 at $1200^{o}C$; CO at $1400^{o}C$	(227)

CHEMICAL (cont.)

    Reactivity: (cont.)

        Maximum operating temperature $2295^{o}C$
        in vacuum or helium   (227)

        Soluble in hot, conc. H_2SO_4 (3)

        Insoluble in HCl, HF, NH_4OH, dil H_2SO_4 (3)

ELECTRICAL

    Resistivity: 5.7×10^{-6} ohm-cm at $20^{o}C$ (29)
                5.2×10^{-6} ohm-cm at $20^{o}C$ **(444)**
    Temperature coefficient: 0.0046 per ^{o}C (29)
    T_c = $1.0^{o}K$ (557)
    Magnetic susceptibility: 0.04×10^{-6} CGS   (29)

    Thermionic work function: 4.15 ev   (29)
        $A = 55$ amp/cm^2/$^{o}K^2$ (29)

MECHANICAL

    Reviewed by Hampel, Ref. 258   (227)

    Strength: Tensile: 120-200,000 psi at rt   (29)
          35- 65,000 psi at $500^{o}C$ (29)
          20- 30,000 psi at $1000^{o}C$ (29)

    Young's modulus: 46×10^6 psi at rt   (29)
        41×10^6 psi at $500^{o}C$ (29)
        39×10^6 psi at $1000^{o}C$ (29)

    Poisson's ratio: 0.321 at $20^{o}C$ (29)

NUCLEAR

    Cross section: 2.4 barns   (29)

OPTICAL

Spectral emissivity: 0.65μ :   0.37 at 1000°C   (29)
                                            0.42 at 0°C   (158)
                                            0.328 at 2610°C   (158)

Total emissivity: 0.071 at 100°C   (polished)   (227)
                               0.096 at 725°C   (filament)   (227)
                               0.202 at 1650°C   (filament)   (227)
                               0.096 at 540°C   (general)   (227)
                               0.189 at 950°C   (general)   (227)
                               0.230 at 1200°C   (general)   (227)
                               0.248 at 1315°C   (general)   (227)

STRUCTURE

Body - centered cubic   (557, 29)

d = 3.1468 A at 20°C   (29)

A hexagonal close packed form is known   (248)

THERMAL

Conductivity: 0.35 CGS at 20°C   (29)

                     0.35 CGS at rt   (227)

Expansion: 4.9×10^{-6} per °C   (29)

                5.35×10^{-6} per °C;   0-20°C   (227)

                5.21×10^{-6} per °C;   100-500°C   (227, 308)

                $5.8 - 6.2 \times 10^{-6}$ per °C;   30-700°C

                5.90×10^{-6} per °C;   25-500°C   (329)

                5.75×10^{-6} per °C;   25-1000°C   (329)

                6.51×10^{-6} per °C;   25-1500°C   (329)

THERMAL (cont.)

    Expansion: (cont.)

$$7.45 \times 10^{-6} \text{ per } {}^{\circ}C; \quad 25\text{-}2000^{\circ}C \qquad (329)$$

$$8.00 \times 10^{-6} \text{ per } {}^{\circ}C; \quad 25\text{-}2500^{\circ}C \qquad (329)$$

$$4.65 \times 10^{-6} + 0.9 \times 10^{-2} t^2 \text{ per } {}^{\circ}C;$$
$$20\text{-}1300^{\circ}C \qquad (227, \ 251)$$

$$6.8 \times 10^{-3} - 5 \times 10^{-6} t + 4.4 \times 10^{-9} t^2 \text{ per } {}^{\circ}C;$$
$$1300\text{-}2250^{\circ}C \qquad (227, \ 251)$$

  Specific heat: 0.061 cal/gm/${}^{\circ}$C at 20${}^{\circ}$C   (29)

NIOBIUM

Formula weight:	92.91 g/mole	
Formula volume:	10.83 cc/mole	
Melting point:	2415°C	(29)
	2415 $\pm$ 15°C	(161, 227, 298)
	2410°C	
Boiling point:	4930°C	(29)
Vapor pressure:	10^{-3} mm of Hg at 2540°C	(227)
Evaporation rate:	1.6 x 10^{-7} g/cm^2/sec at 2195°C	(227)
X-ray density:	8.57 g/cc at 20°C	(29)
Pycnometric density:	8.4 g/cc	(3)

CHEMICAL

Etchant:	HF - NH$_4$F solution	(29)
Reactivity:	Stable to mp in He	(227)
	Reacts with H$_2$O at 500°C	(227)
	Rapid oxidation in O$_2$ above 600°C	(227)
	H$_2$ very soluble in Nb, releases H$_2$ above 1200°C	(227)
	Slight reaction with N$_2$ to 1000°C	(227)
	Soluble in hot H$_2$SO$_4$	(3)
	Slightly soluble in HCl or HNO$_3$	(3)
	Maximum operating temperature, 2345°C in vacuum or helium	(227)

ELECTRICAL

Resistivity: 14.1×10^{-6} ohm-cm at 20°C   (29)

Temperature coefficient: 0.00395 per $^{\circ}$C   (29)
T_c $=$ $8.35 - 8.85^{\circ}$K   (557)

MECHANICAL

Reviews by Hampel, Miller, etc.   (227, 258, 284, 295)

Strength: Tensile: 75-150,000 psi at rt   (29)
35,000 psi at 500°C   (29)
13- 17,000 psi at 1000°C   (29)

Young's modulus: $12-15 \times 10^{6}$ psi at rt   (29)
$6.5 \quad \times 10^{6}$ psi at 500°C   (29)

Poisson's ratio: 0.38 at 20°C   (29)

NUCLEAR

Cross section: 1.1 barns   (29)

OPTICAL

Emissivity: Spectral 0.65μ : 0.49 below mp   (29)

STRUCTURE

Body - centered cubic   (557,(29)

d = 3.3004 A at 20°C   (29)

THERMAL

Conductivity: 0.13 CGS at 20°C   (29)

0.128 CGS at rt   (227)

THERMAL   (cont.)

Expansion:

7.2 x 10^{-6} per $^{\circ}C$		(29)
7.1 x 10^{-6} per $^{\circ}C$ at $100^{\circ}C$		(227)
7.38 x 10^{-6} per $^{\circ}C$ at $300^{\circ}C$		(227, 286)
7.55 x 10^{-6} per $^{\circ}C$ at $400^{\circ}C$		
7.63 x 10^{-6} per $^{\circ}C$ at $500^{\circ}C$		
7.85 x 10^{-6} per $^{\circ}C$ at $600^{\circ}C$		
8.03 x 10^{-6} per $^{\circ}C$ at $700^{\circ}C$		
8.19 x 10^{-6} per $^{\circ}C$ at $800^{\circ}C$		
8.37 x 10^{-6} per $^{\circ}C$ at $900^{\circ}C$		
8.53 x 10^{-6} per $^{\circ}C$ at $1000^{\circ}C$		

7.30 x 10^{-6} per $^{\circ}C$; $\quad$ 20-$300^{\circ}C$		(227, 286)
7.38 x 10^{-6} per $^{\circ}C$; $\quad$ 20-$400^{\circ}C$		
7.47 x 10^{-6} per $^{\circ}C$; $\quad$ 20-$500^{\circ}C$		
7.56 x 10^{-6} per $^{\circ}C$; $\quad$ 20-$600^{\circ}C$		
7.64 x 10^{-6} per $^{\circ}C$; $\quad$ 20-$700^{\circ}C$		
7.73 x 10^{-6} per $^{\circ}C$; $\quad$ 20-$800^{\circ}C$		
7.81 x 10^{-6} per $^{\circ}C$; $\quad$ 20-$900^{\circ}C$		
7.90 x 10^{-6} per $^{\circ}C$; $\quad$ 20-$1000^{\circ}C$		

8.00 x 10^{-6} per $^{\circ}C$; $\quad$ 25-$500^{\circ}C$	(329)
8.31 x 10^{-6} per $^{\circ}C$; $\quad$ 25-$1000^{\circ}C$	(329)
8.68 x 10^{-6} per $^{\circ}C$; $\quad$ 25-$1500^{\circ}C$	(329)
9.03 x 10^{-6} per $^{\circ}C$; $\quad$ 25-$2000^{\circ}C$	(329)

Specific heat: 0.065 cal/gm/$^{\circ}C$ at $20^{\circ}C$ (29)

Formula weight: 144.27 g/mole

Formula volume: 20.9   cc/mole

Melting point: 840°C   (3)

Pycnometric density: 6.9 g/cc   (3)

CHEMICAL

Reactivity: Decomposed by water   (3)

STRUCTURE

"Double hexagonal close packed"   (248)

THERMAL

Expansion: 7.58 x 10^{-6} per $^{\circ}$C;   25-500°C   (329)

                9.03 x 10^{-6} per $^{\circ}$C;   25-800°C   (329)

Formula weight: 58.69 g/mole

Formula volume: 6.4 cc/mole

Melting point: 1455 $\pm$ 1°C   (227, 284, 298)
                       1452°C   (3)

Boiling point: 2900°C

Evaporation rate: 1.18 x 10^{-7} g/cm^2/sec at 1155°C   (227)

Pycnometric density: 8.90 g/cc   (3)

CHEMICAL

Reactivity: Rapidly oxidized in O$_2$ above 800°C   (227)

                   Soluble dilute HNO$_3$ (3)

                   Slightly soluble in H$_2$SO$_4$, HCl   (3)

                   Insoluble NH$_4$OH   (3)

STRUCTURE

Hexagonal close packed and cubic close packed structures
known   (248)

THERMAL

Conductivity: 0.195 CGS at rt   (227, 308)

Expansion: 13.3 x 10^{-6} per $^{\circ}$C at rt   (227, 308)

                   13.9 x 10^{-6} per $^{\circ}$C; 200°C   (227, 308)

                   14.8 x 10^{-6} per $^{\circ}$C; 400°C   (227, 308)

                   16.2 x 10^{-6} per $^{\circ}$C; 900°C   (227, 308)

                   15.75 x 10^{-6} per $^{\circ}$C; 25-500°C   (329)

                   17.10 x 10^{-6} per $^{\circ}$C; 25-1000°C   (329)

<u>OSMIUM</u>

Formula weight: 190.2 g/mole

Formula volume: 8.48 cc/mole

Melting point: 2700 $\pm$ 200°C   (227, 298)

                          2700°C   (3)

Boiling point: >5300°C   (3)

Vapor pressure: 10^{-3} mm of Hg at 2450°C   (227)

Evaporation rate: 1.65 x 10^{-7} g/cm^2/sec at 2100°C   (227)

Pycnometric density: 22.48 g/cc   (3)

CHEMICAL

Reactivity: Slightly soluble in aqua regia, HNO_3 (3)

                        Insoluble NH_4OH (3)

                        Maximum use temperature, 2450°C in
vacuum or helium   (227)

Strong tendency to form volatile OsO_4 (227)

ELECTRICAL

T_c = 0.71°K   (557)

STRUCTURE

Hexagonal   (3)

Hexagonal close packed   (557, 248)

THERMAL

Expansion: 4.57 x 10^{-6} per °C at 50°C   (227, 258)

                   5.47 x 10^{-6} per °C; 25-500°C   (329)

                   5.57 x 10^{-6} per °C; 25-600°C   (329)

<u>ELECTRICAL</u>

Resistivity: T_c = 7.2°K   (557)

<u>MAGNETIC</u>

H_c = 800 Oersteds   (557)

<u>STRUCTURE</u>

Face-centered cubic   (557)

PALLADIUM

Formula weight: 106.70 g/mole

Formula volume: 8.9 cc/mole

Melting point: 1555°C   (227, 284)

Vapor pressure: 1.53 x 10^{-6} mm at Hg at
                       1320°C   (227, 244)

Pycnometric density: 12.0 g/cc at 20°C   (3)
                       11.0 g/cc at 1550°C   (3)

CHEMICAL

Reactivity: Superficially oxidized at 700°C, oxide
               dissociates at 875°C   (227)

               Dissolves large amounts of H_2

               Soluble in aqua regia, hot H_2SO_4 (3)

               Insoluble in NH_4OH (3)

MECHANICAL

Reviewed by Hampel, Ref. 258   (227)

STRUCTURE

Cubic   (3)

Close packed cubic   (248)

THERMAL

Conductivity: 0.174 CGS at rt   (227, 308)

THERMAL (cont.)

Expansion: 11.7×10^{-6} per $^{\circ}$C at rt (227, 258)

11.1×10^{-6} per $^{\circ}$C at rt (227, 308)

12.4×10^{-6} per $^{\circ}$C; 500°C (227, 308)

13.6×10^{-6} per $^{\circ}$C; 1000°C (227, 308)

12.82×10^{-6} per $^{\circ}$C; 25-500°C (329)

13.95×10^{-6} per $^{\circ}$C; 25-1000°C (329)

PLATINUM

Formula weight: 195.23 g/mole

Formula volume: 9.11 cc/mole

Melting point: 1780 $\pm$ 5°C   (227, 284, 295)
                        1755°C   (3)

Boiling point: 4300°C   (3)

Vapor pressure: 1.81 x 10^{-6} mm Hg at 1745°C   (227, 244)
                               10^{-3} mm Hg at 1905°C   (227)

Pycnometric density: 21.45 g/cc at 20°C   (3)
                (liq)   19.00 g/cc at 1755°C   (3)

CHEMICAL

Reactivity: Soluble in aqua regia, fused alkali   (3)

                Stable in He to mp   (227)

                Maximum use 1705°C in He or vacuum   (227)

                Useful to 1400°C in air   (227)

                O_2 insoluble in Pt   (227)

MECHANICAL

Reviewed by Hampel, Ref. 258   (227)

STRUCTURE

Cubic   (3)

Close packed cubic   (248)

THERMAL

Conductivity: 0.168 CGS at rt   (227, 308)

Expansion: 8.9×10^{-6} per $^{\circ}$C at rt   (227, 258)

9.1×10^{-6} per $^{\circ}$C at rt   (227, 308)

9.6×10^{-6} per $^{\circ}$C; 500°C   (227, 308)

10.1×10^{-6} per $^{\circ}$C; 1000°C   (227, 308)

9.70×10^{-6} per $^{\circ}$C; 25-500°C   (329)

10.45×10^{-6} per $^{\circ}$C; 25-1000°C   (329)

11.31×10^{-6} per $^{\circ}$C; 25-1500°C   (329)

11.69×10^{-6} per $^{\circ}$C; 25-1770°C   (329)

Formula weight: 186.31 g/mole

Melting point: 3130 ± 60^oC (227, 298)
                     3440^oC (3)
                     3170^oC

Vapor pressure: 10^{-3} mm Hg at 2775^oC (227)

Evaporation rate: 1.55×10^{-7} g/cm^2/sec at 2380^oC (227)

CHEMICAL

Reactivity: Insoluble in HF, HCl   (3)

                  Soluble HNO_3, H_2SO_4 (3)

                  No reaction with N_2 (227)

                  Metal properties improved by H_2 at
                  1700-2200^oC (227)

                  Maximum use temperature 2775^oC in
                  He or vacuum   (227)

                  Rapidly oxidized in O_2 at 350^oC (227, 258)

                  Forms volatile oxide at 600^oC (227, 302)

                  Water vapor promotes oxidation   (227)

ELECTRICAL

      T_c = 0.9 - 2.42^oK (557)

MECHANICAL

Reviewed by Hampel   (227, 258)

STRUCTURE

Hexagonal   (3)

Hexagonal close packed   (557, 248)

THERMAL

Conductivity:	0.168 CGS at rt	(227, 308)
Expansion:	12.4×10^{-6} per $°C$ at rt, $\parallel$ C axis	(227)
	4.68×10^{-6} per $°C$ at rt, $\perp$ C axis	(227)
	6.74×10^{-6} per $°C$; 25-500°C	(329)
	6.88×10^{-6} per $°C$; 25-1000°C	(329)
	7.05×10^{-6} per $°C$; 25-1500°C	(329)
	7.29×10^{-6} per $°C$; 25-2000°C	(329)

RHODIUM

Formula weight: 102.91 g/mole

Formula volume: 8.23 cc/mole

Melting point: 1965 $\pm$ 4°C   (227, 282, 298)

                          1955°C   (3)

                          1970°C

Boiling point: >2500°C   (3)

Vapor pressure: 10^{-3} mm Hg at 1960°C   (227)

Evaporation rate: 1.34 x 10^{-7} g/cm^2/sec at 1680°C   (227)

Pycnometric density: 12.5 g/cc   (3)

CHEMICAL

Reactivity: Oxidizes superficially at red heat to Rh_2O_3
                  which decomposes at 1100°C   (227)

                  Stable in H_2 and N_2 (227)

                  Slightly soluble in aqua regia, acids   (3)

MECHANICAL

Reviewed by Hampel   (227, 258)

STRUCTURE

Cubic   (3)

Cubic close packed   (248)

THERMAL

Conductivity: 0.190 CGS at rt   (227, 308)

THERMAL (cont.)

Expansion: $8.5 \ \times 10^{-6}$ per $^{\circ}$C at rt   (227, 308)

8.29×10^{-6} per $^{\circ}$C; 23-100°C   (227, 258)

9.80×10^{-6} per $^{\circ}$C at 500°C   (227, 308)

$10.8 \ \times 10^{-6}$ per $^{\circ}$C at 1000°C   (227, 308)

9.48×10^{-6} per $^{\circ}$C; 25-500°C   (329)

10.75×10^{-6} per $^{\circ}$C; 25-1000°C   (329)

12.15×10^{-6} per $^{\circ}$C; 25-1500°C   (329)

RUTHENIUM

Formula weight: 101.70 g/mole

Formula volume: 8.34 cc/mole

Melting point: $2500 \pm 100^{\circ}C$ (227, 298)
                         $2450^{\circ}C$ (3)
                         $2400^{\circ}C$

Boiling point: $>2700^{\circ}C$ (3)

Vapor pressure: 10^{-3} mm Hg at $2230^{\circ}C$ (227)

Evaporation rate: 1.26×10^{-7} g/cm^2/sec at $1915^{\circ}C$ (227)

Pycnometric density: 12.2 g/cc   (3)

CHEMICAL

Reactivity: Maximum use temperature, $2230^{\circ}C$ in
                     vacuum   (227)

                   Ignites in air to stable RuO_2 (227)

ELECTRICAL
       T_c = $0.47^{\circ}K$ (557)

STRUCTURE

Hexagonal   (3)

Hexagonal close packed   (557, 248)

THERMAL

Expansion: 9.1×10^{-6} per $^{\circ}C$ at $23^{\circ}C$ (227)

              7.79×10^{-6} per $^{\circ}C$; $25-500^{\circ}C$ (329)

              7.99×10^{-6} per $^{\circ}C$; $25-600^{\circ}C$ (329)

SILICON

Formula weight: 28.06 g/mole

Melting point: 1430 $\pm$ 20°C   (6, 161, 227, 298)
                     1420°C   (3)

Boiling point: 2600°C   (3)

Vapor pressure: 7.99 x 10^{-6} mm Hg at 1170°C   (227, 244)

X-ray density: 2.33 g/cc   (6)

Pycnometric density: 2.4 g/cc   (3)

CHEMICAL

Reactivity: Oxidized at 1100°C in O_2 (227)

                Burns with difficulty in air   (227)

                Soluble in HNO_3 + HF, silver   (3)

                Insoluble HF   (3)

                Slightly soluble in Pb, Zn   (3)

MECHANICAL

Hardness: Vickers 10 g: 715-960 kg/mm^2 (8)

Young's modulus: 16 x 10^6 psi at 20°C   (6)

Review of mechanical properties by Hampel   (227, 258)

NUCLEAR

Cross section: 0.00633 barns   (6)

OPTICAL

Refractive index   3.736   (3)

- 236 - Si

STRUCTURE

Cubic (3)

Diamond structure, Si-Si = 2.34 A (248)

THERMAL

Conductivity: 0.20 CGS at 20°C (6)
0.194 CGS at rt (185, 227)

Expansion: 3.7×10^{-6} per °C; 0-1000°C (7)

3.99×10^{-6} per °C; 0-1400°C (7)

4.68×10^{-6} per °C; 15-1000°C (185, 227, 258)

5.14×10^{-6} per °C

5×10^{-6} per °C at 20°C (6)

3.54×10^{-6} per °C; 25-500°C (329)

3.93×10^{-6} per °C; 25-1000°C (329)

Heat content: 0.162 cal/gm/°C at 20°C (6)

ELECTRICAL

Resistivity: T_c = 3.702 - 3.752°K   (557)

MAGNETIC

H_c = 304 - 310 Oersteds   (557)

STRUCTURE

Tetragonal   (557)

Formula weight:	180.95 g/mole	(29)
	180.88 g/mole	(3)
Formula volume:	10.9 cc/mole	(29)
	10.6 cc/mole	(436)
Melting point:	2996°C	(29)
	2850°C	(39)
	3000 ± 50°C	(161, 227, 298)
Boiling point:	5430°C	(29)
Vapor pressure:	10^{-3} mm Hg at 2830°C	(227)
Evaporation rate:	1.55×10^{-7} g/cm^2/sec at 2400°C	(227)
X-ray density:	16.6 g/cc at 20°C	(29)

CHEMICAL

Etchant:	HF-NH$_4$F solution	(29)
Reactivity:	Maximum use temperature, 2830°C in He or vacuum	(287)
	Stable to 2850 in pure He	(227, 274)
	Oxidation resistance poor likely to react with O$_2$ and H$_2$O at 500°C	(227)
	H$_2$ very soluble in Ta, liberates H$_2$ above 1200°C	(227)
	Extensive formation of Ta$_2$N in N$_2$ at 2100°C	(227)
	Soluble in fused alkali, HF	(3)
	Insoluble in HCl, HNO$_3$, H$_2$SO$_4$	(3)

ELECTRICAL

| Resistivity: | 13.5×10^{-6} ohm cm at 20°C | (29) |
| T_c = | 4.05 - 4.6°K | (557) |

ELECTRICAL (cont.)

Temperature coefficient: 0.0038 per ^{o}C (29)

Magnetic susceptibility: 0.93 (29)

Thermionic work function: 4.12 ev (29)
A 60 amp/cm^2/^{o}K^2 (29)

MECHANICAL

Reviewed by Hampel (227, 258)

Strength: Tensile: 150-200,000 psi at rt (29)
35- 45,000 psi at 500°C (29)
15- 20,000 psi at 1000°C (29)

Young's modulus: 27 x 10^6 psi at rt (29)
25 x 10^6 psi at 500°C (29)
22 x 10^6 psi at 1000°C (29)

Poisson's ratio: 0.35 (29)

NUCLEAR

Cross section: 21.3 barns (29)

OPTICAL

Emissivity: Spectral 0.65μ : 0.46 at 900°C (29)
0.55 at 27°C to
0.36 at 3027°C (158)

Total: 0.19 at 1325°C to
0.31 at 3000°C (227)

STRUCTURE

Body-centered cubic (29)
d = 3.3026 A at 20°C (29)

THERMAL

Conductivity: 0.130 CGS at 20°C   (29)
0.084 CGS at rt   (227, 251)
0.133 CGS at rt   (227, 258)
0.0442 $\pm$ .025 CGS at 100°C   (227, 251)

Expansion: 6.5×10^{-6} per °C at 20°C   (29)
5.93×10^{-6} per °C at rt   (227, 251)
5.51×10^{-6} per °C at 220°C   (227, 251)
6.41×10^{-6} per °C at 615°C   (227, 251)
6.30×10^{-6} per °C at 825°C   (227, 251)

6.53×10^{-6} per °C; 25-500°C   (329)
7.09×10^{-6} per °C; 25-1000°C   (329)
7.46×10^{-6} per °C; 25-1500°C   (329)
7.75×10^{-6} per °C; 25-2000°C   (329)
8.00×10^{-6} per °C; 25-2500°C   (329)

Specific heat: 0.036 cal/g/°C at 20°C   (29)

Diffusivity: 0.207 at 1300°C   (146)
0.212 at 1400°C   (146)
0.225 at 1500°C   (146)
0.238 at 1600°C   (146)
0.230 at 1700°C   (146)

Formula weight: 159.20 g/mole   (3)

STRUCTURE

Hexagonal close packed   (248)

THERMAL

Expansion: 9.48×10^{-6} per $^{\circ}$C; 25-500°C   (329)

11.7×10^{-6} per $^{\circ}$C; 25-1000°C   (329)

Melting point: 2700°C

ELECTRICAL

Resistivity: $T_c = 11.2°K$ (557)

MAGNETIC

$H_c = 300 - 400$ Oersteds   (557)

STRUCTURE

Close-packed hexagonal   (557)

Formula weight:	232.12 g/mole	
Formula volume:	20.6 cc/mole	
Melting point:	1845°C	(3)
Boiling point:	> 3000°C	(3)
Pycnometric density:	11.2 g/cc	(3)

CHEMICAL

Reactivity:	Soluble in HCl, H_2SO_4	(3)
	Slightly soluble in HNO_3	(3)
	Insoluble HF, alkali	(3)

ELECTRICAL

T_C = 1.37^{o}K (557)

STRUCTURE

Cubic close packed (248)
Face centered cubic (557)

THERMAL

Expansion:	(α form) 12.6 x 10^{-6} per ^{o}C;	25-500°C	(329)
	12.5 x 10^{-6} per ^{o}C;	25-1000°C	(329)
	12.5 x 10^{-6} per ^{o}C;	25-1400°C	(329)
	(β form) 12.3 x 10^{-6} per ^{o}C;	25-1400°C	(329)

<u>TITANIUM</u>

Formula weight: 47.90 g/mole

Formula volume: 10.65 cc/mole
                       11.5 cc/mole   (436)
Melting point: 1725 $\pm$ 10°C   (75)
                       1820 $\pm$ 100°C   (227, 298)
                       1660 °C   (161, 227)
                       1690°C   (227, 310)
                       1800°C   (3)

Boiling point: >3000°C   (3)

Vapor pressure: 10^{-3} mm Hg at 1380°C   (227)

Evaporation rate: 1.08 x 10^{-7} g/cm^2/sec at 1132°C   (227)

Pycnometric density: 4.50 g/cc   (3)

CHEMICAL

Reactivity: Should be useful to 1400°C in He   (227)

               Dissolves O_2, embrittles at 550°C (227, 274)

               Reacts with H_2 and N_2 (227)

               Decomposed by hot water   (3)

               Soluble in acids   (3)

ELECTRICAL

Resistivity: 42 x 10^{-6} ohm-cm at 20°C   (224)
               55 x 10^{-6} ohm-cm at 20°C   (444)
Temperature coefficient: 0.00546 per °C   (224)
T_c =   0.387 - 0.56°K   (557)

MECHANICAL

Reviewed by Hampel   (227, 258)

(Cont.)

OPTICAL

Emissivity: Spectral 0.65μ : 0.419 to 0.533 (226)

STRUCTURE

Hexagonal close packed (222)

c = 4.729 A ; a = 2.953 A (222)

Body-centered cubic above 880°C (223)

a = 3.32 A (223)

THERMAL

Conductivity: 0.037 CGS at rt (185, 227, 255)

Expansion: 8.2 x 10^{-6} per °C; 0-300°C (224)
8.55 x 10^{-6} per °C at 100°C (185, 227)
8.82 x 10^{-6} per °C; 100-800°C (185, 227, 308)
(α form) 10.5 x 10^{-6} per °C; 25-500°C (329)
10.6 x 10^{-6} per °C; 25-882°C (329)

Specific heat: 0.1291 cal/g/°C; 0-219°C (225)
0.1316 cal/g/°C; 0-266°C (225)
0.1334 cal/g/°C; 0-320°C (225)
0.1386 cal/g/°C; 0-500°C (225)

Formula weight:	238.07 g/mole	
Formula volume:	12.88 cc/mole	(3)
Melting point:	1150°C	(3)
	1133 $\pm$ 1	(447, 448)
Boiling point:	3500°C	(3)
X-ray density:	Orthorhombic 19.12 g/cc	(447, 448)
	Tetragonal 18.11 g/cc	(447, 448)
Pycnometric density:	18.485 g/cc	(3)
	Body centered cubic 18.06	(447, 448)

CHEMICAL

Reactivity:	Soluble in acids	(3)
	Insoluble in alkalis	(3)

ELECTRICAL

Resistivity:	29×10^{-6} ohm-cm	(447, 448)
T_c =	$0.8 - 1.3°K$	(557)

STRUCTURE

Body centered cubic (γ form)   (248)

3 polymorphs known, α to 660°C; β to 760°C; γ above 760°C   (248)

α is orthorhombic

β is tetragonal

THERMAL

Conductivity:	0.065 CGS at 100°C	(447, 448)
	0.081 CGS at 400°C	(447, 448)
	0.104 CGS at 700°C	(447, 448)

THERMAL (cont.)

Expansion: (α form) 17.9×10^{-6} per °C;   25 - 500°C   (329)

                      20.0×10^{-6} per °C;   25 - 662°C   (329)

            (β form) 24.7×10^{-6} per °C;   25 - 662°C   (329)

                      23.8×10^{-6} per °C;   25 - 770°C   (329)

            (γ form) 22.8×10^{-6} per °C;   25 - 770°C   (329)

                      23.1×10^{-6} per °C;   25 - 1000°C   (329)

        gamma

            a direction 35.1×10^{-6} per °C;   25 - 625°C (447, 448)

            b direction 8.51×10^{-6} per °C;   25 - 625°C (447, 448)

            c direction 32.6×10^{-6} per °C;   25 - 625°C (447, 448)

Volume expansion: 59.3×10^{-6} per °C;   25 - 625°C   (447, 448)

V

VANADIUM

Formula weight:	50.95 g/mole	
Formula volume:	8.52 cc/mole	
Melting point:	1735 $\pm$ 50°C	(227, 298, 310)
	1710°C	(3)
Boiling point:	3000°C	(3)
Vapor pressure:	10^{-3} mm Hg at 1730°C	(227)
Evaporation rate:	1.0 x 10^{-7} g/cm^2/sec at 1430°C	(227)
Pycnometric density:	5.96 g/cc	(3)

CHEMICAL

Reactivity:	Soluble HNO_3, H_2SO_4	(3)
	Insoluble in aqueous alkalis	(3)
	Maximum use temperature, 1620°C in He or vacuum	(227)
	Embrittled by small amounts of O_2, forms V_2O_5 above 400°C	(227)
	V_2O_5 causes destructive oxidation of other metals	(227)

ELECTRICAL

Resistivity:	26 x 10^{-6} ohm-cm	(444)

STRUCTURE

Cubic	(3)
Body centered cubic	(248)

THERMAL

Conductivity:	0.0735 CGS at 100°C	(227, 258)
	0.128 CGS at 500°C	(227, 258)
Expansion:	8.3 x 10^{-6} per °C; 23- 100°C	(227, 258)
	9.61 x 10^{-6} per °C; 23- 500°C	(227, 258)
	10.41 x 10^{-6} per °C; 23- 900°C	(227, 258)
	10.87 x 10^{-6} per °C; 23-1100°C	(227, 258)
	9.69 x 10^{-6} per °C; 25- 500°C	(329)
	10.7 x 10^{-6} per °C; 25-1000°C	(329)

Formula weight:	183.86 g/mole	
Formula volume:	9.53 cc/mole	
Melting point:	$3390^{\circ}C$	
	$3340^{\circ}C$	(39)
	$3405^{\circ}C$	(403)
	$3410^{\circ}C$	(29)
	$3410 \pm 15^{\circ}C$	(227, 298)
	$3370^{\circ}C$	(3)
	$3400^{\circ}C$	
Boiling point:	$5530^{\circ}C$	(29)
	$5900^{\circ}C$	(3)
Vapor pressure:	10^{-3} mm Hg at $3020^{\circ}C$	(227)
Evaporation rate:	1.47×10^{-7} g/cm^2/sec at $2550^{\circ}C$	(227)
	10^{-6} g/cm^2/sec at $3040^{\circ}C$	(227, 2)
X-ray density:	19.31 g/cc	
	19.3 g/cc	(29)
Pycnometric density:	19.3 g/cc	(3)

CHEMICAL

Etchant:	Hot 6% H_2O_2	(29)
Reactivity:	Maximum use temperature; $3040^{\circ}C$ in neutral atmosphere	(227)
	No failure in pure He at $2850^{\circ}C$	(227, 274)
	Oxidizes easily, WO_3 volatile above $800^{\circ}C$	(227)
	Oxidation rate in O_2 increases with pressure	(227)
	Oxidation rate linear from $600-850^{\circ}C$	(227, 232)
	Oxidizes in steam above $700^{\circ}C$	(227)
	Soluble in hot concentrated KOH	(3)
	Slightly soluble NH_4OH, HNO_3, aqua regia	(3)

(Cont.)

CHEMICAL (cont.)

    Reactivity: (cont.)

        Useful in H_2 to 2500^oC, higher for short times (227)

        Forms nitrides with N_2 above 1500^oC (227)

        No reaction with NH_3 at 1400^oC (227)

        Reacts with CO_2 above 1200^oC (227)

        Reacts with CO above 1400^oC (227)

ELECTRICAL

    Resistivity: 5.5×10^{-6} ohm-cm at 20^oC (29)

    Temperature coefficient: 0.0046 per ^oC (29)

    Magnetic susceptibility: 0.28×10^{-6} (29)

    Thermionic work function: 4.54 ev   (29)
        A   :   60 amps$/cm^2/^oK^2$ (29)

MECHANICAL

    Reviewed by Hampel, Ref. 258   (227)

    Strength: Tensile: 150,000 psi at rt
            100-500,000 psi at rt   (29)
            175-200,000 psi at 500^oC (29)
            50- 75,000 psi at 1000^oC (29)

    Young's modulus: 59×10^6 psi at rt   (29)
        55×10^6 psi at 500^oC (29)
        50×10^6 psi at 1000^oC (29)
        58.59×10^6 psi at rt   (185, 205)
        58.14×10^6 psi at 100^oC (185, 205)
        57.55×10^6 psi at 200^oC (185, 205)
        56.31×10^6 psi at 400^oC (185, 205)
        54.96×10^6 psi at 600^oC
        53.12×10^6 psi at 800^oC
        51.76×10^6 psi at 1000^oC
        50.39×10^6 psi at 1200^oC

MECHANICAL (cont.)

Shear modulus:
22.92×10^6 psi at rt   (185, 205)
22.65×10^6 psi at 100°C   (185, 205)
22.40×10^6 psi at 200°C   (185, 205)
21.87×10^6 psi at 400°C   (185, 205)
21.36×10^6 psi at 600°C   (185, 205)
20.49×10^6 psi at 800°C   (185, 205)
19.94×10^6 psi at 1000°C   (185, 205)
19.38×10^6 psi at 1200°C   (185, 205)

Compressibility:
22.31×10^{-9} in^2/lb at rt   (185, 205)
22.36×10^{-9} in^2/lb at 100°C   (185, 205)
22.45×10^{-9} in^2/lb at 200°C   (185, 205)
22.63×10^{-9} in^2/lb at 400°C   (185, 205)
22.93×10^{-9} in^2/lb at 600°C   (185, 205)
23.04×10^{-9} in^2/lb at 800°C   (185, 205)
23.42×10^{-9} in^2/lb at 1000°C   (185, 205)
23.82×10^{-9} in^2/lb at 1200°C   (185, 205)

Poisson's ratio:
0.284   at rt   (29)
0.28201 at rt   (205)
0.28314 at 100°C   (205)
0.28451 at 200°C   (205)
0.28745 at 400°C   (205)
0.28938 at 600°C   (205)
0.29592 at 800°C   (205)
0.29779 at 1000°C   (205)
0.29978 at 1200°C   (205)

Elastic constants of single crystals:

$C_{11} = 5.2239 \pm .0017$ dynes/cm^2 at 27°C   (192)
$C_{12} = 2.0583 \pm .0008$ dynes/cm^2 at 27°C   (192)
$C_{44} = 1.6011 \pm .0002$ dynes/cm^2 at 27°C   (192)

$C_{11} = 5.22$ dynes/cm^2 at 27°C   (193)
$C_{12} = 2.06$ dynes/cm^2 at 27°C   (193)
$C_{44} = 1.604$ dynes/cm^2 at 27°C   (193)

NUCLEAR

Cross section: 19.2 barns   (29)

OPTICAL

Emissivity: Spectral 0.65μ : 0.45 at 900°C   (29)
       0.47 at 27°C to 0.44 at
          3282°C (158)

       0.66μ : 0.441 at 1325°C (148)
          0.424 at 2125°C (148)

      Total: 0.66 at 100°C; polished (227)
         0.32 at 100°C to 0.35-0.39 at
          3320°C; filament (227)
         0.174 at 750°C; commercial (227)
         0.279 at 1200°C; commercial (227)
         0.311 at 1425°C; commercial (227)

STRUCTURE

Body - centered cubic
a = 3.1585 A at 20°C (29)
   3.1677 A, vacuum annealed at 1400°C (205)
   3.1649 A, H_2 annealed at 1400°C (205)
β form stable at low temperatures is known (248)

THERMAL

Conductivity: 0.40 CGS at 20°C (29)
       0.307 CGS at rt (227, 258)

Expansion: 4.58×10^{-6} per °C; 0- 750°C (206)
      4.75×10^{-6} per °C; 0-1000°C
      4.90×10^{-6} per °C; 0-1275°C
      5.00×10^{-6} per °C; 0-1525°C
      5.18×10^{-6} per °C; 0-1775°C

      5.05×10^{-6} per °C; to 800°C (205)
      5.28×10^{-6} per °C; to 1000°C
      5.50×10^{-6} per °C; to 1200°C
      5.73×10^{-6} per °C; to 1400°C
      6.00×10^{-6} per °C; to 1600°C
      6.27×10^{-6} per °C; to 1800°C
      5.14×10^{-6} per °C; to 1400°C (7)

THERMAL (cont.)

Expansion: (cont.)

4.97×10^{-6} per $^{\circ}$C; to 1200°C

4.3×10^{-6} per $^{\circ}$C; at 20°C (29)

4.5×10^{-6} per $^{\circ}$C; at rt (227, 308)

4.5×10^{-6} per $^{\circ}$C; at 100°C (227, 258)
(annealed at 3000°C)

4.93×10^{-6} per $^{\circ}$C; at 100°C (227, 258)
(worked at 3000°C)

4.52×10^{-6} per $^{\circ}$C; 25- 500°C (329)

4.77×10^{-6} per $^{\circ}$C; 25-1000°C (329)

5.08×10^{-6} per $^{\circ}$C; 25-1500°C (329)

5.47×10^{-6} per $^{\circ}$C; 25-2000°C (329)

5.86×10^{-6} per $^{\circ}$C; 25-2500°C (329)

Specific heat: 0.032 cal/g/$^{\circ}$C at 20°C (29)

Diffusivity: 0.78 cm^2/sec

Formula weight: 173.04 g/mole

Melting point: 1500°C   (161, 227, 258, 298)
                       1450°C

Vapor pressure: 1.36 x 10^{-6} mm Hg at 1418°C   (227, 244)

CHEMICAL

Reactivity: Pyrophoric in air   (227)

               Forms hydrated oxides with H_2O (227)

STRUCTURE

Hexagonal close packed   (248)

ELECTRICAL

    Resistivity: $T_c = 0.84 - 0.96^{o}K$ (557)

MAGNETIC

    $H_c = 42 - 52.5$ Oersteds   (557)

STRUCTURE

    Close-packed hexagonal   (557)

ZIRCONIUM

Formula weight:	91.22 g/mole	
Formula volume:	14.25 cc/mole	
Melting point:	1845°C	(38)
	1830 ± 45°C	(227, 298, 310)
	1700°C	(3)
Boiling point:	2900°C	(3)
Vapor pressure:	10^{-3} mm of Hg at 1815°C	(227)
Evaporation rate:	1.3 x 10^{-7} g/cm^2/sec at 1525°C	(227)
Pycnometric density:	6.38 g/cc	(38)
	6.4 g/cc	(3)

CHEMICAL

Reactivity:*   HF, soluble

HCl + HNO$_3$, soluble

O$_2$, pyrophoric if fine

O$_2$, N$_2$, H$_2$, H$_2$O, etc., reacts if heated 200-800°C

Acids, generally corrosion resistant

*"Materials for Advanced Technology," The Carborundum Co. Lists corrosion rates in common industrial chemicals.

ELECTRICAL

Resistivity:	41-52 x 10^{-6} ohm-cm at 25°C	(38)
	41 x 10^{-6} ohm-cm at 25°C	(444)
Superconductivity:	T_c = 0.546-0.70°K (∝Zr)	(557)

MECHANICAL

Strength: Tensile   : 62,000 psi at rt   (38)
    Yield   : 50,000 psi at rt   (38)
    Elongation : 28%   (38)

Hardness: Rockwell B: 88-92   (38)

Young's modulus: 13.7 x 10^6 psi at rt   (38)

Mechanical properties reviewed by Hampel   (227, 258)

NUCLEAR

Thermal neutron obsorption iron section: 0.180 barns/atom   (38)

STRUCTURE

(α): rt to 860°C: close-packed hexagonal   (38)
  6 neighbors at 1.589 A   (248)
  6 neighbors at 3.223 A   (248)

(β): above 860°C: body-centered cubic   (38)

THERMAL

Conductivity: 0.0393 CGS; 50-100°C   (38)

     0.0348-0.0397 at 125°C   (185, 227, 258)

Expansion: 5.4 x 10^{-6} per °C;   20-200°C   (185, 227)

    6.42 x 10^{-6} per °C; 200-300°C (185, 227, 258)

    6.92 x 10^{-6} per °C; 300-400°C (185, 227, 258)

    7.40 x 10^{-6} per °C; 400-500°C (185, 227, 258)

    8.39 x 10^{-6} per °C; 500-600°C (185, 227, 258)

    8.91 x 10^{-6} per °C; 600-700°C (185, 227, 258)

THERMAL (cont.)

    Expansion: (cont.)

       (α form) 6.95×10^{-6} per °C; 25-500°C   (329)

                   7.34×10^{-6} per °C; 25-870°C   (329)

       (β form) 4.74×10^{-6} per °C; 25-870°C   (329)

                   5.33×10^{-6} per °C; 25-1000°C   (329)

                   6.43×10^{-6} per °C; 25-1500°C   (329)

      c axis 6.16×10^{-6} per °C; at 21°C   (38)

      a axis 5.69×10^{-6} per °C; at 21°C   (38)

    Specific heat: .0659 cal/gm/°C   (38)

DATA TABLES

V NITRIDES

ALUMINUM NITRIDE

Note: In the literature survey by Myers (155) the references 405-430
        are cited without specific details.

Formula weight:	40.98 g/mole	
Formula volume:	12.55 cc/mole	(150)
Melting point:	d 2235°C	
	2230-2240°C	(227, 261)
	2200°C	(38, 39)
	d 2300°C	(150)
	2150 at 4 atmospheres	(3)
Boiling point:	dissociates	
Vapor pressure:	10^{-3} mm Hg at 1190°C	(227)
X-ray density:	3.26 g/cc	(150)
	3.30 g/cc at 20°C	(155)
Pycnometric density:	3.05 g/cc at 25°C	(3)
	3.247 g/cc at 20°C	(155)

CHEMICAL

Theoretical analysis: 34.18% nitrogen
                        65.82% aluminum

Synthesis: Aluminum metal + N_2 at high temperatures
             ($\sim$1800°C or higher)

Reactivity: Decomposed slowly by cold water   (3)

              Soluble with decomposition in aqueous alkalis   (3)

              Corrosion rate of hot pressed AlN   (150)

H_2O	100°C	72 hrs	14 mils/year
HCl (conc)	72	72	320
HCl:H_2O (1:1)	110	72	570
H_2SO_4 (conc)	305	72	180
H_2SO_4:H_2O (1:1)	145	72	550
HNO_3 (conc)	120	72	150
HNO_3:H_2O (1:1)	111	72	200
HF:HNO_3 (1:1)	57	72	160
HF:H_2O (1:1)	57	72	215

CHEMICAL (cont.)

Reactivity: (cont.)

Conversion of 1/4 x 1/2 x 1/2 hot pressed
AlN by gasses: (150)

Air	1000°C	30 hrs	0.3% Al_2O_3
	1400° C	30 hrs	1.3% Al_2O_3
	1700°C	4 hrs	10.6% Al_2O_3
O_2	1400°C	30 hrs	0.9 % Al_2O_3
Steam	1000°C	30 hrs	0.3 % Al_2O_3
Cl_2	500°C	30 hrs	< 0.1 % $AlCl_3$
	700°C	30 hrs	19.2% $AlCl_3$
H_2	1700°C	4 hrs	Nil

No reaction with H_2, C, B or Si   (155)

Reacts slowly at high temperature with P,
O_2, S, and Br_2 (155)

Reacts rapidly with fused alkalis   (155)

Reacts with aqueous acids   (155)

Firing to 1800°C yields a material less
susceptible to decomposition than that
made at 1400°C   (155)

1.2% converted to Al_2O_3 in 50 hr at 1400°C
(1/2" cube)   (156)

No reaction with molten aluminum   (156)

Oxidizes in O_2 above 1000°C   (227)

Unstable in H_2O vapor   (227)

Stable in N_2-H_2 mixtures; 1200-1600°C   (227, 265)

ELECTRICAL

Resistivity: 2×10^{11} ohm-cm at 100 CPS at rt   (156)

Dielectric constant: 8.15 at 25°C; 8.5 x 10^9 CPS   (156)
    8.48 at 550°C; 8.5 x 10^9 CPS   (156)
    8.77 at 800°C; 8.5 x 10^9 CPS   (156)

Dissipation factor: 0.0033 at 25°C; 8.5 x 10^9 CPS   (156)
    0.0043 at 550°C; 8.5 x 10^9 CPS   (156)
    0.027 at 800°C; 8.5 x 10^9 CPS   (156)

Note: For additional data; resistivity, dielectric constant and dissipation factor as a function of frequency see Taylor and Lenie   (150)

FABRICATION

Hot press at 2000°C, 5000 psi to 98% density   (150)

MECHANICAL

Strength: Bending (MOR): 38,500 psi at 25°C (hot pressed)
    (150, 156)
    18,100 psi at 1400°C   (150, 156)
    27,000 psi at 1000°C   (150)

Hardness: Mohs   : 5-5.5   (155)
    7+   (150)
    Knoop 100 g : 1225-1230 kg/mm^2 (150)

Young's modulus: 50 x 10^6 psi at 25°C   (150, 156)
    46 x 10^6 psi at 1000°C   (150)
    40 x 10^6 psi at 1400°C   (150, 156)

OPTICAL

Color: yellow   (3)
    white (when pure) to pale blue

OPTICAL (cont.)

Refractive index:	$\varepsilon = 2.13 \pm 0.02$	(155)
	$\omega = 2.20 \pm 0.02$	(155)

Form:	Hexagonal prisms from $Al + N_2$ at $1850^{\circ}C$	(150)
	Platelets from $AlN + N_2$ above $1900^{\circ}C$	(150)

STRUCTURE

Hexagonal	(3)

Wurtzite structure   (155)
$$a = 3.10 \pm 0.01A$$
$$c = 4.965 \pm 0.010A$$
$$c/a = 1.602$$

$a = 3.10 \pm 0.01A$:	$c = 4.968 \pm 0.010$;	$c/a = 1.602$	(155)
3.113	4.981		(151)
$3.104 \pm .005$	$4.96 \pm .01$		(152)
$3.10 \pm .01$	$4.965 \pm .01$		(153)
3.114	4.968		(154)
3.111	4.980		(150)

THERMAL

Conductivity:	0.072 CGS at $200^{\circ}C$	(150, 156)
	0.060 CGS at $400^{\circ}C$	(150)
	0.053 CGS at $600^{\circ}C$	(150)
	0.048 CGS at $800^{\circ}C$	(150, 156)
	0.0042 CGS at $25^{\circ}C$	(431)

Expansion:	4.03×10^{-6} per $^{\circ}C$; $25\text{-}200^{\circ}C$	(150, 156)
	4.83×10^{-6} per $^{\circ}C$; $25\text{-}600^{\circ}C$	(150)
	5.64×10^{-6} per $^{\circ}C$; $25\text{-}1000^{\circ}C$	(150)
	6.09×10^{-6} per $^{\circ}C$; $25\text{-}1350^{\circ}C$	(150, 156)
	4.84×10^{-6} per $^{\circ}C$; $25\text{-}500^{\circ}C$	(329)
	5.54×10^{-6} per $^{\circ}C$; $25\text{-}1000^{\circ}C$	(329)

THERMAL (cont.)

Heat content: $H_T\text{-}H_{293} = 10.79\,T + 0.73 \times 10^{-3}\,T^2 + 3.85 \times 10^5$
$T^{-1} - 4,574$ cal/mole; $479\text{-}1113^{\circ}K$ (128)

Heat capacity: $10.79 + 1.46 \times 10^{-3}\,T - 3.85 \times 10^5\,T^{-2}$ (128)
cal/gm/$^{\circ}K$; $479\text{-}1113^{\circ}K$

Heat of formation: $57,500 \pm 750$ cal/mole   (155)

Heat of vaporization: 100 kcal/mole   (205)

Thermal diffusivity: 0.16 cm^2/sec

Specific heat: 0-100°C   0.1961 cal/gm/$^{\circ}$C   (155)

0-420   0.2277   (155)

0-598   0.2399   (155)

Note: Unless stated otherwise all data are for hot-pressed material.

Formula weight: 24.83 g/mole

Formula volume: 10.93 cc/mole

Melting point: d. 3000°C   (38)
                        2730°C   (39)
                        2720-2730°C   (161, 227, 261)
                        >3000°C in N_2 (160)

Vapor pressure: 158 mm Hg at 2045°C   (160, 161)
                        10^{-3} mm Hg at 925°C   (227)
                        >4 x 10^{-3} mm Hg at 1800°C (160, 161)

X-ray density: 2.27 g/cc   (160)
                        3.49 g/cc (cubic)   (434)

Pycnometric density: 2.20 g/cc   (129)
                        2.15 g/cc   (159)
                        3.48 g/cc (cubic)   (434)

CHEMICAL

Theoretical analysis: 43.57% boron
                        56.43% nitrogen

Synthesis: a) $B(NH_2)_3$ thermal decomposition   (73)
             b) B + N_2 or NH_3 (73)
             c) Borax + NH_4Cl (73)

   (cubic form)

             a) Boron + Nitrogen containing mixtures
                 >62,000 atmospheres, >1350°C   (434)

Reactivity: Air oxidation becomes severe, 1100-1400°C (227)

                 1.6% weight change; 3 hrs., 1000°C in air (227)

                 Air oxidation in mg/cm^2 (159)

                        700°C   1000°C

CHEMICAL (cont.)

Reactivity: (cont.)

Air oxidation in mg/cm^2 (cont.)   (159)

2 hrs	0.014	0.35
10 hrs	0.062	0.85
30 hrs	0.138	4.8
60 hrs	0.235	10.0

Reaction with chlorine, mg/cm^2 change   (159)

	700°C	1000°C
3 hrs	-	2.7
20 hrs	0.25	17.0
40 hrs	0.55	-

Stable to 3000°C in helium   (227)

Stable to 1000°C in oxygen   (227)

ELECTRICAL

Resistivity: 1.7×10^{13} ohm-cm at   25°C   (159)
                    2.3×10^{10} ohm-cm at   480°C   (159)
                    3.1×10^{4} ohm-cm at 1000°C   (159)

                    10^{12} ohm-cm at 25°C; 20% humidity   (159)
                    7×10^{10} ohm-cm at 25°C; 50% humidity   (159)
                    5×10^{9} ohm-cm at 25°C; 90% humidity   (159)

Dielectric constant:

	25°C	350°C	480°C	
10^{2} CPS	4.15	4.4	4.0	(159)
10^{4} CPS	4.15	-	4.5	(159)
10^{6} CPS	4.15	-	4.25	(159)
10^{8} CPS	4.15	-	-	(159)

ELECTRICAL (cont.)

Dissipation factor:		25°C	350°C	480°C	
10^2	CPS	0.00103	0.032	1.0	(159)
10^4	CPS	0.00042	0.0043	0.1	(159)
10^6	CPS	0.00020	0.0012	0.0056	(159)
10^8	CPS	0.000095	-	-	(159)
10^{10}	CPS	0.0003	0.0004	0.0005	(159)

MECHANICAL

Strength: Bending (MOR):

	(ǁ)	(⊥)	
25°C	15,880 psi	7,280 psi	(159)
300°C	15,140 psi	7,030 psi	(159)
700°C	3,840 psi	1,900 psi	(159)
1000°C	2,180 psi	1,080 psi	(159)
rt		13,200 psi	(149)
1500°C		1,250 psi	(149)
1800°C		1,500 psi	(149)
2000°C		2,450 psi	(149)

Compressive:

(ǁ)	(⊥)	
45,000 psi	34,000 psi	(149, 227)

Tensile:

1000°C	350 psi	(147)
1500°C	350 psi	(147)
1800°C	1,150 psi	(147)
2000°C	2,250 psi	(147)
2400°C	6,800 psi	(147)

Hardness:	Cubic form has approximately the same hardness as diamond	(434)
	(hexagonal form) Mohs: 2	(159)

Young's modulus:

	(ǁ)	(⊥)	
23°C	12.46 x 10^6 psi	4.91 x 10^6 psi	(45)
300°C	8.79 x 10^6 psi	3.47 x 10^6 psi	(45)
700°C	1.54 x 10^6 psi	0.51 x 10^6 psi	(45)
1000°C	1.65 x 10^6 psi	-	(45)

OPTICAL

(Hexagonal form)

Color: white

Refractive index:	1.74	(160)
Birefringence:	0.3	(160)
Optic axial angle:	0°, hexagonal	
Sign:	negative	(160)
Electroluminescence:	between 2500 and 6500 A	(164, 165)

STRUCTURE

Two forms known:

Common type: Hexagonal, graphite-type structure

a = 2.5038 $\pm$ 0.0001 A,   c = 6.60 $\pm$ 0.01 A   (433)

B-N distance   1.45 A   (433)

Cubic form: Zinc blende structure   (434)

a = 3.615 A   (434)

B-N distance   1.57 A   (434)

THERMAL

Conductivity: ($\perp$)   (||)

	$(\perp)$	$(\parallel)$	
300^oC	0.0687 CGS	0.0362 CGS	(159)
700^oC	0.0646 CGS	0.0318 CGS	(159)
1000^oC	0.0637 CGS	0.0295 CGS	(159)

Expansion: 3.8×10^{-6} per ^oC

12.2×10^{-6} per ^oC; 25-500^oC (329)

13.3×10^{-6} per ^oC; 25-1000^oC (329)

THERMAL (cont.)

Expansion: (cont.) $(\parallel)$ $(\perp)$

25-350°C	0.59×10^{-6} per °C	10.15×10^{-6} per °C	(159)
25-700°C	0.89×10^{-6} per °C	8.06×10^{-6} per °C	(159)
25-1000°C	0.77×10^{-6} per °C	7.51×10^{-6} per °C	(159)

Single crystal

41×10^{-6} per °C parallel C axis   (162)

-2.3×10^{-6} per °C parallel A axis   (162)

Note: Frequently on first heating an irreversible expansion is observed.

Heat of formation: $\Delta H^{o}_{298} = -60.7 \pm 2.5$ kcal/mole   (161)

Entropy: $\Delta S^{o}_{298} = 3.67 \pm 0.05$ kcal/mole   (161)

Free energy: $\Delta F_{T} = 26,000 - 9.7\,T$ kcal/mole   (161)

Heat capacity: $C_{p} = 1.8 + 3.62 \times 10^{-3}\,T$ cal/mole/°C   (161)

Be_3N_2

TRIBERYLLIUM DINITRIDE

Formula weight: 55.08 g/mole

Melting point: 2200°C   (227)
                     2205 ± 5°C   (161, 227, 261)

Vapor pressure: 10^{-3} mm of Hg at 1705°C   (227)

CHEMICAL

Theoretical analysis: 50.87% nitrogen
                        49.13% beryllium

Reactivity: Unstable in O_2 at elevated temperatures   (227)

               Poor oxidation resistance at 600°C   (227)

               Unstable in water vapor   (227)

CERIUM MONONITRIDE

Formula weight: 154.14 g/mole

ELECTRICAL

Resistivity: 0.3×10^{-3} ohm-cm   (205)

THERMAL

Conductivity: 0.05 watts/cm$^{\circ}$C   (205)

CHROMIUM MONONITRIDE

Formula weight:	66.02 g/mole	
Formula volume:	10.74 cc/mole	
Melting point:	d. 1500°C	(3)
X-ray density:	6.14 g/cc	(19)
Pycnometric density:	5.8 g/cc at 20°C	(3)
	6.1 g/cc	(520)

CHEMICAL

Theoretical analysis:	21.2% nitrogen	
	78.8% chromium	
Synthesis:	$Cr + NH_3$ at 735°C (pure)	(351)
Reactivity:	Insoluble in acids or alkalis	(3)

OPTICAL

Color: yellow

STRUCTURE

Cubic, NaCl type	(351, 519, 520)
a = 4.150-4.140 A depending on temperature	(351)

THERMAL

Heat of formation:	$\Delta H^o_{298} = -29.5$ kcal/mole	(444)
Thermodynamic data:		(390, 393, 437, 438, 521)

Cr$_2$N

DICHROMIUM MONONITRIDE

Formula weight: 118.03 g/mole

CHEMICAL

Theoretical analysis: 11.87% nitrogen
88.13% chromium

Synthesis: Cr + N$_2$, 1180°C (pure)   (351)

Reactivity: Resistant to acids

MECHANICAL

Hardness: Knoop 100 g : 1200 kg/mm^2 (127)

OPTICAL

Color: yellow

STRUCTURE

Hexagonal   (351, 519, 520)

a = 4.805-4.786 A,   c = 4.480-4.463 A   (351)
depending on temperature

THERMAL

Thermodynamic data: (390, 393, 437, 438, 521)

HAFNIUM MONONITRIDE

Formula weight: 192.61 g/mole

Melting point: 3300°C   (38)
                         3307°C   (39)

CHEMICAL

Theoretical analysis: 7.27% nitrogen
                         92.73% hafnium

Synthesis: See ZrN

Reactivity: Easily oxidized   (227)

ELECTRICAL

Thermionic emission; lower than Hf   (505)

OPTICAL

Color: yellow brown, metallic

STRUCTURE

Face-centered cubic, isomorphous with HfB, HfC,
ZrC, etc.   (44)

Lattice spacing should be somewhat less ($\sim$.05 A) than
that in HfC (4.64 A).

Radius ratio: 0.45   (47)

THERMAL

Heat of formation: -81.4 kcal/mole
                         -88.24 $\pm$ 0.34 kcal/mole   (79)

Thermodynamic properties: See Ref. 438, 393

MOLYBDENUM MONONITRIDE

Formula weight: 109.96 g/mole

Melting point: Dissociates above 725°C   (19)

CHEMICAL

Theoretical analysis: 12.73% nitrogen
                         87.27% molybdenum

Synthesis: $Mo + NH_3$, 700°C, 120 hrs   (19)

ELECTRICAL

Resistivity: Superconductive: 12°K   (340)

STRUCTURE

Hexagonal, isomorphous with MoC & WC   (19)

a = 2.860 A,   c = 2.804 A,   c/a = 0.908   (19)

THERMAL

Heat of formation: ΔH_{298}^{o} = +17.0 kcal/mole   (444)

Thermodynamic data: (437, 390, 438, 393)

DIMOLYBDENUM MONONITRIDE

Formula weight: 205.91 g/mole

Melting point: Dissociates above 725°C   (19)

CHEMICAL

Theoretical analysis: 6.8% nitrogen
                         93.2% molybdenum

Synthesis: (finely divided) Mo + NH_3, 400-725°C   (19)

Reactivity: Very narrow homogeneity range   (19)

ELECTRICAL

Resistivity: Superconductive: 5°K   (340)

STRUCTURE

Face-centered cubic   (19)

a = 4.155-4.160 A   (19)

THERMAL

Thermodynamic data: (437, 438, 390, 393)

NIOBIUM MONONITRIDE

Formula weight:	106.92 g/mole
Formula volume:	14.69 cc/mole

Melting point: 2050°C   (38, 78, 227, 261, 298)
                  2050°C, partial decomposition   (516)

Vapor pressure: 10^{-3} mm of Hg at 1495°C   (227)

X-ray density: 7.28 g/cc   (a = 4.39 A)   (19)

Pycnometric density: 8.26   (127)
                      8.4   (516)

CHEMICAL

Theoretical analysis: 13.1% nitrogen
                       86.9% niobium

Synthesis: $Nb + N_2$, 1000-1200°C   (60, 334, 512)

              $NbCl_5 + N_2 + H_2$, 1350°C   (508)

Reactivity: Corrosion in air becomes severe;
              500-800°C   (227)

              Soluble in HF + HNO_3 (3)

              Insoluble in HNO_3 (3)

              HCl, insoluble   (19)

              Na_2CO_3, decomposes, evolves NH_3 (19)

              H_2SO_4, insoluble, even at boiling temp.   (19)

              Alkali, decomposes, evolves NH_3 (19)

              O_2, air, easily oxidized, liberating N_2 (514)

ELECTRICAL

Resistivity: Superconductive 14.7°K   (340)
                Superconductive 16.8-14.6°K   (323)
                Superconductive 15.2°K   (469)
                Superconductive 20-30°K   (514)
                Superconductive 12.2-16.2°K *   (470)
                450×10^{-6} ohm-cm at 2050°C   (19)
                200×10^{-6} ohm-cm at rt   (19)

* Varies with N content

MECHANICAL

Hardness: Mohs : 8 +   (19)

OPTICAL

Color: Light gray powder, yellowish luster

STRUCTURE

Cubic   (78)

a = 4.41 A   (77, 78)
   NaCl type, cubic, (B 1)   (77, 469, 513, 514, 515)
a = 4.375 A   (469)

Radius ratio: 0.50   (47)

THERMAL

Heat of formation: ΔH^O_{298} = - 59 kcal/mole   (444)

Energy of activation: Diffusion of N_2 into Nb, 25.40 kcal/mole
                                                    (256, 476)
Thermodynamic data: (393, 437, 517)

DINIOBIUM MONONITRIDE

Formula weight: 199.83 g/mole

Pycnometric density: 8.08 g/cc (512)

CHEMICAL

Synthesis: NbN + Nb, 1700°C (19)

Reactivity: Stable to acids

Decomposed by alkali, liberating N$_2$

ELECTRICAL

Resistivity: Not superconductive to 9.5 $^\circ$K (340, 469)

STRUCTURE

Hexagonal close packed (512)

a = 3.058 A, c = 4.961 A, c/a = 1.623 (512)
 3.017 5.580 1.849 (515)

Tetragonal (unconfirmed) (514)

a = 4.35, c/a = 0.98 (514)

NEPTUNIUM MONONITRIDE

Formula weight: 253 g/mole

STRUCTURE

Cubic, NaCl type

a = 4.897 A   (523)

PaN

PROTACTINIUM MONONITRIDE

Formula weight: 245 g/mole

Vapor pressure: 10^{-5} mm of Hg at 2230°C   (227)

PLUTONIUM MONONITRIDE

Formula weight: 252 g/mole

STRUCTURE

Cubic, NaCl type

a = 4.905 A   (523)

SCANDIUM MONONITRIDE

Melting point: 2050°C   (38, 39)

TRISILICON TETRANITRIDE

Note: An excellent review of preparation and properties of Si_3N_4 is presented in Reference 526.

Formula weight:	140.22 g/mole	
Formula volume:	40.0 cc/mole	
Melting point:	d 1900°C	(38, 227)
Vapor pressure:	10^{-3} mm of Hg at 1060°C	(227)
X-ray density:	3.43 g/cc	
	$\propto$ 3.184 g/cc	(526)
	β 3.187 g/cc	(526)

CHEMICAL

Theoretical analysis: 39.9% nitrogen
60.1% silicon

Synthesis $\propto Si_3N_4$: $Si + N_2$ 1200-1300°C

βSi_3N_4: $Si + N_2$ 1450°C and higher

Reactivity: 0.4% weight change in air in 3 hrs at 1100-1400°C (227)

Useful to 1850°C in neutral atmosphere (227, 247)

Stable in air from 1100-1400°C (227)

Useful to 1850°C in reducing atmosphere (227,247)

ELECTRICAL

Resistivity: $>10^{13}$ ohm-cm (526)

MECHANICAL

Strength:	Bending (MOR) :	10,000 psi	(227, 247)
		16-20,000	(526)
	Tensile :	15,800 psi	(38)
Hardness:	Mohs :	9+	(526)
	Rockwell A :	99	(227, 247)

Young's modulus: 13×10^6 psi

STRUCTURE

$\propto$ Hexagonal

a = 7.748 A ;   c = 5.617   (526)

β Hexagonal

a = 7.608 A ;   c = 2.911   (526)

THERMAL

Conductivity: 0.035 CGS   (38, 45)

0.041 CGS, 200-750°C   (227, 247)

0.0037 CGS   (526)

Expansion:

($\propto$ form) 2.11×10^{-6} per °C;   25-500°C   (329)

2.87×10^{-6} per °C;   25-1000°C   (329)

3.66×10^{-6} per °C;   25-1500°C   (329)

(β form) 1.47×10^{-6} per °C;   25-500°C   (329)

2.25×10^{-6} per °C;   25-1000°C   (329)

3.66×10^{-6} per °C;   25-1500°C   (329)

2.5×10^{-6} per °C;   (38, 45, 526)

2.47×10^{-6} per °C;   25-1000°C   (227, 247)

THERMAL (cont.)

Specific heat: 0.17 cal/gm/$^{\circ}$C

Diffusivity: 0.01 cm^2/sec

TANTALUM MONONITRIDE

Formula weight: 194.89 g/mole

Formula volume: 13.8 cc/mole

Melting point: 3360°C   (38, 78)
                          3090°C   (52, 129)
                          3075 $\pm$ 50°C   (161, 227, 298)
                          2980°C   (227, 261)
                        d 2980°C   (516)

Vapor pressure: 10^{-3} mm of Hg at 1460°C   (227)

X-ray density: 14.1 g/cc   (516)

Pycnometric density: 14.1 g/cc   (516)

CHEMICAL

Theoretical analysis: 7.19% nitrogen
                          92.81% tantalum

Synthesis: Ta + N_2, 1100-1200°C   (19)

Reactivity: Soluble in HF + HNO_3 (3)

                  Insoluble in simple aqueous acids   (3)

                  Poor oxidation resistance at 650°C   (227)

ELECTRICAL

Resistivity: 135×10^{-6} ohm-cm at rt   (52, 127)

                  103.7×10^{-6} ohm-cm at 1480°C   (52)

                  116.3×10^{-6} ohm-cm at 2840°C   (52)

                  Not superconductive to 1.88°K   (340, 469)

Thermionic emmision: Smaller than that of Ta metal   (505)

MECHANICAL

    Hardness: Mohs   : 8 +   (19)

                Knoop 30 g : 3236 kg/mm^2 (8)

STRUCTURE

Hexagonal   (127)

Hexagonal close packed (Wurtzite type)   (77, 436)

$a = 3.05 A$, $c = 4.95 A$, $c/a = 1.62$ (19)

It has been suggested that this structure is actually due
to Ta_2N.   (100, 518)

Radius ratio: 0.50   (47)

THERMAL

Heat of formation: $\Delta H^0_{298} = -58$ kcal/mole   (444)

Activation energy: Diffusion of N_2 into Ta,
39.40 kcal/mole   (256, 476)

Thermodynamic data: (437, 393, 438, 390)

Ta$_2$N

DITANTALUM MONONITRIDE

Formula weight: 375.77 g/mole

CHEMICAL

Theoretical analysis: 3.73% nitrogen
96.27% tantalum

Synthesis: Ta + NH$_3$, 18 hrs, 900°C then
2000°C, in vacuum   (100)

ELECTRICAL

Resistivity: Superconductive: 9.5^{o}K   (340)

THORIUM MONONITRIDE

Formula weight: 246.13 g/mole

Melting point: $2630^{\circ}C$ (19, 227)
$2630 \pm 50^{\circ}C$ (100)

Vapor pressure: 10^{-5} mm Hg at $2230^{\circ}C$ (227)

CHEMICAL

Theoretical analysis: 5.7% nitrogen
94.3% thorium

Synthesis: Decompose Th_2N_3 (Th_3N_4) at $1750^{\circ}C$

STRUCTURE

Cubic, NaCl type   (518)

$a = 5.2 A$ (100)

Radius ratio: 0.39   (47)

DITHORIUM TRINITRIDE

Note: See Th_3N_4

    Formula weight: 506.27 g/mole

    Melting point: d to ThN 1750°C   (100)

CHEMICAL

    Theoretical analysis: 5.54% nitrogen
                          94.46% thorium

    Synthesis: $Th + N_2$ below 1500°C

STRUCTURE

    Hexagonal   (100)
    a = 3.87 A;   c = 6.16 A   (100)

    Radius ratio: 0.39   (47)

TRITHORIUM TETRANITRIDE

Note: "It appears possible that nitride preparations previously identified as Th_3N_4 are identical with the phase ..... Th_2N_3."   (19)

Formula weight: 752.40 g/mole

Melting point: 2100°C   (38)
                       2360°C   (227)

Vapor pressure: 10^{-3} mm of Hg at 1990°C   (227)

CHEMICAL

Theoretical analysis: 5.58% nitrogen
                         94.42% thorium

Synthesis: $Th + N_2$

Reactivity: Burns in air when ignited   (227)

                  Decomposes in water, liberating NH_3 (227)

THERMAL

Thermodynamic data: (524, 437, 393, 390, 438)

TITANIUM MONONITRIDE

Formula weight: 61.91 g/mole

Formula volume: 11.87 cc/mole

Melting point: $2900^{\circ}C$ (45)
$2945 \pm 50^{\circ}C$ (227, 261, 298)
$3205^{\circ}C$ (227)
$3220^{\circ}C$ (78)
$2950^{\circ}C$ (129)

Vapor pressure: 10^{-3} mm of Hg at $2060^{\circ}C$ (227)

X-ray density: 5.43 g/cc   (129)

Pycnometric density: 5.213 g/cc   (197)

CHEMICAL

Theoretical analysis: 22.6% nitrogen
77.4% titanium

Synthesis: $TiO_2 + N_2 + C$, $1250^{\circ}C$

$TiO_2 + NH_3$

$Ti + N_2$, $1200^{\circ}C$

$TiCl_4 + NH_3$, below $1400^{\circ}C$

$TiCl_4 + N_2 + H_2$, hot wire

Reactivity: Oxidizes rapidly in air at $1200^{\circ}C$ (143)

0.8% weight change at $1100^{\circ}C$,
3 hrs in air   (227)

16.2% weight change at $1400^{\circ}C$,
3 hrs in air   (227)

Rapid oxidation by CO_2 at $1200^{\circ}C$ (227)

Stable in H_2 at $1450^{\circ}C$ (227)

O_2, poor resistance at $600^{\circ}C$ (227)

ELECTRICAL

Resistivity: 22×10^{-6} ohm-cm

72×10^{-6} ohm-cm (127)

130×10^{-6} at rt (53)

340×10^{-6} at melting point (53)

21.7×10^{-6} at rt (52, 440)

8.13×10^{-6} at liquid air (52, 440)

11.07×10^{-6} ohm-cm at rt (441)

Superconductive $1.2-1.6^{o}$K (333)

Superconductive $1.2-2.6^{o}$K (283)

Superconductive $1.1-5.5^{o}$K (333, 465)

Superconductive 1.34^{o}K (460)

Magnetic susceptibility: $+0.8 \times 10^{-6}$ (19, 394)

Thermionic emission: See Ref. 337 (19)

MECHANICAL

Strength: Bending (MOR) : 34,000 psi (45, 127)

Compression : 141,000 psi (127)

Hardness: Mohs : 8-10 (227, 19)

Knoop 30 g : 2160 kg/mm^2 (8)

100 g : 1770 kg/mm^2 (8)

Young's modulus: 36.3×10^6 psi (166)

11.47×10^6 psi; ρ = 3.03 g/cc (175)

OPTICAL

Color: Yellow - bronze

STRUCTURE

Cubic   (127)

a = 4.23 A   (19, 197, 78, 436)

Homogeneity range: $TiN_{0.42}$ - $TiN_{1.16}$ yields   (19)

a = 4.213 to ~4.24 A   (19)

a = 4.234A at $TiN_{1.0}$ (19)

Radius ratio: 0.49   (47)

THERMAL

Conductivity: 0.07 CGS   (45)

0.069 CGS at $25^{\circ}C$ (227)

0.040 CGS at $200^{\circ}C$ (227)

0.027 CGS at $650^{\circ}C$ (227)

0.020 CGS at $1000^{\circ}C$ (227)

Expansion: 9.35×10^{-6} per $^{\circ}C$ (166)

Thermodynamic data: See Refs. 145, 390, 391, 392, 393, 437, 438, 439)

Heat of formation: ΔH^{o}_{298} = - 80.3 kcal/mole   (444)

Energy of activation: diffusion of N_2 into Ti, 33.8 kcal/mole (256, 476)

URANIUM MONONITRIDE

Formula weight: 252 08 g/mole

Formula volume: 17 6   cc/mole

Melting point: $2850 \pm 30^{\circ}C$ (2.9 atm. N_2)   (435)

                     $2650^{\circ}C$ (141, 142)

                     $2630-2650^{\circ}C$ (161, 227, 261)

Vapor pressure: $10^{-4}-10^{-5}$ mm of Hg at $2230^{\circ}C$ (227)

                     Stable in vacuum at $1700^{\circ}C$ (141)

X-ray density: 14. 31-14. 32 g/cc   (61, 138)

Pycnometric density: 11. 35-12. 0 g/cc   (136)

CHEMICAL

Theoretical analysis: 5. 56% nitrogen
                        94. 44% uranium

Synthesis: a) By decomposition of other uranium nitrides
                 above $1300^{\circ}C$ (141)

              U (UH) + N_2 or NH_3 (61)

              UN_2, U_2N_3 + H_2 or vac.   (61)

Reactivity: Slowly decomposed by water at $100^{\circ}C$ (136)

              Rapidly decomposed by water at $200^{\circ}C$ (136)

              Powder is pyrophoric   (140)

              Reacts with O_2, and H_2O (195)

              No reaction with H_2 (195)

              Complete oxidation at $500^{\circ}C$ (227)

              Thermodynamically stable in H_2 (227)

              Reacts with carbon, $2230^{\circ}C$ (227)

              Reacts with UO_2, $2300^{\circ}C$ (227)

ELECTRICAL

Resistivity: 208×10^{-6} ohm-cm ; $\rho = 11.83$ g/cc   (136)

MECHANICAL

Strength: Bending (MOR) :   10,500 psi at 25°C   (136)

11,200 psi at 800°C   (136)

19,300 psi at 1200°C   (136)

13,700 psi at 1400°C   (136)

Hardness: Knoop 100 g   :   455 kg/mm^2 (142)

Young's modulus: 21.6×10^6 psi   (136)

Shear modulus: 8.7×10^6 psi   (136)

Poisson's ratio: 0.24   (136)

Creep: 1600°C   (136)

STRUCTURE

Face-centered cubic, isomorphous with UC   (61)

a = 4.880 A   (61)

4.8899   (138)

THERMAL

Conductivity: 0.13 CGS at 800°C   (136)

0.011 CGS at 25°C   (227, 245)

Expansion: 8.00×10^{-6} per °C;   25-500°C   (329)

8.62×10^{-6} per °C;   25-1000°C   (329)

8.5×10^{-6} per °C;   25-200°C   (136)

8.7×10^{-6} per °C: 25-400°C   (136)

THERMAL (cont.)

    Expansion: (cont.)

                8.9×10^{-6} per $^{\circ}C$; 25-600$^{\circ}C$ (136)

                9.1×10^{-6} per $^{\circ}C$; 25-800$^{\circ}C$ (136)

                9.3×10^{-6} per $^{\circ}C$; 25-1000$^{\circ}C$ (136)

                9.7×10^{-6} per $^{\circ}C$; 25-1200$^{\circ}C$ (136)

URANIUM DINITRIDE

Formula weight: 266.09 g/mole

Formula volume: 22.65 cc/mole

Melting point: d. to UN at 1300°C   (141)

X-ray density: 11.73 g/cc   (140)

CHEMICAL

Theoretical analysis: 10.5% nitrogen
                        89.5% uranium

Synthesis: U + N$_2$, 126 atm.   (61)

Reactivity: Reduced to UN by H$_2$ (227)

STRUCTURE

Face-centered cubic, fluorite type   (61)

a = 5.31 A   (61)

U_2N_3

DIURANIUM TRINITRIDE

Formula weight: 518.17 g/mole

Formula volume: 46.1 cc/mole

Melting point: d. to UN

Vapor pressure: Decomposes in vacuum; 7-800°C   (227)

X-ray density: 11.24 g/cc   (61)

CHEMICAL

Theoretical analysis: 8.11% nitrogen
91.89% uranium

Synthesis: UH or U + N_2 or NH_3 (61)

Reactivity: Reduced to UN by H_2 (227)

STRUCTURE

Body-centered cubic, Mn_2O_3 type (D 5_3)   (61)

a = 10.678 A   (61)

U_2N_3 - UN_2 form continuous solid solution series   (61)

Hexagonal   (139)

a = 3.70 A,   c = 5.80 A   (139)

VANADIUM MONONITRIDE

Formula weight: 64.96 g/mole

Formula volume: 10.65 cc/mole

Melting point: 2050°C   (3, 38, 53, 161, 227, 298)
                       2320°C   (39, 78)
                       2360°C   (227)

Vapor pressure: 10^{-3} mm of Hg at 930°C   (227)
                       Pn_2 = <0.5 mm of Hg at 1270°C   (510)

X-ray density: 6.102 g/cc   (19)

Pycnometric density: 6.13 g/cc   (127)
                       5.63 g/cc   (3, 77)
                       5.91 g/cc   (53)

CHEMICAL

Theoretical analysis: 21.55% nitrogen
                       78.45% vanadium

Synthesis: a. NH_4VO_3 + NH_3,
                       900-1100°C   (496, 497, 498, 499, 500)

                 b. NH_4VO_3 + N_2 + H_2,
                       600-1400°C   (496, 497, 498, 499, 500)

                 c.   V + N_2 (500)

                 d. Cl_4 + N_2 + H_2,
                       1000-1600°C   (52, 334, 508)

Homogeneity range: $VN_{0.71}$ - $Vn_{1.0}$ (500)

Reactivity: Easily oxidized   (227)

                 Converted by C to VC at 1200°C   (227)

                 Not expected to react with H_2, 1450°C   (227)

                 HNO_3, soluble

                 HCl, insoluble

CHEMICAL (cont.)

Reactivity: (cont,)

H_2SO_4, insoluble

H_2SO_4, soluble on prolonged boiling, liberating N_2.

Strong alkali, decomposes, liberating NH_3

ELECTRICAL

Resistivity:	86×10^{-6} ohm-cm	(127)
	Superconductive $1.3^{\circ}K$	(283, 460)
	Superconductive $1.5 - 3.2^{\circ}K$	(465)
	200×10^{-6} ohm-cm at rt	(53)
	850×10^{-6} ohm-cm at mp	(53)
	85.9×10^{-6} ohm-cm at rt	(42)
	59.9×10^{-6} ohm-cm at liq. air temperature	(42)
	332×10^{-6} ohm-cm at rt	(499)

OPTICAL

Color:	gray brown, with violet luster	(19)
	Metallic bronze if high in N_2	(500)

STRUCTURE

Cubic, NaCl type	(127)
a = 4.13 A	(78)
4.126 A	(500)
4.28 A	(77)
4.129 A (most probable)	(19, 497, 498, 499, 509)

STRUCTURE (cont.)

    Radius ratio: 0.55   (47)

THERMAL

    Heat of formation: ΔH^{O}_{298} = -60.0 kcal/mole   (444)

    Thermodynamic data: 390, 438, 511, 393

TRIVANADIUM MONONITRIDE

Formula weight: 166. 86 g/mole

CHEMICAL

Theoretical analysis: 8. 40% nitrogen
                       91. 60% vanadium

Synthesis: VN + V, sealed tube, $1000\text{-}1100^{\circ}C$ (500)

Homogeneity range: $VN_{0.43} - VN_{0.37}$ (500)

Reactivity: less stable than VN   (500)

STRUCTURE

Hexagonal

a = 2. 831 A,   c = 4. 533 A   (500)

DITUNGSTEN MONONITRIDE

Formula weight: 381.85 g/mole

CHEMICAL

Theoretical analysis: 3.67% nitrogen
96.33% tungsten

STRUCTURE

β phase: face-centered cubic
a = 4.118   (522)

γ phase: cubic
a = 4.122-4.133 depending on temperature   (351)

THERMAL

Heat of formation: ΔH^o_{298} = +17.0 kcal/mole   (444)

Thermodynamic data: (393, 437)

<u>ZIRCONIUM MONONITRIDE</u>

Formula weight:	105.23 g/mole	
Formula volume:	14.32 cc/mole	
Melting point:	$2985 \pm 50^\circ C$	(227, 261, 298)
	$2980^\circ C$	(52, 78, 129)
	$2982^\circ C$	(39)
	$2950^\circ C$	(38, 45)
	$2965^\circ C$	
	$2930^\circ C$	(53)
Vapor pressure:	10^{-3} mm of Hg at $2045^\circ C$	(227)
X-ray density:	7.349 g/cc	(19)
Pycnometric density:	7.09 g/cc	(127)
	6.93 g/cc	(53)

CHEMICAL

Theoretical analysis: 13.31% nitrogen
                               86.69% Zirconium

Synthesis:	(a)	$ZrO_2 + C + N_2$; $1300^\circ C$	(53)
	(b)	$Zr + NH_3$	(100)
	(c)	$ZrH + NH_3$, $1000^\circ C$	(442, 443)
	(d)	$ZrCl_4$ vapor phase decomposition in N_2	(504)

Reactivity:	Melts without decomposition	(19)
	Corrosion in air becomes severe; 1100-$1400^\circ C$	(227)
	32% weight change, 3 hr; $1400^\circ C$ in air	(227)
	H_2O, slow hydrolysis	(227)
	O_2, poor resistance at $600^\circ C$	(227)
	HNO_3, insoluble	

CHEMICAL (cont.)

    Reactivity: (cont.)

        HCl, difficulty soluble when dilute

        H_2SO_4, difficulty soluble when dilute

        H_2SO_4, readily soluble when concentrated

        Alkali, d. on boiling, evolving NH_3

        Na_2CO_3, d. evolves NH_3

ELECTRICAL

Resistivity:	14×10^{-6} ohm-cm	(127)
	11.52×10^{-6} ohm-cm at rt	(441)
	13.6×10^{-6} ohm-cm at rt	(52)
	3.97×10^{-6} at liquid air temperature	(52)
	160×10^{-6} at rt	(53)
	320×10^{-6} at mp	(53)
	Superconductive 3.2-$7.8^\circ K$	(333, 465)
	Superconductive $9.45^\circ K$	(333, 460)
	Superconductive $3.2^\circ K$	(283)
Magnetic susceptibility:	$+0.6 \times 10^{-6}$	(394)
Thermionic emission		(505, 337)

MECHANICAL

Hardness:	Mohs	:	8 +	(227)
	Knoop 30 g	:	1983 kg/mm^2	(8)
	100 g	:	1510 kg/mm^2	(8)

OPTICAL

    Color: Yellow brown, brassy when dense

STRUCTURE

Cubic, NaCl type, B 1   (19, 436, 77, 60)

a = 4.62 A   (78)
   4.63 A   (77)
   4.567 A   (60)

Radius ratio: 0.46   (47)

THERMAL

Conductivity: 0.040 CGS   (129)

                0.040 CGS at 200°C   (227)

                0.025 CGS at 425°C   (227)

                0.018 CGS at 650°C   (227)

                0.016 CGS at 875°C   (227)

                0.015 CGS at 1100°C   (227)

Expansion: 6.13×10^{-6} per$^{\circ}$C; $20\text{-}450^{\circ}$C   (227, 232)

             7.03×10^{-6} per$^{\circ}$C; $20\text{-}680^{\circ}$C   (227, 232)

Activation energy: diffusion of N_2 into Zr, 39.20
                     kcal/mole   (256, 476)

Thermodynamic data: (437, 390, 438, 506, 507, 393)

Heat of formation: ΔH°_{298} = - 79.53 kcal/mole   (41)
                                - 82.2 kcal/mole   (444)

DATA TABLES

VI OXIDES

Al_2O_3

ALUMINUM OXIDE

Formula weight:	101.94 g/mole	
Formula volume:	25.5 cc/mole	
Melting point:	2015°C	(129)
	2040°C	
	2015-2050°C	(227, 261, 298)
Vapor pressure:	10^{-7} mm Hg at 1780°C	(227)
X-ray density:	3.986 g/cc	(126)
	3.97 g/cc	(4, 5)

CHEMICAL

Theoretical analysis: 47.1% oxygen
 52.9% aluminum

Reactivity:	Limit of usefulness; air, 1980°C	(227)
	vacuum, 1800°C	(227)
	reducing atm, 1925°C	(227)
	H_2O, limited effect at 1500°C	(227)
	N_2, resistant to at least 1700°C	(227)
	F_2, reacts at 1750°C	(227)
	C, attacked at mp	(227)
	$C_3H_8 + O_2$, stable to 1980°C	(227)

ELECTRICAL

Resistivity:	$>10^{14}$ ohm-cm at 25°C	(157)
	1×10^{13} ohm-cm at 300°C	(157)
	6.3×10^{10} ohm-cm at 500°C	(157)
	5.0×10^{8} ohm-cm at 700°C	(157)
	2×10^{6} ohm-cm at 1000°C	(157)

Al_2O_3

(Cont.)

ELECTRICAL (cont.)

Dielectric constant:		25°C	500°C	800°C	(157)
	1 KC	9.30	11.25	-	
	1 MC	9.30	10.09	-	
	100 MC	9.30	9.88	-	
	8.5 KMC	9.30	9.88	-	
	14 KMC	9.30	9.85	10.41	
	24 KMC	9.30	9.85	10.41	
	50 KMC	9.30	-	-	

Loss tangent:	(tan b)	25°C	500°C	800°C	(157)
	1 KC	0.0042	0.14	-	
	1 MC	0.0015	0.0052	-	
	100 MC	0.00006	-	-	
	8.5 KMC	0.00013	0.00017	0.00057	
	14 KMC	-	0.0002	0.00047	
	24 KMC	0.00024	0.00025	0.00034	
	50 KMC	0.00052	-	-	

Loss factor:		at 25°C	at 500°C	at 800°C	(157)
	1 KC	0.039	1.6	-	
	1 MC	0.014	0.052	-	
	100 MC	0.0006	-	-	
	8.5 KMC	0.0012	0.0017	-	
	14 KMC	-	0.002	0.0049	

ELECTRICAL (cont.)

  Loss factor: (cont.)

           24 KMC   0.0022   0.0025   0.0035

           50 KMC   0.0048   -   -

MECHANICAL

  Strength: Bending (MOR):

        60×10^3 psi at rt   (157)

        56×10^3 psi at 20°C, 98% dense, 3μ grain size   (191)

        30×10^3 psi at 20°C, 98% dense, 20μ grain size   (191)

        62×10^3 psi at 600°C, 98% dense, 3μ grain size   (191)

        28×10^3 psi at 600°C, 98% dense, 20μ grain size (191)

        58×10^3 psi at 900°C, 98% dense, 3μ grain size   (191)

        31×10^3 psi at 900°C, 98% dense, 20μ grain size (191)

        31×10^3 psi at 1090°C,   (157)

        42×10^3 psi at 1100°C, 98% dense, 3μ grain size (191)

        30×10^3 psi at 1100°C, 98% dense, 20μ grain size (191)

        Tensile: 37.8×10^3 psi at rt   (2, 227, 297)

                    37×10^3 psi at rt   (4)

                    33.6×10^3 psi at 300°C   (227)

                    40×10^3 psi at 500°C   (190)

                    34.6×10^3 psi at 800°C   (227)

                    35×10^3 psi at 1000°C   (190)

MECHANICAL (cont.)

 Strength: Tensile: (cont.)

	33.9×10^3 psi at 1050^oC	(227)
	31.4×10^3 psi at 1140^oC	(227)
	18.5×10^3 psi at 1200^oC	(227)
	20×10^3 psi at 1200^oC	(190)
	6.4×10^3 psi at 1300^oC	(227)
	4.3×10^3 psi at 1400^oC	(227)
	15×10^3 psi at 1460^oC	(4)
	1.5×10^3 psi at 1460^oC	(227)
Shear:	29.4×10^3 psi at rt	(2, 227)
	29.3×10^3 psi at 500^oC	(227)
	23.6×10^3 psi at 1000^oC	(227)
	19.8×10^3 psi at 1100^oC	(227)
	13.2×10^3 psi at 1200^oC	(227)
	11.4×10^3 psi at 1300^oC	(227)
	6.5×10^3 psi at 1400^oC	(227)
	3.35×10^3 psi at 1500^oC	(227)

 Compression:

427×10^3 psi at rt	(2, 227)
214×10^3 psi at 400^oC	(227)
199×10^3 psi at 600^oC	(227)
183×10^3 psi at 800^oC	(227)

MECHANICAL (cont.)

    Strength: Compression: (cont.)

128×10^3 psi at 1000°C	(227)
85×10^3 psi at 1100°C	(227)
71×10^3 psi at 1200°C	(227)
35.6×10^3 psi at 1400°C	(227)
14×10^3 psi at 1500°C	(227)
7×10^3 psi at 1600°C	(227)

Hardness: Mohs　　　: 9
    Vickers 20 g : 2600 kg/mm^2 (8)
          50 g : 2720 kg/mm^2 (8)
    Knoop 100 g : 2000-2050 kg/mm^2 (8)

Young's modulus:　　$E = 59,495 \times 10^3 \, e^{-3.95\,P}$ (215)
        where P is porosity

57.41×10^6 psi at rt, $\rho = 3.942$	(126)
52×10^6 psi at rt	(7)
50×10^6 psi at 500°C	(7)
32×10^6 psi at 1250°C	(7)
50×10^6 psi at rt	(157)
53×10^6 psi at 500°C	(190)
50×10^6 psi at 1000°C	(190)
48×10^6 psi at 1200°C	(190)
59.3×10^6 psi at rt	(191, 185)
57.275×10^6 psi at 500°C	(191, 185)

<u>MECHANICAL</u> (cont.)

    Young's modulus: (cont.)

54.890×10^6 psi at 1000^oC	(191, 185)
53.650×10^6 psi at 1200^oC	(191, 185)
52.4×10^6 psi at rt	(227, 297)
55.5×10^6 psi at rt	(2, 227)
51.2×10^6 psi at 800^oC	(227)
45.5×10^6 psi at 1000^oC	(227)
39.8×10^6 psi at 1200^oC	(227)
32.7×10^6 psi at 1400^oC	(227)
25.6×10^6 psi at 1500^oC	(227)

      (Single Crystal)

$$E = 67.248 \times 10^6 - 6.395\, T\, e^{-T_o/T} \quad (185, 187)$$

    $T_o = 100^oC$

    T = temperature in question

Shear modulus:	22.7×10^6 psi at rt	(126)
Bulk modulus:	38.36×10^6 psi at rt	(126)
Rigidity modulus:	50.6×10^6 psi at rt	(2, 227)
	40×10^6 psi at 500^oC	(227)
	42.5×10^6 psi at 1000^oC	(227)
	32.9×10^6 psi at 1100^oC	(227)
	29.5×10^6 psi at 1200^oC	(227)
	17.4×10^6 psi at 1300^oC	(227)
	2.3×10^6 psi at 1400^oC	(227)
	1.8×10^6 psi at 1500^oC	(227)

MECHANICAL (cont.)

    Poisson's ratio: 0.254   (126)

    Speed of sound: 32,870 fps   (126)

    Creep: See Ref. 237   (227)

NUCLEAR

    Neutron capture cross section: 0.0101 barns   (5)

OPTICAL

    Colorless if pure

    Refractive index: ω 1.7676-1.7682

    Emissivity: Spectral 0.65μ : 0.15; 1025-1625°C   (158)

        Total: 0.44 at 500°C   (227)

                0.39 at 600°C   (227)

                0.30 at 800°C   (227)

                0.29 at 1000°C   (227)

                0.29 at 1200°C   (227)

                0.33 at 1400°C   (227)

                0.43 at 1600°C   (227)

STRUCTURE

    Hexagonal

    a = 4.758 A;   c = 12.99 A   (126)

THERMAL

Conductivity: 0.069 CGS at $100^{\circ}C$ (9, 157, 185)

0.021 CGS at $600^{\circ}C$ (9)

0.013 CGS at $1200^{\circ}C$ (9)

0.037 CGS at $315^{\circ}C$ (157, 185)

0.0147 CGS at $1000^{\circ}C$ (129)

0.035 CGS at $500^{\circ}C$ (190)

0.016 CGS at $1000^{\circ}C$ (190)

0.015 CGS at $1200^{\circ}C$ (190)

0.06 CGS at rt   (328)

0.04 CGS at $100^{\circ}C$ (328)

0.03 CGS at $200^{\circ}C$ (328)

0.02 CGS at $400^{\circ}C$ (328)

0.015 CGS at $800^{\circ}C$ (328)

0.014 CGS at $1000^{\circ}C$ (328)

0.064 CGS at $200^{\circ}C$ (227)

0.031 CGS at $400^{\circ}C$ (227)

0.022 CGS at $600^{\circ}C$ (227)

0.017 CGS at $800^{\circ}C$ (227)

0.015 CGS at 1000° (227)

0.013 CGS at $1200^{\circ}C$ (227)

0.013 CGS at $1400^{\circ}C$ (227)

0.014 CGS at $1600^{\circ}C$ (227)

0.017 CGS at $1800^{\circ}C$ (227)

THERMAL (cont.)

Diffusivity:	0.091 cm^2/sec at rt	(328)
	0.059 cm^2/sec at 100°C	(328)
	0.043 cm^2/sec at 200°C	(328)
	0.024 cm^2/sec at 400°C	(328)
	0.018 cm^2/sec at 800°C	(328)
	0.015 cm^2/sec at 1000°C	(328)
Expansion:	2.7 x 10^{-6} per °C; -150 to 20°C	(157)
	6.8 x 10^{-6} per °C; 20-260°C	(157)
	8.3 x 10^{-6} per °C; 260-540°C	(157)
	9.2 x 10^{-6} per °C; 540-980°C	(157)
	8.33 x 10^{-6} per °C; 20-980°C	(157)
	9.0 x 10^{-6} per °C	(129)
	7.0 x 10^{-6} per °C; to 500°C	(190)
	9.5 x 10^{-6} per °C; to 1000°C	(190)
	10.0 x 10^{-6} per °C; to 1200°C	(190)
	7.58 x 10^{-6} per °C; 25-500°C	(329)
	8.52 x 10^{-6} per °C; 25-1000°C	(329)
	9.29 x 10^{-6} per °C; 25-1500°C	(329)
	8.5 x 10^{-6} per °C; 30-800°C	(297)
	8.0 x 10^{-6} per °C; 25-1850°C	(2)

THERMAL (cont.)

    Expansion: (cont.)

Single Crystal		Polycrystalline (329)	
Parallel c axis	Perpendicular c axis		
1.95×10^{-6} per $^{\circ}$C	1.65×10^{-6} per $^{\circ}$C	1.89×10^{-6} per $^{\circ}$C;	0 to -273°C
3.01×10^{-6} per $^{\circ}$C	2.55×10^{-6} per $^{\circ}$C	2.91×10^{-6} per $^{\circ}$C;	-173°C
4.39×10^{-6} per $^{\circ}$C	3.75×10^{-6} per $^{\circ}$C	4.10×10^{-6} per $^{\circ}$C;	-73°C
5.31×10^{-6} per $^{\circ}$C	4.78×10^{-6} per $^{\circ}$C	5.60×10^{-6} per $^{\circ}$C;	$+27^{\circ}$C
6.26×10^{-6} per $^{\circ}$C	5.51×10^{-6} per $^{\circ}$C	6.03×10^{-6} per $^{\circ}$C;	$+127^{\circ}$C
6.86×10^{-6} per $^{\circ}$C	6.10×10^{-6} per $^{\circ}$C	6.55×10^{-6} per $^{\circ}$C;	227°C
7.31×10^{-6} per $^{\circ}$C	6.52×10^{-6} per $^{\circ}$C	6.93×10^{-6} per $^{\circ}$C;	327°C
7.68×10^{-6} per $^{\circ}$C	6.88×10^{-6} per $^{\circ}$C	7.24×10^{-6} per $^{\circ}$C;	427°C
7.96×10^{-6} per $^{\circ}$C	7.15×10^{-6} per $^{\circ}$C	7.50×10^{-6} per $^{\circ}$C;	527°C
8.19×10^{-6} per $^{\circ}$C	7.35×10^{-6} per $^{\circ}$C	7.69×10^{-6} per $^{\circ}$C;	627°C
8.38×10^{-6} per $^{\circ}$C	7.53×10^{-6} per $^{\circ}$C	7.83×10^{-6} per $^{\circ}$C;	727°C
8.52×10^{-6} per $^{\circ}$C	7.67×10^{-6} per $^{\circ}$C	7.97×10^{-6} per $^{\circ}$C;	827°C *
8.65×10^{-6} per $^{\circ}$C	7.80×10^{-6} per $^{\circ}$C	8.08×10^{-6} per $^{\circ}$C;	927°C
8.75×10^{-6} per $^{\circ}$C	7.88×10^{-6} per $^{\circ}$C	8.18×10^{-6} per $^{\circ}$C;	1027°C
8.84×10^{-6} per $^{\circ}$C	7.96×10^{-6} per $^{\circ}$C	8.25×10^{-6} per $^{\circ}$C;	1127°C
8.92×10^{-6} per $^{\circ}$C	8.05×10^{-6} per $^{\circ}$C	8.32×10^{-6} per $^{\circ}$C;	1227°C
8.98×10^{-6} per $^{\circ}$C	8.12×10^{-6} per $^{\circ}$C	8.39×10^{-6} per $^{\circ}$C;	1327°C

*Data above 827°C obtained by extrapolation.

THERMAL (cont.)

Expansion: (cont.)

Single Crystal		Polycrystalline	(329)
Parallel c axis	Perpendicular c axis		
9.02×10^{-6} per °C	8.16×10^{-6} per °C	8.45×10^{-6} per °C;	1427°C
9.08×10^{-6} per °C	8.20×10^{-6} per °C	8.49×10^{-6} per °C;	1527°C
9.13×10^{-6} per °C	8.26×10^{-6} per °C	8.53×10^{-6} per °C;	1627°C
9.18×10^{-6} per °C	8.30×10^{-6} per °C	8.58×10^{-6} per °C;	1727°C

BaO

BARIUM OXIDE

Formula weight: 153.36 g/mole

Melting point: 1923°C   (3)
                       1920°C   (161, 227, 261)

Vapor pressure: 10^{-7} mm Hg at 1540°C   (227)

Boiling point: Approximately 2000°C   (3)

X-ray density: 5.72 g/cc

CHEMICAL

Reactivity: O_2, can form less stable BaO_2

                   H_2O, hydrates

                   H_2O, soluble 1.5 parts per 100 parts at 0°C   (3)

                   Aqueous acids, soluble   (3)

                   Absolute alcohol, soluble   (3)

                   NH_4OH, insoluble   (3)

OPTICAL

Color - colorless

Refractive index: 1.98   (3)

STRUCTURE

Cubic   (3)

BERYLLIUM OXIDE

Formula weight:	25.02 g/mole	
Formula volume:	8.33 cc/mole	
Melting point:	2520°C	(129)
	2530°C	(127)
	2520-2550°C	(227, 261)
Vapor pressure:	$\log P = -3.22 \times 10^4/T + 10.93$; 1775-1975°C	(227)
Pycnometric density:	3.03 g/cc	(129)
	3.01 g/cc	(127)

CHEMICAL

Theoretical analysis:	64% oxygen 36% beryllium	
Reactivity:	Useful to 2400°C in air	(227)
	Useful to 2000°C in vacuum	(227)
	Becomes appreciably volatile, 2100°C	(227)
	Volatilizes in H_2O vapor at 1200°C	(2)
	Uneffected by H_2O vapor at 950°C	(236)
	Dissociates in H_2O vapor at 1600°C	(236)
	H_2, excellent resistance	
	N_2, not attacked	
	NH_3, not attacked	
	CO, significant reaction at high temp.	
	CO_2, no reaction	

ELECTRICAL

Resistivity: 18 ohm-cm at 1000°C   (127)

MECHANICAL

Strength: Bending (MOR) : 24 x 10^3 psi   (127)
                         29 x 10^3 psi   (227)

Tensile: 35 x 10^3 at 500°C, 99.5% dense   (190)

30 x 10^3 at 1000°C, 99.5% dense   (190)

22 x 10^3 at 1200°C, 99.5% dense   (190)

18.5 x 10^3 at rt   (88, 227, 319)

13.8 x 10^3 at rt   (227, 297)

11.1 x 10^3 at 500°C   (2, 227)

7.0 x 10^3 at 900°C   (2, 227)

2.0 x 10^3 at 1140°C   (227)

0.6 x 10^3 at 1300°C   (227)

Compressive: 231 x 10^3 psi   (127)

200 x 10^3 at rt   (88, 227, 319)

114 x 10^3 at rt   (2, 227)

71 x 10^3 at 500°C   (227)

64 x 10^3 at 800°C   (227)

35.5 x 10^3 at 1000°C   (227)

28.5 x 10^3 at 1145°C   (227)

24 x 10^3 at 1400°C   (227)

MECHANICAL (cont.)

Strength: Compressive: (cont.)

$$17 \times 10^3 \text{ at } 1500^{\circ}C \qquad (227)$$

$$7 \times 10^3 \text{ at } 1600^{\circ}C \qquad (227)$$

Hardness: Mohs : 9

    Knoop 100 g : 1300 kg/mm^2 (127)

Young's modulus: 55.1 $\times 10^6$ psi   (127)

        53 $\times 10^6$ at 500°C, 99.5% dense (190)

        50 $\times 10^6$ at 1000°C, 99.5% dense (190)

        48 $\times 10^6$ at 1200°C, 99.5% dense (190)

        45.5 $\times 10^6$ at rt   (2)

        42.8 $\times 10^6$ at rt   (227, 297)

        40 $\times 10^6$ at 800°C   (2, 227)

        33 $\times 10^6$ at 1000°C   (227)

        20 $\times 10^6$ at 1145°C   (227)

OPTICAL

Colorless if pure

Refractive index: 1.719   (3)

Emissivity: Spectral 0.65μ : 0.212 at 927°C (158)

           0.235 at 1627°C (158)

    Total: 0.36 at 1000°C (227)

        0.42 at 1200°C (227)

        0.46 at 1400°C (227)

        0.49 at 1600°C (227)

        0.50 at 1800°C (227)

STRUCTURE

Hexagonal (127)

a = 2.698 A,   c = 4.380 A (527)

THERMAL

Conductivity: 0.0485 CGS at 1000°C (129)

0.11 CGS at 800°C (127)

0.035 CGS at 550°C (127)

0.22 CGS at 500°C, 99.5% dense (190)

0.068 CGS at 1000°C, 99.5% dense (190)

0.053 CGS at 1200°C, 99.5% dense (190)

0.496 CGS, 100°C, 5-10% pores (227)

0.521 CGS, 100°C, 0 pores (227)

0.42 CGS, 200°C, 0 pores (2)

0.22 CGS, 400°C, 0 pores (2)

0.13 CGS at 600°C (2)

0.07 CGS at 800°C (2)

0.045 CGS at 1000°C (2)

0.040 CGS at 1200°C (2)

0.035 CGS at 1600°C (2)

0.184 CGS at 400°C (531)

0.160 CGS at 400°C (534)

0.122 CGS at 600°C (531)

0.089 CGS at 600°C (534)

THERMAL (cont.)

Conductivity: (cont.)

0.137 CGS at 600°C	(532)
0.093 CGS at 800°C	(532)
0.062 CGS at 800°C	(534)
0.045 CGS at 1000°C	(534)
0.043 CGS at 1100°C	(533)
0.041 CGS at 1200°C	(534)
0.054 CGS at 1200°C	(532)
0.038 CGS at 1300°C	(533)
0.036 CGS at 1400°C	(533)
0.034 CGS at 1500°C	(533)
0.033 CGS at 1600°C	(533)
0.039 CGS at 1600°C	(532)
0.033 CGS at 1700°C	(533)
0.036 CGS at 1800°C	(531, 532, 533)
0.036 CGS at 1900°C	(531, 532, 533)
0.036 CGS at 2000°C	(531, 532, 533)

Expansion:

7.59×10^{-6} per °C;	25-500°C	(329)
9.03×10^{-6} per °C;	25-1000°C	(329)
10.3×10^{-6} per °C;	25-1500°C	(329)
11.1×10^{-6} per °C;	25-2000°C	(329)
9.5×10^{-6} per °C;	--	(129)

THERMAL (cont.)

Expansion: (cont.)

8.9×10^{-6} per $^{\circ}$C;   --   (127)

9.4×10^{-6} per $^{\circ}$C;   500-1200°C   (190)

9.18×10^{-6} per $^{\circ}$C;   25-1250°C   (297)

9.54×10^{-6} per $^{\circ}$C;   20-1400°C   (2, 227)

9.54×10^{-6} per $^{\circ}$C;   20-1000°C   (227, 280)

Parallel a axis	Parallel c axis	Average*		
7.1	6.3	6.83	; 28-252°C	(325)
7.8	6.7	7.43	; 28-474°C	(325)
8.5	7.8	8.27	; 28-749°C	(325)
9.2	8.2	8.87	; 28-872°C	(325)
9.9	8.9	9.57	; 28-1132°C	(325)

* $\dfrac{2a + c}{3}$

Specific heat: 0.52   (129)

CALCIUM OXIDE

Formula weight:	56.08 g/mole	
Melting point:	2600°C	(2)
	2572°C	(39)
	2580°C	(127)
	2570°C	(129)
	2545-2570°C	(227, 261)
Vapor pressure:	10^{-7} mm Hg at 2055°C	(227)
Pycnometric density:	3.32 g/cc	(127, 129)

CHEMICAL

Reactivity:	Useful in air to 2400°C	(227)
	Useful in reducing atmosphere to $>$1370°C	(227)
	H_2, not expected to react, 1400°C	(227)

MECHANICAL

Hardness: Mohs   :   4.5

Knoop 100 g :   560 kg/mm^2 (127)

STRUCTURE

Cubic, NaCl type	(127)
a = 4.8105 A	(527)

THERMAL

Conductivity:	0.0186 CGS at 1000°C	(129)
	0.037 CGS at 100°C, 0% porosity	(227)
	0.027 CGS at 200°C, 0% porosity	(227)
	0.022 CGS at 400°C, 0% porosity	(227)

THERMAL (cont.)

    Conductivity: (cont.)

                0.020 CGS at 600°C, 0% porosity   (227)

                0.019 CGS at 800°C, 0% porosity   (227)

                0.019 CGS at 1000°C, 0% porosity   (227)

    Expansion: 11.79×10^{-6} per $^{\circ}$C; 25-500°C   (329)

                13.12×10^{-6} per $^{\circ}$C; 25-1000°C   (329)

                15.25×10^{-6} per $^{\circ}$C; 25-1500°C   (329)

                13.6×10^{-6} per $^{\circ}$C; 20-1200°C

                14×10^{-6} per $^{\circ}$C;   -   (127)

                14.5×10^{-6} per $^{\circ}$C;   -   (129)

                13.7×10^{-6} per $^{\circ}$C; 20-1400°C   (2)

                12×10^{-6} per $^{\circ}$C; 28-255°C   (325)

                12.9×10^{-6} per $^{\circ}$C; 28-550°C   (325)

                13.1×10^{-6} per $^{\circ}$C; 28-722°C   (325)

                13.3×10^{-6} per $^{\circ}$C; 28-945°C   (325)

                13.9×10^{-6} per $^{\circ}$C; 28-1125°C   (325)

    Specific heat: 0.19 cal/gm/$^{\circ}$C   (129)

CeO_2

CERIUM DIOXIDE

Formula weight:	172.13 g/mole	
Formula volume:	24.14 cc/mole	
Melting point:	2600°C	(161, 227)
Vapor pressure:	Volatilizes at 1880°C	(227)
Pycnometric density:	7.13 g/cc	(2)

CHEMICAL

Reactivity:	Useful in air to 2400°C	(227)
	Not useful in reducing atmospheres	(227)
	Cannot be used in non-oxidizing atmospheres	(2)
	Subject to hydration	(227)
	H_2, easily reduced to Ce_2O_3, mp, 1690°C	(227)
	N_2, reacts at elevated temperature	

MECHANICAL

Hardness: Mohs:	6	(2)

OPTICAL

Color: white to pale yellow

STRUCTURE

Cubic   (2)

THERMAL

Expansion: 8.22×10^{-6} per °C;   25-500°C   (329)

THERMAL (cont.)

    Expansion: (cont.)

$$8.92 \times 10^{-6} \text{ per } {}^{\circ}C; \quad 25\text{-}1000{}^{\circ}C \tag{329}$$

$$8.65 \times 10^{-6} \text{ per } {}^{\circ}C; \quad \text{at } 500{}^{\circ}C \tag{32}$$

$$8.5 + 0.54\,T \text{ per } {}^{\circ}C; \quad 0\text{-}1000{}^{\circ}C \tag{2}$$

Heat content: $H_T - H_{293} = 14.24\,T + 2.8 \times 10^{-3}T^2$
$$-4,496 \text{ cal/mole}; \ 491\text{-}1140{}^{\circ}K \tag{128}$$

Specific heat: $C_p = 14.24\,T + 5.62 \times 10^{-3}T \text{ cal/mole/}{}^{\circ}K$
$$491 - 1140{}^{\circ}K \tag{128}$$

Formula weight:	165.88 g/mole	
Formula volume:	32.02 cc/mole	
Melting point:	d. 900°C	(3)
Pycnometric density:	5.18 g/cc	(3)

CHEMICAL

Reactivity: Soluble in aqueous acids

THERMAL

Expansion: 11.58×10^{-6} per °C; 25-500°C   (329)

11.80×10^{-6} per °C; 25-635°C   (329)

Cr_2O_3

DICHROMIUM TRIOXIDE

Formula weight:	152.02 g/mole	
Formula volume:	29.18 cc/mole	
Melting point:	d. 1995°C	(227, 265, 298)
	2265-2275°C	(227, 261)
	2400°C	(161, 227)
	2440°C	(227, 265)
	2265°C	(2)
	2275°C	(127)
Pycnometric density:	5.21 g/cc	(2)

CHEMICAL

Reactivity: O_2, forms less stable oxides   (227)

STRUCTURE

Trigonal

Rhombic   (127)

THERMAL

Expansion: 8.43×10^{-6} per °C;   25-500°C   (329)

8.62×10^{-6} per °C;   25-1000°C   (329)

8.82×10^{-6} per °C;   25-1500°C   (329)

9.55×10^{-6} per °C;   20-1400°C   (2, 227)

6.83×10^{-6} per °C;   at 200°C   (314)

7.38×10^{-6} per °C;   at 1200°C   (314)

12×10^{-6} per °C;   (127)

Dy_2O_3

DIDYSPROSIUM TRIOXIDE

Formula weight: 372.92 g/mole

Pycnometric density: 7.81 g/cc

CHEMICAL

Reactivity: Soluble in aqueous acids

THERMAL

Expansion: (C - form) 7.58×10^{-6} per ^{o}C; 25-500°C   (329)

9.75×10^{-6} per ^{o}C; 25-1000°C   (329)

Eu_2O_3

DIEUROPIUM TRIOXIDE

Formula weight: 352.00 g/mole

Pycnometric density: 7.42 g/cc

THERMAL

Expansion: (B - form) 9.47×10^{-6} per $^{\circ}C$; 25-500$^{\circ}C$ (329)

11.09×10^{-6} per $^{\circ}C$; 25-1000$^{\circ}C$ (329)

FeO

IRON MONOXIDE

Formula weight: 71.85 g/mole

Melting point: 1370°C   (161, 227)

FeO_{1-x}

IRON SUBOXIDE

THERMAL

Expansion: 11.79×10^{-6} per $^{\circ}C$; $25\text{-}500^{\circ}C$ (329)

14.57×10^{-6} per $^{\circ}C$; $25\text{-}1000^{\circ}C$ (329)

Fe_2O_3

DIIRON TRIOXIDE
(Ferric Oxide)

Formula weight:	159.70 g/mole	
Melting point:	1455°C	(161, 227)
	d. 1560°C	(3)
Pycnometric density:	5.12 g/cc	

STRUCTURE

Trigonal (3)

THERMAL

Expansion: (α form) 11.36×10^{-6} per °C; 25-500°C (329)

12.31×10^{-6} per °C; 25-1000°C (329)

Fe_3O_4

TRI IRON TETROXIDE
(Magnetite)

Formula weight: 231.55 g/mole

Melting point: 1595°C   (161, 227)

1565°C   (227, 261)

MAGNETIC

Strongly magnetic   (558)

OPTICAL

Color: black, opaque, metallic luster   (558)

Form: octahedra   (558)

STRUCTURE

Cubic   (558)

DIGADOLINIUM TRIOXIDE

Formula weight: 361.80 g/mole

Pycnometric density: 7.407 g/cc at 18^oC (3)

MECHANICAL

Young's modulus: 18.0×10^6 psi at 20^oC

STRUCTURE

Body-centered cubic   (83)

THERMAL

Conductivity: 0.005 CGS at 1000^oC

Expansion: (C - form) 9.05×10^{-6} per ^oC; 25-500^oC (329)

10.45×10^{-6} per ^oC; 25-1000^oC (329)

HAFNIUM DIOXIDE

Formula weight:	210.60 g/mole
Formula volume:	21.7 cc/mole

Melting point:
    2810°C
    2777°C   (2)
    2812°C   (39)
    2780°C   (227, 261)

Vapor pressure: $10^{-7.2}$ mm Hg at 1720°C (227, 285)

X-ray density: 9.68 g/cc

Pycnometric density: 9.68 g/cc   (2)

CHEMICAL

Reactivity: Limit of usefulness: air, 2400°C   (227)

reducing atmosphere, >1925°C (227)

H$_2$, stable to 1925°C (227)

N$_2$, probably nitrides at high temperature (227)

MECHANICAL

Strength: Bending (MOR): 10,000 psi

Young's modulus: 8.2 x 10^6 psi   (5)

STRUCTURE

Monoclinic to 1700°C

Tetragonal above 1700°C

Monoclinic at rt, 4 molecules per cell   (527)

a = 5.12 A; b = 5.18 A; c = 5.25 A (527)

β = 98° (527)

THERMAL

Conductivity: 0.00273 CGS; 25-425°C

Diffusivity: 0.01 cm^2/sec

Expansion: (Monoclinic)

5.47 x 10^{-6} per °C;	25-500°C	(329)
5.85 x 10^{-6} per °C;	25-1000°C	(329)
6.30 x 10^{-6} per °C;	25-1500°C	(329)
6.45 x 10^{-6} per °C;	25-1700°C	(329)
5.8 x 10^{-6} per °C;	25-1300°C	(5, 80)
6.45 x 10^{-6} per °C;	20-1700°C	(2)

Parallel a axis	Parallel b axis	Parallel c axis		
6.8	0	11	28-262°C	(325)
6.2	0.9	11.4	28-494°C	(325)
6.7	1.3	10.8	28-697°C	(325)
7.5	1.4	11.9	28-903°C	(325)
7.9	2.1	12.1	28-1098°C	(325)

(tetragonal)

1.31 x 10^{-6} per °C;	25-1700°C	(329)
3.03 x 10^{-6} per °C;	25-2000°C	(329)

Heat of formation: -251.8 kcal/mole

-266.06 ± 0.28 kcal/mole   (79)

Specific heat: 0.067 cal/gm/°C

La_2O_3

DILANTHANUM TRIOXIDE

Formula weight:	325.84 g/mole	
Melting point:	2305°C	(2)
	2316°C	(39)
	2310°C	
	2320°C	(227)
	2000°C	(3)
Boiling point:	4200°C	
X-ray density:	6.51 g/cc	(2)

CHEMICAL

Reactivity:	Useful in reducing atmospheres to 1370°C	(227)
	Reacts with H_2O	(227)
	Resistant to O_2	(227)
	H_2O, slight solubility, forms hydroxides	(3)
	Aqueous acids, soluble	(3)
	Aqueous alkali, soluble	(3)
	NH_4Cl solution, soluble	(3)
	Acetone, insoluble	(3)

STRUCTURE

Hexagonal   (83)

MAGNESIUM OXIDE

Formula weight: 40.32 g/mole

Formula volume: 11.2 cc/mole

Melting point: 2800°C   (2, 129, 227)

Vapor pressure: $\log P = -2.732 \times 10^4/T + 13.13$
                     1540 - 1940°C   (227)

X-ray density: 3.60 g/cc   (167)

Pycnometric density: 3.65 g/cc   (3)
                     3.58 g/cc   (129)

CHEMICAL

Theoretical analysis: 39.7% oxygen
                        60.3% magnesium

Reactivity: Limit of usefulness: air, 2400°C   (227)
                                      vacuum, 1600°C   (227)
                                      reducing atmosphere,
                                      1700-1980°C   (227)

               Melts at 2680 in O_2 free He   (227)

               Not volatile to 1800°C   (227)

               No volatility in H_2O vapor but crystalline
               form hydrates easily.   (236)

               H_2, not expected to react at 1400°C   (227)

               C, stable in contact to 1800°C

               CO, poor resistance

               CO_2, absorbed

ELECTRICAL

Resistivity: 10^8 ohm-cm at 1000°C   (127)

MECHANICAL

Strength: Tensile : 14×10^3 psi at rt   (2, 227)

14×10^3 psi at 200°C   (227)

15.2×10^3 psi at 400°C   (227)

16×10^3 psi at 800°C   (227)

11.5×10^3 psi at 1000°C   (227)

3.1×10^3 psi at 1000°C   (847)

10×10^3 psi at 1100°C   (227)

8×10^3 psi at 1200°C   (227)

6×10^3 psi at 1300°C   (2, 227)

Shear   : 12.1×10^3 psi at rt   (2, 227)

8.5×10^3 psi at 500°C   (227)

7.7×10^3 psi at 1000°C   (227)

5.7×10^3 psi at 1300°C   (227)

Compressive: 112,000 psi   (127)

Hardness: Mohs : 6   (2)

5.5 - 6   (3)

Young's modulus: 36.3×10^6 psi at rt   (127)

30.5×10^6 psi at rt   (2, 227)

12×10^6 psi at 20°C

12.4×10^6 psi at   (227, 297)

29.5×10^6 psi at 600°C   (227)

21×10^6 psi at 1000°C   (227)

MECHANICAL (cont.)

Young's modulus: (cont.)

10×10^6 psi at 1200°C   (227)

4×10^6 psi at 1300°C   (227)

42.74×10^6 psi at rt, $\rho = 3.506$ g/cc (167)

Modulus of rigidity: 16.7×10^6 psi at rt   (2, 227)

13.7×10^6 psi at 500°C

10.3×10^6 psi at 1100°C   (227)

8.4×10^6 psi at 1200°C   (227)

7.7×10^6 psi at 1300°C   (227)

5.2×10^6 psi at 1400°C   (227)

Shear modulus: 18.03×10^6 psi at rt, $\rho = 3.506$ g/cc (167)

Bulk modulus: 22.68×10^6 psi at rt, $\rho = 3.506$ g/cc (167)

Poisson's ratio: 0.163 at rt; $\rho = 3.506$ g/cc   (167)

Speed of Sound: 30,090 fps at rt; $\rho = 3.506$ g/cc   (167)

Creep rate:

Hot pressed: 0.15×10^{-6} in/cm/hr at 1100°C, 1200 psi   (2)

Sintered: 0.7×10^{-6} in/cm/hr at 1100°C, 1200 psi   (2)

NUCLEAR

Thermal neutron cross section: 0.0032 cm^2/cm^3

OPTICAL

Refractive index: 1,736

Emissivity: Spectral 0.65μ :   0.20 at 627°C to
0.45 at 1427°C   (158)

Total :   0.49 at 500°C   (227)

0.42 at 600°C   (227)

0.34 at 800°C   (227)

0.32 at 1000°C   (227)

0.30 at 1200°C   (227)

0.22 at 1400°C   (227)

STRUCTURE

Cubic

THERMAL

Conductivity: 0.097 CGS at rt   (328)

0.078 CGS at 100°C   (328)

0.064 CGS at 200°C   (328)

0.045 CGS at 400°C   (328)

0.026 CGS at 800°C   (328)

0.020 CGS at 1000°C   (328)

0% porosity: 0.082 CGS at 100°C   (125, 227)

0.065 CGS at 200°C   (227)

0.038 CGS at 400°C   (125, 227)

THERMAL (cont.)

    Conductivity: 0% porosity: (cont.)

0.027 CGS at 600°C	(125, 227)
0.020 CGS at 800°C	(227)
0.016 CGS at 1000°C	(227)
0.014 CGS at 1200°C	(125, 227)
0.014 CGS at 1400°C	(227)
0.016 CGS at 1600°C	(227)
0.020 CGS at 1800°C	(227)
0.0198 CGS at 800°C	(533, 536)
0.0167 CGS at 1000°C	(533, 536)
0.0148 CGS at 1200°C	(533, 536)
0.0139 CGS at 1200°C	(537)
0.0136 CGS at 1400°C	(533, 536)
0.0120 CGS at 1400°C	(537)
0.0153 CGS at 1600°C	(533, 536)
0.0108 CGS at 1600°C	(537)
0.0191 CGS at 1800°C	(533, 536)
0.0096 CGS at 1800°C	(537)

Diffusivity:

0.11 cm^2/sec at rt	(328)
0.096 cm^2/sec at 100°C	(328)
0.077 cm^2/sec at 200°C	(328)

THERMAL (cont.)

    Diffusivity: (cont.)

               0.051 cm^2/sec at 400°C   (328)

               0.028 cm^2/sec at 800°C   (328)

               0.023 cm^2/sec at 1000°C   (328)

    Expansion: 12.83 x 10^{-6} per °C; 25-500°C   (329)

               13.63 x 10^{-6} per °C; 25-1000°C   (329)

               15.11 x 10^{-6} per °C; 25-1500°C   (329)

               15.89 x 10^{-6} per °C; 25-1800°C   (329)

               14.0 x 10^{-6} per °C; 20-1400°C   (2)

               14.2-14.9 x 10^{-6} per °C; 20-1700°C   (227)

               13.3 x 10^{-6} per °C; 20-1800°C   (297)

               13.90 x 10^{-6} per °C; 0-1000°C   (7)

               14.46 x 10^{-6} per °C; 0-1200°C   (7)

               15.06 x 10^{-6} per °C; 0-1400°C   (7)

               11 x 10^{-6} per °C at 100°C   (5)

               15 x 10^{-6} per °C at 1000°C   (5)

    Specific heat: 0.29 cal/gm/°C   (129)

MANGANESE MONOXIDE

Formula weight: 70.93 g/mole

Melting point: 1775°C (227)

CHEMICAL

Reactivity: He, stable (227)

O_2, forms lower melting oxide (227)

Nb_2O_3

DINIOBIUM TRIOXIDE

Formula weight: 233.82 g/mole

Melting point: 1780°C
                    1775°C   (2, 227, 261)

Nb_2O_5

DINIOBIUM PENTOXIDE

Formula weight: 265.82 g/mole

Melting point: 1520°C   (3)
                       1460°C   (161, 227)

Pycnometric density: 4.6 g/cc

CHEMICAL

Reactivity: H_2SO_4, soluble

                  HF, soluble

NICKEL MONOXIDE

Formula weight:	74.69 g/mole	
Melting point:	1950°C	(2)
	2090°C	
	1950-1960°C	(161, 227, 261)
Vapor pressure:	10^{-7} mm Hg at 1585°C	(227)
Pycnometric density:	6.8 g/cc	(2)
	7.45 g/cc	(3)

CHEMICAL

Reactivity:	Easily reduced	(227)
	Oxidized in O_2, dissociates in higher temp.	(227)
	O_2, oxidizes at 400°C to Ni_2O_3	(3)
	Aqueous acids, soluble	(3)
	NH_4OH, soluble	(3)

MECHANICAL

Hardness: Mohs : 5.5

OPTICAL

Color: green to black	(3)
Refractive index: 2.37	(3)

STRUCTURE

Cubic	(3)
NaCl type	(2)

THERMAL

Conductivity: 0% porosity:

0.029 CGS at 100°C	(2, 227)
0.024 CGS at 200°C	(2, 227)
0.017 CGS at 400°C	(2, 227)
0.012 CGS at 800°C	(2, 227)
0.011 CGS at 1000°C	(2, 227)

$$Pr_4O_7$$

TETRAPRASEODYMIUM HEPTOXIDE

Formula weight: 675.68 g/mole

STRUCTURE

Face-centered cubic   (83)

SiO_2

SILICON DIOXIDE

Formula weight: 60.06 g/mole

Melting point: 1710°C   (227, 261, 298)

Vapor pressure: 10^{-7} mm Hg at 1725°C   (227)
                       Volatile at 1800°C   (227)

CHEMICAL

Reactivity: Vitreous SiO_2 devitrifies at 1100°C   (227)

                May be used in air to 1675°C   (227)

                He, loses O_2 at 1300°C   (227)

                H_2O, volatilizes at 1425-1475°C   (227)

                O_2, crystalline form may be used to
                1675°C   (227)

MECHANICAL

Hardness: Vickers 50 gm : 1103 ($\perp$ Opt. axis)   (8)
                             1260 (// Opt. axis)   (8)
                             1120   (8)

           10 μ diagonal : 1120 (1010 face)   (8)
                             1130 (1011 face)   (8)
                             1230 (1010 face)   (8)
                             1040 (1011 face)   (8)
                             1300 (Polished 1010 face)   (8)

           Knoop 100 g   :   710 (// Opt. axis)   (8)
                             790 ($\perp$ Opt. axis)   (8)

OPTICAL

Color: colorless

OPTICAL (cont.)

Refractive index: $\propto$ quartz: $\mathcal{W}$ = 1.5442   (558)

$\qquad\qquad\qquad\qquad\qquad$ ϵ = 1.5533   (558)

Sign: +   (558)

STRUCTURE

Hexagonal

THERMAL

Conductivity: 0.0025 at 200°C   (2, 227)

$\qquad\qquad\qquad$ 0.003 at 400°C   (2, 227)

$\qquad\qquad\qquad$ 0.004 at 800°C   (2, 227)

$\qquad\qquad\qquad$ 0.005 at 1200°C   (2, 227)

$\qquad\qquad\qquad$ 0.006 at 1600°C   (2, 227)

Expansion: ($\propto$ quartz)

$\qquad$ 19.35×10^{-6} per °C; 25-500°C   (329)

$\qquad$ $22.2 \ \times 10^{-6}$ per °C; 25-575°C   (329)

$\qquad$ (β quartz)

$\qquad$ $27.8 \ \times 10^{-6}$ per °C; 25-575°C   (329)

$\qquad$ 14.58×10^{-6} per °C; 25-1000°C   (329)

$\qquad$ ($\propto$ tridymite)

$\qquad$ $18.5 \ \times 10^{-6}$ per °C; 25-117°C   (329)

$\qquad$ (β_1 tridymite)

$\qquad$ $25.0 \ \times 10^{-6}$ per °C; 25-117°C   (329)

$\qquad$ $27.5 \ \times 10^{-6}$ per °C; 25-163°C   (329)

THERMAL (cont.)

    Expansion: (cont.)

        (β_2 tridymite)

           31.9 x 10^{-6} per ^{o}C; 25-163°C (329)

           19.35 x 10^{-6} per ^{o}C; 25-500°C (329)

           10.45 x 10^{-6} per ^{o}C; 25-1000°C (329)

        (Vitreous)

           0.527 x 10^{-6} per ^{o}C; 25-500°C (329)

           0.564 x 10^{-6} per ^{o}C; 25-1000°C (329)

           0.5   x 10^{-6} per ^{o}C; 20-1250°C (2)

        (Crystalline)

           43.0 x 10^{-6} per ^{o}C;   20-300°C (2, 227)

           3.0 x 10^{-6} per ^{o}C;   300-1100°C (2, 227)

DISAMARIUM TRIOXIDE

Formula weight: 348.86 g/mole

Pycnometric density: 7.43 g/cc at 15°C   (3)

CHEMICAL

Reactivity: H_2O, insoluble   (3)

Aqueous acids, very soluble   (3)

MECHANICAL

Young's modulus: 26.5×10^6 psi at 20°C

STRUCTURE

Face-centered cubic   (83)

THERMAL

Conductivity: 0.005 CGS at 1000°C

Expansion: (B - form) 9.06×10^{-6} per °C; 25-500°C   (329)

10.45×10^{-6} per °C; 25-1000°C   (329)

TIN DIOXIDE

Formula weight: 150.70 g/mole

Melting point: > 1900°C
                       1127°C   (3)

Pycnometric density: 7.00 g/cc   (3)

CHEMICAL

Reactivity: H$_2$SO$_4$, soluble; hot, concentrated   (3)

               Alkali, insoluble   (3)

OPTICAL

Color: colorless when pure

Refractive index: ω = 1.996   (558)
                     ϵ = 2.093   (558)

Birefringence: 0.097   (558)

Sign: +   (558)

STRUCTURE

Tetragonal   (558)

THERMAL

Expansion: 3.37 x 10^{-6} per °C; 25-500°C   (329)

              4.19 x 10^{-6} per °C; 25-1000°C   (329)

STRONTIUM OXIDE

Formula weight: 103.63 g/mole

Melting point: 2415°C   (2)
                                  2430°C
                                  $2415-2460^{\circ}$C   (161, 227, 261)

Vapor pressure: $\text{Log } P = 3.07 \times 10^{4}/T + 13.12$
                                  $1135-1375^{\circ}$C   (227)

CHEMICAL

Reactivity: Useful in reducing atmosphere to 1370°C   (227)

                      Vaporizes in He at 1600°C   (227)

                      Unstable in moist air   (2)

                      Good resistance to O_2 (227)

                      Forms hydrates with H_2O (227)

                      H_2, not expected to react 1400°C   (227)

MECHANICAL

Hardness: Mohs : 3.5

STRUCTURE

Cubic

THERMAL

Expansion: 12.40×10^{-6} per $^{\circ}$C; $25-500^{\circ}$C   (329)

                    13.53×10^{-6} per $^{\circ}$C; $25-1000^{\circ}$C   (329)

                    13.95×10^{-6} per $^{\circ}$C; $25-1200^{\circ}$C   (329)

Ta_2O_5

DITANTALUM PENTOXIDE

Formula weight:	441.76 g/mole	
Melting point:	1880°C	
	1875-1890°C	
	1890°C	(161, 227, 261)
	d. 1470°C	(3)
Pycnometric density:	8.735 g/cc	(3)

CHEMICAL

Reactivity:	Air; stable to melting point	(227)
	$KHSO_4$, soluble fused	(3)
	Acids, insoluble	(3)

OPTICAL

Form: Rhombic needles

THORIUM DIOXIDE

Formula weight:	264.12 g/mole	
Formula volume:	26.378 $\pm$ .005 cc/mole at 298.16°K	(553)
Melting point:	3205°C	(227, 275)
	3300°C	(129)
	3050°C	(127)
	3220 $\pm$ 50°C	(6)
Vapor pressure:	Log P = -3.71 x 10^4/T + 11.53 1775 - 1975°C	(227)
X-ray density:	9.821 g/cc	(126)
Pycnometric density:	10.03 g/cc	(127)
	10.01 g/cc	(129)
	9.69 g/cc	(2)

CHEMICAL

Reactivity:	Useful in air to 2700°C	(227)
	Becomes appreciably volatile in He at 2300°C	(227)
	Useful to 2150°C in oxygen	(227)
	H_2, good resistance	(227)
	C, reduced at high temperature	(2)

MECHANICAL

Strength: Tensile:	14 x 10^3 psi at rt	(10, 227, 252)
Compression:	214 x 10^3 psi at rt	(2, 227)
	146 x 10^3 psi at rt	(127)
	156 x 10^3 psi at 400°C	(227)
	85 x 10^3 psi at 600°C	(227)

MECHANICAL (cont.)

    Strength: (cont.)

        Compression: (cont.)

	71 x 10^3 psi at 800°C	(227)
	51 x 10^3 psi at 1000°C	(227)
	28.5 x 10^3 psi at 1200°C	(227)
	5.7 x 10^3 psi at 1400°C	(227)
	1.5 x 10^3 psi at 1500°C	(227)
Shear:	3.5 x 10^3 psi at 1100°C	(2, 227)
	1.2 x 10^3 psi at 1300°C	(227)

Hardness: Mohs   : 6.5   (2)

           Knoop 100 g : 945 kg/mm^2 (127)

Young's modulus: 34.87 x 10^6 psi at rt, density

                                 9.722 g/cc   (126)

	17.9 x 10^6 psi at rt	(227, 297)
	31.9 x 10^6 psi at rt	(127)
	21.3 x 10^6 psi at rt	(2, 227)
	21 x 10^6 psi at 20	(5)
	18.5 x 10^6 psi at 800°C	(227)
	18 x 10^6 psi at 800°C	(5)
	17.1 x 10^6 psi at 1000°C	(227)
	12.8 x 10^6 psi at 1200°C	(227)

Shear modulus: 13.66 x 10^6 psi at rt, density

                                 9.722 g/cc   (126)

Bulk modulus: 25.89 x 10^6 psi at rt, density

                                 9.722 g/cc   (126)

MECHANICAL (cont.)

Modulus of rigidity: 14.3 x 10^6 psi at rt   (227)

11.4 x 10^6 psi at 700°C   (227)

8.6 x 10^6 psi at 1100°C   (227)

5.6 x 10^6 psi at 1300°C   (227)

Poisson's ratio: 0.275 at rt, density 9.722 g/cc   (126)

Speed of sound: 16,310 fps at rt, density 9.722 g/cc (126)

Creep rate: 2.06 x 10^{-6} in/in/hr at 1200 psi at 1100°C   (2, 227)

NUCLEAR

Thermal neutron capture cross section: 0.160 cm^2/cm^3 (6)

STRUCTURE

Cubic, fluorite type   (527)

4 molecules per unit cell   (527)

a = 5.5997 at 26°C   (527)

    5.5968 at rt   (126)

    5.59525 $\pm$ 0.0001A at 298.16°K   (553)

THERMAL

Conductivity: 0.024 CGS at rt, 0% porosity   (227)

0.020 CGS at 100°C, 0% porosity   (5)

0.019 CGS at 200°C, 0% porosity   (227)

0.014 CGS at 400°C, 0% porosity   (227)

THERMAL (cont.)

    Conductivity: (cont.)

0.010 CGS at 600°C, 0% porosity	(227)
0.008 CGS at 800°C, 0% porosity	(227)
0.007 CGS at 1000°C, 0% porosity	(227)
0.0074 CGS at 1000°C, 0% porosity	(129)
0.006 CGS at 1200°C, 0% porosity	(227)
0.0076 CGS at 1200°C, 0% porosity	(5)
0.006 CGS at 1400°C, 0% porosity	(227)

Diffusivity: $0.04 \ cm^2/sec$

Expansion:		
8.63×10^{-6} per °C;	25-500°C	(329)
9.44×10^{-6} per °C;	25-1000°C	(329)
10.17×10^{-6} per °C;	25-1500°C	(329)
10.43×10^{-6} per °C;	25-1700°C	(329)
9.55×10^{-6} per °C;	20-1400°C	(2)
9.55×10^{-6} per °C;	20-800°C	(297)
3.67×10^{-6} per °C;	0 to -273°C	(327)
5.32×10^{-6} per °C;	0 to -173°C	(327)
6.47×10^{-6} per °C;	0 to -73°C	(327)
8.10×10^{-6} per °C;	0 to +27°C	(327)
8.06×10^{-6} per °C;	0-127°C	(327)
8.31×10^{-6} per °C;	0-227°C	(327)
8.53×10^{-6} per °C;	0-327°C	(327)

THERMAL (cont.)

    Expansion: (cont.)

                8.71×10^{-6} per $^{\circ}C$; $0-427^{\circ}C$ (327)

                8.87×10^{-6} per $^{\circ}C$; $0-527^{\circ}C$ (327)

                9.00×10^{-6} per $^{\circ}C$; $0-627^{\circ}C$ (327)

                9.14×10^{-6} per $^{\circ}C$; $0-727^{\circ}C$ (327)

                9.24×10^{-6} per $^{\circ}C$; $0-827^{\circ}C$*   (327)

                9.34×10^{-6} per $^{\circ}C$; $0-927^{\circ}C$ (327)

                9.42×10^{-6} per $^{\circ}C$; $0-1027^{\circ}C$ (327)

                9.53×10^{-6} per $^{\circ}C$; $0-1127^{\circ}C$ (327)

                9.60×10^{-6} per $^{\circ}C$; $0-1227^{\circ}C$ (327)

                9.68×10^{-6} per $^{\circ}C$; $0-1327^{\circ}C$ (327)

                9.76×10^{-6} per $^{\circ}C$; $0-1427^{\circ}C$ (327)

                9.83×10^{-6} per $^{\circ}C$; $0-1527^{\circ}C$ (327)

                9.91×10^{-6} per $^{\circ}C$; $0-1627^{\circ}C$ (327)

                9.97×10^{-6} per $^{\circ}C$; $0-1727^{\circ}C$ (327)

*Data above $827^{\circ}C$ obtained by extrapolation

                7.8×10^{-6} per $^{\circ}C$; $27-223^{\circ}C$ (325)

                8.7×10^{-6} per $^{\circ}C$; $27-498^{\circ}C$ (325)

                8.9×10^{-6} per $^{\circ}C$; $27-755^{\circ}C$ (325)

                9.2×10^{-6} per $^{\circ}C$; $27-994^{\circ}C$ (325)

                9.1×10^{-6} per $^{\circ}C$; $27-1087^{\circ}C$ (325)

THERMAL (cont.)

Expansion: (cont.)

8.96×10^{-6} per $^{\circ}$C; 0-1000°C

9.35×10^{-6} per $^{\circ}$C; 0-1200°C

9.84×10^{-6} per $^{\circ}$C; 0-1400°C

α_a = (Linear thermal expansion coefficient) =
$0.6216 \times 10^{-5} + 3.541 \times 10^{-9}T - 0.1125^{-2}$
from 298-1073°K   (553)

α_v = (Volume expansion coefficient) = $1.85 \times 10^{-5} + 10.96 \times 10^{-9}T$
$-0.3375\ T^{-2}$

from 298-1073°K   (553)

TITANIUM MONOXIDE

Formula weight: 63.90 g/mole

Melting point: 1760°C   (161, 227)

Vapor pressure: $10^{-3.3}$ at 1720°C   (227, 285)

THERMAL

Expansion: 9.06×10^{-6} per $^{\circ}$C; 25-500°C   (329)

            12.31×10^{-6} per $^{\circ}$C; 25-1000°C   (329)

            9.01×10^{-6} per $^{\circ}$C;   0-800°C   (7)

TITANIUM DIOXIDE

Formula weight:	79.90 g/mole	
Melting point:	1830°C	
	1920°C	(161, 227, 261)
d.	1640°C	(3)
Vapor pressure:	10$^{-4.7}$ at 1720°C	(227, 285)
Pycnometric density:	3.84 g/cc (Anatase)	(3)
	4.17 g/cc (Brookite)	(3)
	4.26 g/cc (Rutile)	(3)

CHEMICAL

Reactivity:	Useful only in oxidizing atmospheres	
	H_2SO_4, soluble	(3)
	Aqueous alkali, soluble	(3)

OPTICAL

Refractive index:	Anatase: 2.534 - 2.564	(3)
	Brookite: 2.586	(3)
	Rutile: ω = 2.603 - 2.616	(558)
	ϵ = 2.889 - 2.903	(558)
Birefringence	= 0.286 - 0.287	(558)
Emissivity: Total:	0.19 at 500°C	(227)
	0.19 at 600°C	(227)
	0.27 at 800°C	(227)

STRUCTURE

(Rutile) Tetragonal		
a = 4.594 A; c = 2.958 A at 26°C		(527)
(Anatase) Tetragonal		(3)
(Brookite) Rhombic		(3)

THERMAL

Conductivity: 0.016 CGS at 100°C, 0% porosity   (2, 227)

0.012 CGS at 200°C, 0% porosity   (2, 227)

0.009 CGS at 400°C, 0% porosity   (2, 227)

0.008 CGS at 600°C, 0% porosity   (2, 227)

0.008 CGS at 800°C, 0% porosity   (2, 227)

0.008 CGS at 1000°C, 0% porosity   (2, 227)

0.008 CGS at 1200°C, 0% porosity   (2, 227)

Expansion: 8.22×10^{-6} per $^{\circ}$C; 25-500°C   (329)

8.83×10^{-6} per $^{\circ}$C; 25-1000°C   (329)

9.50×10^{-6} per $^{\circ}$C; 25-1500°C   (329)

7.8×10^{-6} per $^{\circ}$C; 20-600°C   (2)

8.98×10^{-6} per $^{\circ}$C;   0-1000°C   (7)

(Rutile)

Parallel a axis	Parallel c axis	Average*		
7.9	9.8	8.53	26-240°C	(325)
8.2	10.5	8.97	26-455°C	(325)
8.1	10.6	8.93	26-670°C	(325)
8.2	10.5	8.97	26-940°C	(325)
8.3	10.8	9.13	26-1110°C	(325)

*$\dfrac{2a + c}{3}$ = average

Ti_2O_3

DITITANIUM TRIOXIDE

Formula weight: 143.80 g/mole

Melting point: 2130°C   (39, 161, 227)

Pycnometric density: 4.6 g/cc   (3)

CHEMICAL

Reactivity: In O_2 forms lower melting oxide   (227)

H_2SO_4, soluble   (3)

HCl, insoluble   (3)

HNO_3, insoluble   (3)

<u>URANIUM DIOXIDE</u>

Formula weight:	270.07 g/mole	
Formula volume:	24.5 cc/mole	
Melting point:	$2750 \pm 40^{o}C$	(448)
	$2870^{o}C$	(161, 227)
	$2880^{o}C$	(129)
	$2878 \pm 22^{o}C$	(6)
X-ray density:	10.97 g/cc	(448)
	10.949 g/cc	(126)
Pycnometric density:	10.88 g/cc	(127)
	10.96 g/cc	(2, 129)
	10.02 g/cc	(6)
	10.37 g/cc	(126)

<u>CHEMICAL</u>

Reactivity:	Easily oxidized to U_3O_8	(227)
	Useful in neutral atmospheres	(227)
	Resistant to air to $1400^{o}C$	(259)
	Forms hydrates with H_2O	(227)
	H_2, good resistance	(227)
	N_2, softens at $2180^{o}C$	(227)
	C, reacts to UC_x at $1600^{o}C$	(227)

<u>ELECTRICAL</u>

Resistivity:	$1-10^{-4}$ ohm-cm	(448)
	(Strongly influenced by surface effects; decreases with increasing temperature)	

<u>MECHANICAL</u>

Hardness:	Mohs	: 6-7	(276)
	Knoop 100 g	: 600 kg/mm^2	(127)

MECHANICAL (cont.)

Young's modulus:	21×10^6 psi, 0-1000°C	(227, 276)
	25×10^6 psi, at 20°C	(5)
	27.98×10^6 psi at rt, density 10.37 g/cc	(126)
Shear modulus:	10.75×10^6 psi at rt, density 10.37 g/cc	(126)
Bulk modulus:	23.50×10^6 psi at rt, density 10.37 g/cc	(126)
Poisson's ratio:	0.302×10^6 psi at rt, density 10.37 g/cc	(126)
Speed of sound:	14,150 fps, density 10.37 g/cc	(126)

NUCLEAR

Thermal neutron capture cross section: 0.165 cm^2/cm^3 (6)

STRUCTURE

Cubic, fluorite type	(448)
a = 5.471 A	(126)

THERMAL

Conductivity:	0.018 CGS at 100°C	(5, 6, 448)
	0.012 CGS at 400°C	(448)
	0.008 CGS at 600°C	(6)
	0.008 CGS at 700°C	(448)
	0.006 CGS at 1000°C	(5)
	0.006 CGS at 1200°C	(6)

THERMAL (cont.)

    Conductivity: (cont.)

            0.009   (127)

            0.025 CGS at 100°C, 0% porosity   (227)

            0.020 CGS at 200°C, 0% porosity   (227)

            0.015 CGS at 400°C, 0% porosity   (227)

            0.010 CGS at 600°C, 0% porosity   (227)

            0.009 CGS at 800°C, 0% porosity   (227)

            0.008 CGS at 1000°C, 0% porosity   (227)

            0.0082 CGS   -   -   (129)

    Expansion: 9.18×10^{-6} per °C; 27-400°C   (448)

               $10.8 \ \times 10^{-6}$ per °C; 400-800°C   (448)

               $12.6 \ \times 10^{-6}$ per °C; 800-1250°C   (448)

               9.47×10^{-6} per °C; 25-500°C   (329)

               11.19×10^{-6} per °C; 25-1000°C   (329)

               12.19×10^{-6} per °C; 25-1200°C   (329)

               11.15×10^{-6} per °C; 25-1750°C   (2, 227)

               $11.2 \ \times 10^{-6}$ per °C; 27-1260°C   (5)

Heating	Cooling		
9.07×10^{-6} per °C	9.28×10^{-6} per °C	27-400°C	(552)
11.1×10^{-6} per °C	10.8×10^{-6} per °C	400-800°C	(552)
13.0×10^{-6} per °C	12.9×10^{-6} per °C	800-1200°C	(552)

THERMAL (cont.)

    Expansion: (cont.)

$$\% \text{ expansion} = y = at^2 + bt + c \tag{552}$$

$$a = 2.1481 \times 10^{-7}$$

$$b = 8.4217 \times 10^{-4}$$

$$c = 3.0289 \times 10^{-2}$$

(27°C, unit length)

U_2O_3

DIURANIUM TRIOXIDE

Formula weight: 524.14 g/mole

Melting point: 1977°C

X-ray density: 4.87 g/cc

STRUCTURE

Hexagonal, Corundum type

VO_2

(V_2O_4)

<u>VANADIUM DIOXIDE</u>

Formula weight:	82.95 g/mole	
Melting point:	1550°C	(161, 227)
	1967°C	(3)
Vapor pressure:	$10^{-7.8}$ at 1720°C	(227, 285)
Pycnometric density:	4.399 g/cc	(3)

CHEMICAL

Reactivity:	Aqueous acids, soluble	(3)
	Aqueous alkali, soluble	(3)

V_2O_3

DIVANADIUM TRIOXIDE

Formula weight:	149.90 g/mole	
Melting point:	1977°C	(2)
	2410°C	
	>2000°C	(161, 227)
	1980°C	(227, 261)
	1970°C	(3)
Pycnometric density:	4.87 g/cc	(2)

CHEMICAL

Reactivity:	Easily oxidized	(227)
	H_2O, slightly soluble cold	(3)
	H_2O, soluble hot	(3)
	HNO_3, soluble	(3)
	HF, soluble	(3)
	Alkali, soluble	(3)

TUNGSTEN TRIOXIDE

Formula weight: 231.92 g/mole

Melting point: 1470°C   (161, 227, 261)
                      > 2130°C   (3)

Pycnometric density: 7.16 g/cc   (3)

CHEMICAL

Reactivity: H_2O, insoluble   (3)

                Alkali, soluble   (3)

                Acids, insoluble   (3)

OPTICAL

Color - yellow

STRUCTURE

Rhombic   (3)

THERMAL

Expansion: (monoclinic)
            13.75×10^{-6} per °C; 25-330°C   (329)

           (orthorhombic)
            13.75×10^{-6} per °C; 25-330°C   (329)

            12.20×10^{-6} per °C; 25-500°C   (329)

            12.20×10^{-6} per °C; 25-730°C   (329)

           (tetragonal)
            11.49×10^{-6} per °C; 25-730°C   (329)

            13.55×10^{-6} per °C; 25-1000°C   (329)

Y_2O_3

DIYTTRIUM TRIOXIDE

Formula weight:	225,84 g/mole	
Melting point:	$2410^{o}C$	(2)
	$2410-2415^{o}C$	(161, 227, 261)
Boiling point:	$4300^{o}C$ (760 mm)	(3)
Pycnometric density:	4,84 g/cc	(2)

CHEMICAL

Reactivity:	Useful in reducing atmospheres to $1370^{o}C$	(227)
	Loses weight in air up to $200^{o}C$	(227)
	Highly subject to hydration	(227)
	H_2, not expected to react at $1400^{o}C$	(227)
	CO_2, absorbed	(227)
	H_2O, insoluble	(3)
	Acids, soluble	(3)
	Alkali, insoluble	(3)

THERMAL

Conductivity:	Like Al_2O_3	(227)

Yb_2O_7

DIYTTERBIUM HEPTOXIDE

Formula weight: 289.84 g/mole

STRUCTURE

Body-centered cubic   (83)

ZINC MONOXIDE

Formula weight: 81.38 g/mole

Melting point: $1975^{o}C$ (2)
                        $>1800^{o}C$ (3)

X-ray density: 5.66 g/cc   (2)

Pycnometric density: 5.606 g/cc   (3)

CHEMICAL

Reactivity: H_2O, insoluble   (3)

                Acids, soluble   (3)

                Alkalis, soluble   (3)

                NH_4Cl, soluble   (3)

                NH_3, insoluble   (3)

MECHANICAL

Hardness: Mohs : 4-4.5   (2)

STRUCTURE

Hexagonal   (2)

THERMAL

Expansion: 5.90×10^{-6} per ^{o}C; $25\text{-}500^{o}C$ (329)

              6.77×10^{-6} per ^{o}C; $25\text{-}1000^{o}C$ (329)

              6.99×10^{-6} per ^{o}C; $25\text{-}1200^{o}C$ (329)

ZIRCONIUM OXIDE

Note: (s) signifies stabilized zirconia

Formula weight: 123.22 g/mole

Formula volume: 21.3 cc/mole

Melting point: 2710°C
                         2690°C
                         2677°C   (2, 39)
                         2680-2705°C   (161, 227, 261, 298)
                    (s) 2600°C   (6)
                    (s) 2650°C   (129)

Vapor pressure: Highly volatile at 2300°C   (227)

X-ray density: (s) 5.754 g/cc   (126)

Pycnometric density: 5.56 g/cc   (2)
                         5.59 g/cc
                    (s) 5.32 g/cc   (6)
                    (s) 5.7 g/cc   (129)
                    (s) 5.78 g/cc   (127)

CHEMICAL

Reactivity: ·   Limit of usefulness: air, 2400°C   (227)
                                        vacuum, 2200°C   (227)
                                        reducing atmosphere
                                        1955-2205°C   (227)

              Becomes volatile in He above 2300°C   (227)

              H$_2$, stable to 2200°C

              N$_2$, nitrides at high temperature

ELECTRICAL

Resistivity: (s) 2300 ohm-cm at 700°C   (129)

             (s) 77 ohm-cm at 1200°C   (129)

ELECTRICAL (cont.)

Resistivity: (cont.)

(s) 9.4 ohm-cm at 1300°C		(129)
(s) 1.6 ohm-cm at 1700°C		(129)
(s) 0.59 ohm-cm at 2000°C		(129)
(s) 0.37 ohm-cm at 2200°C		(129)

MECHANICAL

Strength: Bending (MOR):   30,000 psi   (127)

Tensile:		
16.4×10^3 psi		(45)
17.9×10^3 at rt		(227, 297)
20×10^3 at rt		(2, 227)
16.8×10^3 at 200°C		(2)
17.5×10^3 at 400°C		(227)
20.0×10^3 at 500°C		(190)
17.6×10^3 at 600°C		(227)
16.0×10^3 at 800°C		(227)
6.75×10^3 at 1000°C		(124, 185)
17.0×10^3 at 1000°C		(190)
14.8×10^3 at 1000°C		(227)
13.5×10^3 at 1100°C		(227)
13.0×10^3 at 1200°C		(190)
12.1×10^3 at 1200°C		(227)
10.2×10^3 at 1300°C		(227)

MECHANICAL (cont.)

    Strength: (cont.)

        Compression: 205×10^3 psi at rt   (127)

                        300×10^3 psi at rt   (2, 227)

                        228×10^3 psi at $500^{\circ}C$ (227)

                        171×10^3 psi at $1000^{\circ}C$ (227)

                        114×10^3 psi at $1200^{\circ}C$ (227)

                        18.5×10^3 psi at $1400^{\circ}C$ (227)

                        2.8×10^3 psi at $1500^{\circ}C$ (227)

    Hardness: Mohs   : 6.5   (2)

             Knoop 100 g : 1200 kg/mm^2 (127)

    Young's modulus: (s) 36×10^6 psi at $20^{\circ}C$.   (5)

                        24.8×10^6 psi at rt   (227, 297)

                        27×10^6 psi at rt   (2, 227)

                        22×10^6 psi at $500^{\circ}C$ (190)

                        18.9×10^6 psi at $800^{\circ}C$ (2)

                        18.5×10^6 psi at $1000^{\circ}C$ (227)

                        20.0×10^6 psi at $1000^{\circ}C$ (120)

                        25×10^6 psi at $1000^{\circ}C$ (124, 185)

                        17.1×10^6 psi at $1200^{\circ}C$ (227)

                        18.0×10^6 psi at $1200^{\circ}C$ (190)

                        14.2×10^6 psi at $1400^{\circ}C$ (227)

                        12.8×10^6 psi at $1500^{\circ}C$ (227)

             (s) 19.96×10^6 psi at rt, density
                                    5.634 g/cc   (126)

MECHANICAL (cont.)

      Modulus of rigidity: 14.1×10^6 psi at 1300°C   (2)

      Shear modulus: 7.40×10^6 psi at rt, density
                          5.634 g/cc   (126)

      Bulk modulus: 11.75×10^6 psi at rt, density
                          5.634 g/cc   (126)

      Poisson's ratio: 0.337 at rt, density 5.634 g/cc   (126)

      Speed of sound: (s) 16,210 fps, density 5.634 g/cc   (126)

NUCLEAR

      Cross section: (s) 0.0571 cm^2/cm^3

OPTICAL

      Emissivity: Spectral 0.65μ: 0.40 at 727-1727°C   (158)

                Total : 0.65 at 500°C   (227)

                        0.50 at 600°C   (227)

                        0.36 at 800°C   (227)

                        0.33 at 1000°C   (227)

                        0.34 at 1200°C   (227)

                        0.38 at 1400°C   (227)

STRUCTURE

      Monoclinic to 1050°C

      Tetragonal above 1050°C

STRUCTURE (cont.)

Monoclinic   a = 5.169 A,   b = 5.232 A,   c = 5.341   (527)

β = 99° 15' at rt   (527)

Stabilized: Cubic: a = 5.1195 A   (126)

THERMAL

Conductivity:(s) 0.004 CGS at 100°C   (5)

(s) 0.0049 CGS at 1200°C   (5)

(s) 0.0055 CGS at 1000°C   (129)

(s) 0.005 CGS at 100°C, 0% porosity   (2, 227)

(s) 0.005 CGS at 200°C, 0% porosity   (2, 227)

(s) 0.005 CGS at 400°C, 0% porosity   (2, 227)

(s) 0.0055 CGS at 800°C, 0% porosity   (2, 227)

(s) 0.006 CGS at 1200°C, 0% porosity   (2, 227)

(s) 0.0065 CGS at 1400°C, 0% porosity   (2, 227)

(s) 0.0044 CGS at 500°C   (190)

(s) .0048 CGS at 1000°C   (190)

(s) .0050 CGS at 1200°C   (190)

(s) .002 CGS   -   (127)

Expansion: (monoclinic)

6.53×10^{-6} per °C ;   25-500°C   (329)

7.59×10^{-6} per °C ;   25-1000°C   (329)

7.72×10^{-6} per °C ;   25-1050°C   (329)

8.0×10^{-6} per °C ;   200-1080°C   (80)

THERMAL (cont.)

    Expansion: (cont.)

        (tetragonal)

            -21.7×10^{-6} per $^{\circ}$C; 25-1050°C   (329)

            -11.11×10^{-6} per $^{\circ}$C; 25-1500°C   (329)

            -9.53×10^{-6} per $^{\circ}$C; 25-1600°C   (329)

            5.5×10^{-6} per $^{\circ}$C; 20-1200°C

            5.0×10^{-6} per $^{\circ}$C;   0-1400°C   (297, 227)

            5.58×10^{-6} per $^{\circ}$C; 20-1200°C   (227, 2)

            10.0×10^{-6} per $^{\circ}$C   -   (4)

            7.2×10^{-6} per $^{\circ}$C; -10-1000°C   (5)

            10.6×10^{-6} per $^{\circ}$C;   0-1200°C (CaO)   (7)

            10.52×10^{-6} per $^{\circ}$C;   0-1000°C (MgO)   (7)

            8.64×10^{-6} per $^{\circ}$C;-20-600°C   (84, 185)

            8.7×10^{-6} per $^{\circ}$C;   -   (129)

            4.0×10^{-6} per $^{\circ}$C;   0-500°C   (190)

            10.5×10^{-6} per $^{\circ}$C;   0-1000°C   (190)

            11.0×10^{-6} per $^{\circ}$C;   0-1500°C   (190)

            7.0×10^{-6} per $^{\circ}$C;   -   (127)

Parallel a axis	Parallel b axis	Parallel c axis		
8.4	3	14	27-264°C	(325)
7.5	2	13	27-504°C	(325)
6.8	1.1	11.9	27-759°C	(325)

THERMAL (cont.)

Expansion: (tetragonal) (cont.)

Parallel a axis	Parallel b axis	Parallel c axis		
7.8	1.5	12.8	27-964°C	(325)
8.7	1.9	13.6	27-1110°C	(325)

Specific heat: 0.17 cal/gm °C (129)

DATA TABLES

VII MIXED OXIDES

$Al_2O_3 \cdot BaO$

Formula weight: 255.30 g/mole

Melting point: 2000^oC (2, 227)

Pycnometric density: 3.99 g/cc (2)

THERMAL

Expansion: 6.74×10^{-6} per ^oC; 25-500^oC (329)

              7.48×10^{-6} per ^oC; 25-1000^oC (329)

$Al_2O_3 \cdot BeO$

Formula weight: 126.96 g/mole

Melting point: 1870°C   (2)
                       1910°C   (227)

CHEMICAL

Reactivity: H_2O, hygroscopic   (227)

THERMAL

Expansion: 7.37×10^{-6} per °C; 25-500°C   (329)

                  8.00×10^{-6} per °C; 25-1000°C   (329)

$Al_2O_3 \cdot CaO$

Formula weight: 158,.02 g/mole

Melting point: $>1590^oC$ (227, 314)
                        1960^oC

$Al_2O_3 \cdot CoO$

Formula weight: 176.88 g/mole

Melting point: 1955°C   (2)

Pycnometric density: 4.37 g/cc   (2)

THERMAL

Expansion: 8.00×10^{-6} per ^{o}C;   25-500°C   (329)

8.52×10^{-6} per ^{o}C;   25-1000°C   (329)

$Al_2O_3 \cdot Co_2O_3$

Formula weight: 267.82 g/mole

THERMAL

   Expansion: 7.79×10^{-6} per $^{\circ}C$;   25-500°C   (329)

                   8.52×10^{-6} per $^{\circ}C$;   25-1000°C   (329)

$Al_2O_3 \cdot FeO$

Formula weight: 173.79 g/mole

THERMAL

Expansion: 8.00×10^{-6} per $^{\circ}C$;   25-500°C   (329)

8.62×10^{-6} per $^{\circ}C$;   25-1000°C   (329)

$Al_2O_3 \cdot Li_2O$

Formula weight: 131.82 g/mole

THERMAL

    Expansion: 11.57×10^{-6} per $^\circ$C; 25-500°C   (329)

                     12.42×10^{-6} per $^\circ$C; 25-1000°C   (329)

$Al_2O_3 \cdot MgO$

SPINEL

Formula weight:	142.26 g/mole	
Formula volume:	39.73 cc/mole	
Melting point:	d. 2135°C 2130°C	(2, 227, 265)
X-ray density:	3.580 g/cc	(126)
Pycnometric density:	3.510 g/cc	(126)

CHEMICAL

Reactivity:	Limit of usefulness in air, 1900°C	(265)
	H_2, stable to 1400°C	(227)

MECHANICAL

Strength: Tensile   : 19.2×10^3 psi at rt   (2)

                                    13.7×10^3 psi at 550°C   (227)

                                    10.8×10^3 psi at 900°C   (227)

                                    6.1×10^3 psi at 1150°C   (227)

                                    1.1×10^3 psi at 1300°C   (227)

          Compression : 270×10^3 psi at rt   (2)

                                    199×10^3 psi at 500°C   (227)

                                    171×10^3 psi at 800°C   (227)

                                    85.5×10^3 psi at 1100°C   (227)

                                    $71 \quad \times 10^3$ psi at 1200°C   (227)

                                    21.4×10^3 psi at 1400°C   (227)

                                    8.5×10^3 psi at 1600°C   (227)

          Shear   : 9.4×10^3 psi at rt   (2, 227)

                                    9.3×10^3 psi at 200°C   (227)

                                    9.1×10^3 psi at 400°C   (227)

                                    8.7×10^3 psi at 600°C   (227)

                                    8.2×10^3 psi at 800°C   (227)

                                    7.5×10^3 psi at 1000°C   (227)

MECHANICAL (cont.)

Strength: Shear (cont.) : 6.4×10^3 psi at 1200°C   (227)

5.3×10^3 psi at 1300°C   (227)

Young's modulus: 34.5×10^6 psi at rt   (2, 227)

34.4×10^6 psi at 200°C   (227)

34.3×10^6 psi at 400°C   (227)

$34 \quad \times 10^6$ psi at 600°C   (227)

32.9×10^6 psi at 800°C   (227)

30.4×10^6 psi at 1000°C   (227)

25.0×10^6 psi at 1200°C   (227)

20.1×10^6 psi at 1300°C   (227)

38.23×10^6 psi at rt, density
3.510 g/cc   (126)

Shear modulus: 14.78×10^6 psi at rt, density
3.510 g/cc   (126)

Bulk modulus: 31.52×10^6 psi at rt, density
3.510 g/cc   (126)

Poisson's ratio: 0.294×10^6 psi at rt, density
3.510 g/cc   (126)

Speed of sound: 28,440 fps at rt, density 3.510 g/cc (126)

Modulus of rigidity: 13.2×10^6 psi at rt   (2, 227)

13.1×10^6 psi at 200°C   (227)

12.3×10^6 psi at 400°C   (227)

12.4×10^6 psi at 600°C   (227)

11.6×10^6 psi at 800°C   (227)

MECHANICAL (cont.)

Modulus of rigidity: (cont.)

10.3×10^6 psi at $1000^{\circ}C$ (227)

8.5×10^6 psi at $1200^{\circ}C$ (227)

7.2×10^6 psi at $1300^{\circ}C$ (227)

THERMAL

Conductivity: 0% porosity:

0.035 CGS at $100^{\circ}C$ (2, 227, 287)

0.031 CGS at $200^{\circ}C$ (2, 227, 287)

0.024 CGS at $400^{\circ}C$ (2, 227, 287)

0.019 CGS at $600^{\circ}C$ (2, 227, 287)

0.015 CGS at $800^{\circ}C$ (2, 227, 287)

0.013 CGS at $1000^{\circ}C$ (2, 227, 287)

0.0138 CGS at $1000^{\circ}C$ (129)

0.013 CGS at $1200^{\circ}C$ (2, 227, 287)

Expansion: 7.79×10^{-6} per $^{\circ}C$; 25-$500^{\circ}C$ (329)

8.41×10^{-6} per $^{\circ}C$; 25-$1000^{\circ}C$ (329)

9.17×10^{-6} per $^{\circ}C$; 25-$1500^{\circ}C$ (329)

9.0×10^{-6} per $^{\circ}C$; 20-$1250^{\circ}C$ (2, 227)

9.6×10^{-6} per $^{\circ}C$;   -   (129)

Specific heat: 0.29 cal/gm/$^{\circ}C$ (129)

$Al_2O_3 \cdot MnO$

Formula weight: 172.87 g/mole

THERMAL

    Expansion: 6.52×10^{-6} per $^{\circ}$C ;   25-500°C   (329)

                      7.18×10^{-6} per $^{\circ}$C ;   25-1000°C   (329)

$Al_2O_3 \cdot NiO$

Formula weight: 176.63 g/mole

Melting point: 2020°C   (2, 227)

Pycnometric density: 4.45 g/cc   (2)

THERMAL

    Expansion: 7.79×10^{-6} per °C; 25-500°C   (329)

                           8.41×10^{-6} per °C; 25-1000°C   (329)

$Al_2O_3 \cdot SiO_2$

SILLIMANITE

Formula weight:	162.00 g/mole	
Melting point:	d. 1620°C	(227)
Pycnometric density:	3.23-3.24 g/cc	(3)

CHEMICAL

Limit of usefulness:	Air, 1800°C	(227)
	Vacuum, 1285-1500°C	(227)

MECHANICAL

Hardness: Mohs: 6-7	(3)

OPTICAL

Refractive index: 1.660	(3)

STRUCTURE

Orthorhombic	(3)

THERMAL

Conductivity: 0% porosity:

0.0042 CGS at 100°C	(2, 227, 287)
0.004 CGS at 400°C	(2, 227, 287)
0.0035 CGS at 800°C	(2, 227, 287)
0.0035 CGS at 1200°C	(2, 227, 287)
0.003 CGS at 1500°C	(2, 227, 287)

Expansion: 6.58×10^{-6} per °C at 20°C	(227)

$Al_2O_3 \cdot SrO$

Formula weight:	205.57 g/mole	
Melting point:	2110°C	(2, 227)
	2020°C	

$Al_2O_3 \cdot TiO_2$

Formula weight: 181.84 g/mole

THERMAL

Expansion: 8.63×10^{-6} per ^{o}C; 25-500°C (329)

9.53×10^{-6} per ^{o}C; 25-1000°C (329)

$Al_2O_3 \cdot ZnO$

Formula weight: 183.32 g/mole

Melting point: 1950°C

THERMAL

Expansion: 8.00×10^{-6} per °C; 25-500°C (329)

8.72×10^{-6} per °C; 25-1000°C (329)

8.94×10^{-6} per °C; 25-1200°C (329)

$Al_2O_3 \cdot ZrO_2$

Formula weight: 225.16 g/mole

Melting point: $\sim$2000°C   (227, 298)

- 405 - $2\ Al_2O_3 \cdot 2\ MgO \cdot 5\ SiO_2$

<u>CORDIERITE</u> (Iolite)

Formula weight: 584.82 g/mole

Pycnometric density: 2.6 - 2.7 g/cc (3)

CHEMICAL

Reactivity: Maximum temperature for continuous
service, 1275°C (280)

MECHANICAL

Hardness: Mohs : 7 ¬ 7.5 (3)

OPTICAL

Color: Strongly pleochroic (558)

Refractive index: N_a = 1.532 - 1.552 (558)

N_b = 1.536 - 1.562 (558)

N_c = 1.539 - 1.570 (558)

Birefringence: Weak, 0.007 - 0.011 (558)

2 V 40-80° (558)

Sign: (+) or (-) (558)

STRUCTURE

Orthorhombic, pseudo-hexagonal

THERMAL

Conductivity: 0.0076 at 35-43°C (309)

0.003 - .004 at elevated temperature (227, 288)

Expansion: 6.5 - 10.0 x 10^{-6} per °C (227, 288)

$3 \text{ Al}_2\text{O}_3 \cdot 2 \text{ CeO}_2$

Formula weight: 650.08 g/mole

THERMAL

Expansion: 8.84×10^{-6} per $^{\circ}C$;   25-500°C   (329)

9.54×10^{-6} per $^{\circ}C$;   25-1000°C   (329)

$3\ Al_2O_3 \cdot 2\ SiO_2$

MULLITE

Formula weight:	425.94 g/mole	
Melting point:	1810°C	(227)
	1830°C	(2)
Pycnometric density:	2.779 g/cc	(126)

CHEMICAL

Limit of usefulness:	Air, 1800°C	(227)
	Vacuum, 1500-1700°C	(227)
	Reducing atmosphere, fair	(227)

MECHANICAL

Strength: Shear:	$> 2.4 \times 10^3$ psi at 1100°C	(227)
Young's modulus:	20.75×10^6 psi at rt, density 2.779 g/cc	(126)
Modulus of rigidity:	8.5×10^6 psi at rt	(2)
Shear modulus:	8.38×10^6 psi at rt, density 2.779 g/cc	(126)
Bulk modulus:	13.20×10^6 psi at rt, density 2.779 g/cc	(126)
Poisson's ratio:	0.238 psi at rt, density 2.779 g/cc	(126)
Speed of sound:	23,540 fps at rt	(126)
Creep:	0.4×10^{-6} in/in/hr at 1200 psi, 1100°C	(227)

STRUCTURE

Orthorhombic

THERMAL

    Conductivity: 0% porosity:

        0.0145 CGS at 100^oC (2, 227, 287)

        0.013 CGS at 200^oC (2, 227, 287)

        0.011 CGS at 400^oC (2, 227, 287)

        0.010 CGS at 600^oC (2, 227, 287)

        0.0095 CGS at 800^oC (2, 227, 287)

        0.009 CGS at 1000^oC (2, 227, 287)

        0.009 CGS at 1200^oC (2, 227, 287)

        0.009 CGS at 1400^oC (2, 227, 287)

    Expansion: 4.5×10^{-6} per ^oC; 20-1325^oC (2, 227)

        4.63×10^{-6} per ^oC; 25-500^oC (329)

        5.13×10^{-6} per ^oC; 25-1000^oC (329)

        5.62×10^{-6} per ^oC; 25-1500^oC (329)

$3 \ Al_2O_3 \cdot 2 \ SnO_2$

Formula weight: 607.22 g/mole

THERMAL

Expansion: 7.16×10^{-6} per $^{\circ}$C;   25-500°C   (329)

7.69×10^{-6} per $^{\circ}$C;   25-1000°C   (329)

$3\ Al_2O_3 \cdot 2\ ThO_2$

Formula weight: 834.06 g/mole

THERMAL

    Expansion: 7.79×10^{-6} per $^{\circ}$C;   25-500°C   (329)

                      8.52×10^{-6} per $^{\circ}$C;   25-1000°C   (329)

Formula weight: 677.94 g/mole

THERMAL

Expansion: 1.89×10^{-6} per ^oC; 25-500^oC (329)

2.26×10^{-6} per ^oC; 25-1000^oC (329)

4.17×10^{-6} per ^oC; 25-1200^oC (329)

Formula weight: 539.58 g/mole

THERMAL

Expansion: 7.16×10^{-6} per °C;   25-500°C   (329)

8.31×10^{-6} per °C;   25-1000°C   (329)

$9\ Al_2O_3 \cdot 2\ B_2O_3$

Formula weight: 1056.74 g/mole

THERMAL

    Expansion: 3.89×10^{-6} per $^{\circ}C$;   25-500°C   (329)

                    4.20×10^{-6} per $^{\circ}C$;   25-1000°C   (329)

$ZrO_2 \cdot BaO$

Formula weight: 276.58 g/mole

Melting point: 2700°C   (2)
                       2620°C   (227)

Pycnometric density: 6.26 g/cc   (2)

CHEMICAL

Reactivity: O_2, may form lower melting BaO_2 (227)

                 H_2O, may hydrate   (227)

                 H_2, probably stable at 1400°C   (227)

THERMAL

Expansion: 7.79×10^{-6} per °C; 25-500°C   (329)

                 8.52×10^{-6} per °C; 25-1000°C   (329)

$Ta_2O_5 \cdot 6\ BaO$

$(Ba_3\ Ta\ O_{5.5})$

Formula weight:	1361.92 g/mole
X-ray density:	6.906 g/cc (556)
Pycnometric density:	6.79 g/cc (556)

CHEMICAL

Synthesis: Sinter equivalent amounts of BaO and Ta_2O_5 at 1100°C in oxygen   (556)

ELECTRICAL

Dielectric constant: 13.18   (556)

STRUCTURE

Face-centered cubic   (556)

a = 8.69 A   (556)

$Cr_2O_3 \cdot BeO$

Formula weight: 177.04 g/mole

THERMAL

Expansion: 6.74×10^{-6} per °C;   25-500°C   (329)

7.48×10^{-6} per °C;   25-1000°C   (329)

Formula weight: 295.09 g/mole

THERMAL

Conductivity: 0.10 CGS at 200°C   (227)

0.06 CGS at 425°C   (227)

0.055 CGS at 1000°C   (227)

$SiO_2 \cdot 2\ BeO$

PHENACITE

Formula weight: 110.10 g/mole

Melting point: $> 1750^{\circ}C$ (2)

                         $2000^{\circ}C$ (227)

THERMAL

Expansion: 5.68×10^{-6} per $^{\circ}C$; 25-$500^{\circ}C$ (329)

              6.36×10^{-6} per $^{\circ}C$; 25-$1000^{\circ}C$ (329)

$UO_2 \cdot 2\ BeO$

Formula weight: 320.11 g/mole

Melting point: 2150°C   (227)

2 $ZrO_2 \cdot$ 3 BeO

Formula weight: 321.50 g/mole

Melting point: 2530°C   (2, 227)
                       2535°C

$Cr_2O_3 \cdot CaO$

Formula weight:	208.10 g/mole	
Melting point:	2175°C	(227)
	2100°C	(227)
	2170°C	(2)
Pycnometric density:	4.8 g/cc	(2)

CHEMICAL

Limit of usefulness:	Air, $>980°C$	(227)
	Reducing atmosphere $>980°C$	(227)
	May hydrate	(227)
	H_2, stable to 1000°C	(227)

Formula weight:	266.68 g/mole	
Melting point:	$2470^\circ C \pm 20^\circ C$	(76)
X-ray density:	5.73 g/cc	(76)

CHEMICAL

Synthesis: 1:1 mole ratio, HFO_2 and CaO; sinter
1700°C   (76)

OPTICAL

Biaxial Negative   (76)

Low Birefringence   (76)

Refractive index   >1.957   (76)

STRUCTURE

Orthorhombic

a = 11.08 A; b = 7.94 A; c = 1142 A   (76)

THERMAL

Expansion: 6.12×10^{-6} per °C; 25-500°C   (329)

7.28×10^{-6} per °C; 25-1000°C   (329)

3.6×10^{-6} per °C; 100-600°C   (76)

12.1×10^{-6} per °C; 600-1300°C   (76)

7.08×10^{-6} per °C; 1000-1300°C   (76)

$SiO_2 \cdot CaO$

Formula weight: 116.14 g/cc

Melting point: 1540°C   (227, 265)

THERMAL

Expansion: (α form) 9.90×10^{-6} per °C; 25-500°C   (329)

11.18×10^{-6} per °C; 25-1000°C   (329)

(β form) 5.68×10^{-6} per °C; 25-500°C   (329)

5.93×10^{-6} per °C; 25-700°C   (329)

$$TiO_2 \cdot CaO$$

Formula weight: 135.98 g/mole

Melting point: 1980°C
                       1975°C   (2)

Pycnometric density: 4.10 g/cc   (2)

STRUCTURE

Cubic, perovskite type

THERMAL

Expansion: 13.04×10^{-6} per °C: 25-500°C   (329)

              14.05×10^{-6} per °C; 25-1000°C   (329)

$ZrO_2 \cdot CaO$

Formula weight:	179.30 g/mole	
Melting point:	2330°C	(227)
	2345°C	(2)
	2350°C	
	2550°C	(39)
Pycnometric density:	4.78 g/cc	(2)

STRUCTURE

Monoclinic

THERMAL

Expansion: 9.05×10^{-6} per °C; 25-500°C (329)

10.46×10^{-6} per °C; 25-1000°C (329)

11.14×10^{-6} per °C; 25-1200°C (329)

11.5×10^{-6} per °C; 25-1300°C (546)

Formula weight: 172.22 g/mole

Melting point: 2120°C   (2, 227)
                       2130°C   (39)

Pycnometric density: 3.28 g/cc   (2)

STRUCTURE

Monoclinic

THERMAL

Expansion: (β form) 13.25 x 10^{-6} per °C; 25-500°C   (329)

14.36 x 10^{-6} per °C; 25-1000°C   (329)

$SiO_2 \cdot 3CaO$

Formula weight: 228.30 g/mole

Melting point: 1900°C   (2, 227)

$TiO_2 \cdot 3\ CaO$

Formula weight: 248.14 g/mole

Melting point: $2135^{\circ}C$ (2)

$Cr_2O_3 \cdot FeO$

CHROMITE

Formula weight: 223.87 g/mole

Melting point: 2180°C   (227)

CHEMICAL

Reactivity: Good resistance to O_2 (227)

OPTICAL

Color: black, translucent in thin sections   (558)

Refractive index: 2.07 - 2.16   (558)

STRUCTURE

Cubic   (558)

THERMAL

Expansion: 8.22×10^{-6} per °C;   25-500°C   (329)

8.93×10^{-6} per °C;   25-1000°C   (329)

9.37×10^{-6} per °C;   25-1200°C   (329)

4.32×10^{-6} per °C;   100-200°C   (227, 314)

6.12×10^{-6} per °C;   100-1000°C   (227, 314)

$Cr_2O_3 \cdot MgO$

<u>PICHROMITE</u>

Formula weight: 192.34 g/mole

Melting point: 2000°C   **(2)**

Pycnometric density: 4.39 g/cc   **(2)**

<u>THERMAL</u>

     Expansion: 7.16×10^{-6} per °C;   25-500°C   **(329)**

                       7.90×10^{-6} per °C;   25-1000°C   **(329)**

                       5.76×10^{-6} per °C;   100-200°C   **(227, 314)**

                       8.19×10^{-6} per °C;   100-1200°C   **(227, 314)**

$Cr_2O_3 \cdot MnO$

Formula weight: 222.95 g/mole

THERMAL

    Expansion: 8.63×10^{-6} per $^{\circ}C$; 25-500$^{\circ}C$ (329)

                          9.34×10^{-6} per $^{\circ}C$; 25-1000$^{\circ}C$ (329)

Formula weight: 233.40 g/mole

THERMAL

Expansion: 8.22×10^{-6} per °C; 25-500°C   (329)

8.72×10^{-6} per °C; 25-1000°C   (329)

8.17×10^{-6} per °C; 25-1200°C   (329)

$$Cr_2O_3 \cdot ZrO_2$$

Formula weight: 275.24 g/mole

THERMAL

Expansion: 4.84×10^{-6} per $^{\circ}$C; 25-500°C   (329)

5.64×10^{-6} per $^{\circ}$C; 25-1000°C   (329)

$2Cr_2O_3 \cdot Fe_2O_3$

Formula weight: 463.74 g/mole

THERMAL

Expansion: 6.83×10^{-6} per °C; 100-200°C (227, 314)

8.28×10^{-6} per °C; 100-1200°C (227, 314)

$4Cr_2O_3 \cdot MgO$

Formula weight: 648.40 g/mole

Melting point: 2260°C (227)

$SiO_2 \cdot 2\ FeO$

Formula weight: 203.76 g/mole

THERMAL

Expansion: 9.48×10^{-6} per $^{\circ}$C; 25-500°C   (329)

9.95×10^{-6} per $^{\circ}$C; 25-1000°C   (329)

Formula weight: 200.02 g/mole

THERMAL

Expansion: 10.31×10^{-6} per $^{\circ}$C; 25-500°C (329)

12.11×10^{-6} per $^{\circ}$C; 25-1000°C (329)

$Fe_2O_3 \cdot ZnO$

Formula weight: 241.08 g/mole

THERMAL

  Expansion: 9.05×10^{-6} per $^{\circ}C$; 25-500$^{\circ}C$ (329)

                  9.85×10^{-6} per $^{\circ}C$; 25-1000$^{\circ}C$ (329)

$SiO_2 \cdot HfO_2$

Formula weight: 270.66 g/mole

THERMAL

Expansion: 2.95×10^{-6} per °C; 25-500°C (329)

3.24×10^{-6} per °C; 25-1000°C (329)

3.45×10^{-6} per °C; 25-1300°C (329)

Formula weight: 314.23 g/mole

STRUCTURE

Cubic, perovskite type

a = 4.069 ± 003 A   (547)

Formula weight: 366.16 g/mole

Melting point: 2030°C   (2)

SiO$_2$·MgO

STEATITE

Formula weight: 100.38 g/mole

CHEMICAL

    Reactivity: Maximum temperature of continuous service,
980-1090°C   (227)

THERMAL

    Conductivity: .006 CGS at 35-43°C   (227, 309)

                      0.0022-.006 CGS at elevated temperature(227, 288)

    Expansion: 10.51 x 10^{-6} per ^{o}C; 25-500°C   (329)

                      10.77 x 10^{-6} per ^{o}C; 25-1000°C   (329)

                      10.83 x 10^{-6} per ^{o}C; 25-1500°C   (329)

Formula weight: 120.22 g/mole

THERMAL

Expansion: 7.16×10^{-6} per °C; 25-500°C   (329)

7.90×10^{-6} per °C; 25-1000°C   (329)

$2\ TiO_2 \cdot MgO$

Formula weight: 200.12 g/mole

THERMAL

Expansion: 8.22×10^{-6} per ^{o}C; 25-500°C (329)

9.03×10^{-6} per ^{o}C; 25-1000°C (329)

$ZrO_2 \cdot MgO$

Formula weight: 163.54 g/mole

Melting point: 2110^oC (2, 227)
                       2150^oC

CHEMICAL

Reactivity: O_2, possibly useful to 2400^oC in
                   oxidizing atmosphere   (227)

                   H_2, probably stable at 1400^oC (227)

THERMAL

Expansion: 10.31×10^{-6} per ^oC; 25-500^oC (329)

                  1200×10^{-6} per ^oC; 25-1000^oC (329)

                  13.41×10^{-6} per ^oC; 25-1500^oC (329)

$SiO_2 \cdot 2\, MgO$

FORSTERITE

Formula weight: 140.70 g/mole

Melting point: $1885^{\circ}C$ (2)

CHEMICAL

Reactivity: Maximum temperature of continuous service
980-1090°C   (227)

THERMAL

Conductivity: 0% porosity:

0.0125 CGS at 100°C   (2, 227, 287)

0.011 CGS at 200°C   (2, 227, 287)

0.008 CGS at 400°C   (2, 227, 287)

0.006 CGS at 600°C   (2, 227, 287)

0.005 CGS at 800°C   (2, 227, 287)

Expansion: 10.31×10^{-6} per $^{\circ}$C; 25-500°C   (329)

11.70×10^{-6} per $^{\circ}$C; 25-1000°C   (329)

11.0×10^{-6} per $^{\circ}$C; 0-1425°C   (227, 2)

$TiO_2 \cdot 2\ MgO$

Formula weight: 160.54 g/mole

THERMAL

Expansion: 10.10×10^{-6} per $^{\circ}C$; 25-500$^{\circ}C$ (329)

10.97×10^{-6} per $^{\circ}C$; 25-1000$^{\circ}C$ (329)

2 $UO_2 \cdot$ 3 MgO

Formula weight: 661.10 g/mole

Melting point: 2160°C   (227)

$SiO_2 \cdot ZnO \cdot ZrO_2$

Formula weight: 264.66 g/mole

Melting point: $2080^{o}C$

$SiO_2 \cdot 2\ ZnO$

Formula weight: 222,82 g/mole

THERMAL

Expansion: 2.95×10^{-6} per $^{\circ}$C; 25-500°C   (329)

3.18×10^{-6} per $^{\circ}$C; 25-1000°C   (329)

$SiO_2 \cdot ZrO_2$

ZIRCON

Formula weight: 183.26 g/mole

Melting point: $2550^{O}C$ (227)
                     $2390^{O}C$ (2)
                     $2500^{O}C$
                     $2420^{O}C$ (Incongruent)   (2)
                 d.>$1740^{O}C$ (551)

Pycnometric density: 4.6 g/cc   (2)

CHEMICAL

Limit of usefulness: Air, $1870^{O}C$ (227, 265)

                      Vacuum, $1700^{O}C$ (227)

                      Reducing atmosphere, fair   (227)

                      Continuous service, $1500-1775^{O}C$,
                      decomp. to $ZrO_2 + SiO_2$ (227)

                      Dissociates at $1775^{O}C$, $1500^{O}C$ in
                      presence of impurities

Reactivity: H_2, fair resistance; decomposes at
                      $1900^{O}C$, possible failure at $1400^{O}C$ (227)

                      C, forms carbides at high temperature (2)

MECHANICAL

Strength: Tensile : 12.7×10^3 psi at rt   (227, 297)
                      8.7×10^3 psi at $1050^{O}C$ (2, 227)
                      3.6×10^3 psi at $1200^{O}C$ (227)

          Shear   : 8.7×10^3 psi at rt   (2, 227)
                      6.3×10^3 psi at $1100^{O}C$ (227)
                      2.3×10^3 psi at $1300^{O}C$ (227)

Hardness: Mohs   : 7.5

Young's modulus: 24×10^6 psi at rt   (84, 227, 297)

MECHANICAL (cont.)

Modulus of rigidity:	11.5×10^6 psi at rt	(2, 227)
	9.1×10^6 psi at 900°C	(227)
	6.6×10^6 psi at 1000°C	(227)
	4.5×10^6 psi at 1100°C	(227)
	0.4×10^6 psi at 1300°C	(227)
Creep rate:	435×10^{-6} in/in/hr at 1200 psi, 1100°C	(2)

THERMAL

Conductivity: 0% porosity:

0.0145 CGS at 100°C	(2, 227, 287)
.0135 CGS at 200°C	(2, 227, 287)
.012 CGS at 400°C	(2, 227, 287)
.010 CGS at 800°C	(2, 227, 287)
.0095 CGS at 1200°C	(2, 227, 287)
.0095 CGS at 1400°C	(2, 227, 287)

Expansion:	3.79×10^{-6} per °C;	25-500°C	(329)
	4.62×10^{-6} per °C;	25-1000°C	(329)
	5.63×10^{-6} per °C;	25-1500°C	(329)
	5.5×10^{-6} per °C;	20-1200°C	(2, 227)
	4.03×10^{-6} per °C;	to 600°C	(84)
	4.2×10^{-6} per °C;	-	(4)
	6.37×10^{-6} per °C;	to 600°C	(555)
	8.27×10^{-6} per °C;	to 1100°C	(555)
	10.17×10^{-6} per °C;	to 1300°C	(555)

Formula weight: 183.53 g/mole

Melting point: 2080°C

STRUCTURE

Cubic, perovskite type

THERMAL

Expansion: 8.63 x 10^{-6} per °C; 25-500°C   (329)

9.43 x 10^{-6} per °C; 25-1000°C   (329)

$ZrO_2 \cdot SrO$

Formula weight: 226.85 g/mole

Melting point: 2800°C   (227)
                       >2700°C   (2)
                       >2800°C

Pycnometric density: 5.48 g/cc   (2)

CHEMICAL

Synthesis: Calcine $SrCO_3 + ZrO_2$, then sinter
                  12-1300°C   (549)

STRUCTURE

Cubic, perovskite type,   d = 4.089 $\pm$ .003 A   (548, 547)

THERMAL

Expansion: 8.84×10^{-6} per °C;   25-500°C   (329)

                  9.64×10^{-6} per °C;   25-1000°C   (329)

$Ta_2O_5 \cdot 6\ SrO$

$(Sr_3\ Ta\ O_{5.5})$

Formula weight: 1063.54 g/mole

X-ray density: 6.088 g/cc   (556)

Pycnometric density: 5.93 g/cc   (556)

CHEMICAL

Synthesis: Sinter equivalent amount of SrO
               and Ta_2O_5 at $1100^{\circ}C$ in   (556)

ELECTRICAL

Dielectric constant: 14.3   (556)

STRUCTURE

Face centered cubic   (556)

a = 8.34 A   (556)

$ZrO_2 \cdot ThO_2$

Formula weight: 387.34 g/mole

Melting point: $>2800^{\circ}C$ (2)

$TiO_2 \cdot ZrO_2$

Formula weight: 203.12 g/mole

THERMAL

    Expansion: 7.37×10^{-6} per $^{\circ}$C; 25-500°C   (329)

                       7.90×10^{-6} per $^{\circ}$C; 25-1000°C   (329)

                       8.77×10^{-6} per $^{\circ}$C; 25-1200°C   (329)

DATA TABLES

VIII SILICIDES

B_3Si

TRIBORON MONOSILICIDE

Formula weight: 60.52 g/mole

| Pycnometric density: | 2.52 g/cc | (673, 674) |
| | 2.41 - 2.49 g/cc | (677) |

CHEMICAL

Theoretical analysis: 46.4% silicon
53.6% boron

Synthesis:	B + Si	(673)
	B + Si at 1725°C, no reaction	(87)
	B_4C + Si, no BSi compounds	(675)
	B + Si, hot press, 1600-1800°C, B_3Si only	(676)
	B_2O_3 + SiO_2 + Mg, B_3Si only	(676)

ELECTRICAL

Resistivity: Conductor   (673, 674)

MECHANICAL

| Hardness: Approaches SiC | (673, 674) |
| Micro 30 g : 5350 ± 167 kg/mm^2 | (677) |

OPTICAL

Color - black, translucent in thin layers   (673, 674)

Form - rhombic plates   (673, 674)

STRUCTURE

Tetragonal, $Z = 1$ (677)
$a = 2.829 \pm 0.007$ A. $c = 4.765 \pm 0.013$ A
$c/a = 1.63$ (677)

B_4Si

$(B_{12}Si_3)$

TETRABORON MONOSILICIDE

STRUCTURE

Rhombohedral

$a = 5.602\,A$, $\alpha = 68^\circ\ 49'$

Hexagonal cell

$a = 6.330\,A$, $c = 12.736\,A$, $c/a = 2.015$

B_6Si

HEXABORON MONOSILICIDE

Formula weight:	92.98 g/mole	
X-ray density:	2.18 g/cc	(678)
Pycnometric density:	2.47 g/cc	(673, 674)
	2.15 g/cc	(678)
	2.42 g/cc	(444)

CHEMICAL

Theoretical analysis: 30.2% silicon
69.8% boron

Synthesis:		
	B + Si	(673)
	B + Si at $1725^{\circ}C$, no reaction	(87)
	B_4C + Si, no BSi compounds	(675)
	SiO_2 + B, vacuum, $1300-1400^{\circ}C$	(444)
	Si + excess B, hot press, $2150^{\circ}C$, 1400 psi	(444)

ELECTRICAL

Resistivity: Conductor   (673, 674)

MECHANICAL

Hardness: Approaches SiC   (673, 674)

OPTICAL

Color - opaque   (673, 674)

Form - irregular   (673, 674)

STRUCTURE

Cubic, CaB_6 type, O_h^1 -P/m3m space group   (678)

STRUCTURE (cont.)

Similar to $(RE)B_6$ types   (678)

$Z = 1$ (678)

$a = 4.142 \pm 0.002$ A   (678)

Orthorhombic

$a = 14.392$ A, $b = 18.267$ A, $c = 9.885$ A   (680)

$B_{12}Si$

DODECABORON MONOSILICIDE

Existence of $B_{>10}Si$ reported (679)

Probably $B_{12}Si$ (444)

BARIUM MONOSILICIDE

Formula weight: 165.42 g/mole

CHEMICAL

Theoretical analysis: 16.96% silicon
                        83.04% barium

Synthesis: Ba + Si, 1100-1200°C   (485)

Reactivity: Absorbs H_2 at 500°C, releasing it again
                at 650°C   (444)

                Excess silicon gives $BaSi_3$, etc.   (444)

THERMAL

Heat of formation: 181.5 kcal/mole   (444)

$BaSi_2$

BARIUM DISILICIDE

Formula weight: 193.48 g/mole

CHEMICAL

Theoretical analysis: 29.01% silicon
                        70.99% barium

Synthesis: BaH_2 + Si   (444)

Reactivity: Excess silicon gives $BaSi_3$, etc.   (444)

THERMAL

Heat of formation: 399.2 kcal/mole   (444)

Formula weight: 221.54 g/mole

CHEMICAL

Theoretical analysis: 37.94% silicon
                         62.06% barium

Synthesis: (BaSi, BaSi$_2$) + Si   (444)

                        BaH$_2$ + Si   (444)

$BaSi_4$

BARIUM TETRASILICIDE

Formula weight: 249.60 g/mole

CHEMICAL

Theoretical analysis: 44.94% silicon
                        55.06% barium

Synthesis: Suggested by Wohler   (485)

Ba_2Si_7

DIBARIUM HEPTASILICIDE

Formula weight: 471.14

CHEMICAL

Theoretical analysis: 41.64% silicon
58.36% barium

Synthesis: Suggested by Wohler   (485)

BeSi$_x$

BERYLLIUM SILICIDE

No compounds form and the solubility of Si in Be is negligible.   (488)

The eutectic temperature 1090°C corresponds to a composition of

61% Si by weight.   (489)

CALCIUM MONOSILICIDE

Formula weight: 68.14 g/mole

Formula volume: 21.23 cc/mole

Melting point: 1245°C   (480, 481)

X-ray density: 3.21 g/cc   (444)

CHEMICAL

Theoretical analysis: 41.19% silicon
                        58.81% calcium

Synthesis: 2 Ca + 3-4 Si at 1000-1100°C   (444)

Reactivity: Forms calcium hydride in H_2
                       at 400-450°C   (444)

STRUCTURE

Rhombic, TaB type, D_{2h}^{17} space group   (444, 483, 484)

2 molecules per unit cell

Corrugated layers of Si atoms alternate with Ca atom
layers. Si atom layers alternate ABCA...CABCBC...
AB to give 12 layer CP system   (444)

a = 3.91 A,   b = 4.59 A,   c = 10.79 A   (445)

THERMAL

Heat of formation: 36 ± 2 kcal/mole   (482)

$CaSi_2$

CALCIUM DISILICIDE

Formula weight: 96.20 g/mole

Melting point: 1020°C incongruent   (480, 481)

CHEMICAL

Theoretical analysis: 58.34% silicon
                         41.66% calcium

Synthesis: CaSi in vac, above 1000°C   (444)

STRUCTURE

Rhombic cell,   a = 10.4 A, α = 21°.42' (?)   (445)

THERMAL

Heat of formation: 36 ± 2 kcal/mole   (482)

DICALCIUM MONOSILICIDE

Formula weight: 108.22 g/mole

Formula volume: 51.05 cc/mole

Melting point: 910°C, incongruent   (480, 481)

X-ray density: 2.12 g/cc   (479)

CHEMICAL

Theoretical analysis: 25.93% silicon
                        74.07% calcium

STRUCTURE

F.C. cubic,   Cu$_3$Au type, space group O_h^1 (444)

a = 4.73 A   (445)

Rhombic,   Z = 4, Pbnm, PbCl$_2$ type   (479)

a = 9.002 A,   b = 2.667 A,   c = 4.799 A   (479)

THERMAL

Heat of formation: 50 $\pm$ 3 kcal/mole   (482)

$CdSi_x$

CADMIUM SILICIDE

No compounds formed and elements are mutually insoluble in solid

state. (487)

CERIUM MONOSILICIDE

Formula weight: 168.19 g/mole

Melting point: 1525°C   (444)

CHEMICAL

Theoretical analysis: 16.65% silicon
                         83.35% cerium

Synthesis: Has been reported   (444)

$CeSi_2$

CERIUM DISILICIDE

Formula weight:	196.25 g/mole	
Melting point:	1440°C	(227, 298)
X-ray density:	5.45 g/cc	(567)
Pycnometric density:	5.31 ± 0.02 g/cc	(567)

CHEMICAL

Theoretical analysis:		28.57% silicon	(567)
		71.43% cerium	(567)
Synthesis:	(a)	Electrolysis of fused fluorides	(562)
	(b)	Reaction of elements	(561)
	(c)	Electrolysis of metal oxide, SiO_2, CaO fusion	(563, 564)
	(d)	Metal oxide plus excess silicon, in vacuum, 1500°C	(444, 565)
Reactivity:		Reacts with N_2 to form nitrides	(227, 272)
		Air, O_2; oxidizes to silicates, 1500-1600°C	(444)
		Aq HF : decomposed	(444)
		Aq HCl : decomposed	(444)
		K_2CO_3 fused : decomposed	(444)
		Na_2CO_3 fused : decomposed	(444)

STRUCTURE

Tetragonal, $ThSi_2$ type, D_{4h}^{19} (444)

a = 4.175 ± 0.002 A,	c = 13.848 ± 0.006 A	(108)
4.148	13.81	c/a = 3.32 (567, 568)
4.15 ± 0.03	13.87 ± 0.07	(332)

STRUCTURE (cont.)

    See LaSi$_2$

THERMAL

    Heat of formation: ΔH^O_{298} = -50 kcal/mole   (444)

                                            -33 to -68.8 kcal/mole   (108)

Ce_2Si

DICERIUM MONOSILICIDE

Formula weight: 308.32 g/mole

CHEMICAL

Theoretical analysis: 9.09% silicon
                         90.91% cerium

Synthesis: Present in Ce - Si alloys containing

33 - 37 atom percent Si   (108)

CeSi$_{0.35}$

(Ce$_3$Si)

TRICERIUM MONOSILICIDE

Formula weight: 448.45 g/mole

CHEMICAL

Theoretical analysis: 6.25% silicon
93.75% cerium

Synthesis: Always present in alloys of Ce + Si

containing CeO$_2$, at 1430°C   (108)

$CeSi_{0.75}$

(Ce_4Si_3)

TETRACERIUM TRISILICIDE

Formula weight: 644.70 g/mole

CHEMICAL

Theoretical analysis: 13.08% silicon

                                 86.92% cerium

Synthesis: Present in Ce - Si alloys containing

                                 40 - 45 atom percent Si   (108)

COBALT MONOSILICIDE

Formula weight: 87.00 g/mole

Melting point: 1395°C   (227, 298)
                        1415°C   (444)

CHEMICAL

Theoretical analysis: 32.20% silicon
                        67.80% cobalt

ELECTRICAL

Thermal EMF: - 46 μv/°C   (456, 457)

STRUCTURE

Rhombic, FeSi type, B 20   (444)

Cubic, T^4, 8 atoms per unit cell,   a = 4.438   (445)

Radius ratio: 0.94   (444)

THERMAL

Activation energy: Diffusion of Si into Co

        metal: 13.09 kcal/mole   (473)

Heat of formation: ΔH = -24.0 kcal/mole   (636)

$CoSi_2$

COBALT DISILICIDE

Formula weight:	113.06 g/mole	
Melting point:	1277°C	(444)
Pycnometric density:	4.94 g/cc	(641)

CHEMICAL

Theoretical analysis: 48.77% silicon
51.23% cobalt

ELECTRICAL

Resistivity:	64.8×10^{-6} ohm-cm	(444)
Thermal EMF:	$-8 \, \mu v/°C$	(456, 457)

Temperature coefficient of Thermal EMF:
$+5.42 \pm 0.0 \, \mu v/°C$ (454)

MAGNETIC

Susceptibility: $K = +130 \times 10^{-6}$ (458)

MECHANICAL

Hardness: Micro 50 g : 889 kg/mm^2 (444)

STRUCTURE

Cubic, fluorite type, O_h^5 -Fm3m space group   (641)
a = 5.365 A   (641)

Radius ratio: 0.94   (444)

THERMAL

    Activation energy: Diffusion of Si into Co:
                                 13.09 kcal/mole   (473)

    Heat of formation: ΔH = -24.6 kcal/mole   (636)

- 482 -

CoSi_3$CoSi_3$

COBALT TRISILICIDE

Formula weight:	143.12 g/mole	
Melting point:	1306°C	(444)

CHEMICAL

Theoretical analysis:	58.7% silicon 41.3% cobalt	
Existence doubtful		(639, 642)

ELECTRICAL

Thermal EMF:	+14 μv/°C	(456, 457)

THERMAL

Activation energy:	Diffusion of Si into Co: +13.09 kcal/mole	(473)

Co$_2$Si

DICOBALT MONOSILICIDE

Formula weight: 145.94 g/mole

Melting point: 1332°C   (444)

CHEMICAL

Theoretical analysis: 19.23% silicon
80.77% cobalt

ELECTRICAL

Thermal EMF: -8 μv/^{o}C   (456, 457)

STRUCTURE

Rhombic, D$_{4h}^{16}$ space group   (634, 635)

4 Co$_2$Si per unit cell, severely distorted ccp structure (634, 635)

a = 7.095 A,   b = 4.908 A,   c = 3.730 A

Si atoms form zigzag chains parallel Z axis   (634, 635)

12 atoms/unit cell   (634, 635)

Radius ratio: 0.94   (444)

THERMAL

Activation energy: Diffusion of Si into Co:
13.09 kcal/mole   (473)

Heat of formation: ΔH = -18.4 kcal/mole   (640)

Formula weight: 204. 88 g/mole

Melting point: 1210°C, peritectic   (486)
                  d. below 1160°C to Co_2Si (486)

CHEMICAL

Theoretical analysis: 13.67% silicon
                        86.33% cobalt

Stability range: 1160-1210°C   (486)

STRUCTURE

Cubic, β W type   (444)

See Fe_3Si for details

Radius ratio: 0.94   (444)

THERMAL

Activation energy: Diffusion of Si into Co:
                        17. 09 kcal/mole   (473)

CHROMIUM MONOSILICIDE

Formula weight: 80.07 g/mole

Melting point: 1545 $\pm$ 50°C   (444)

Pycnometric density: 5.43 g/cc   (444)

CHEMICAL

Theoretical analysis: 35.05% silicon
                        64.95% chromium

Synthesis: $Cr_2O_3 + SiO_2$ (SiC) + C   (603, 604, 605)

Reactivity: Acids, very stable   (25)

               HF, easily soluble   (25)

               Fused alkali, easily soluble   (25)

ELECTRICAL

Resistivity: 143×10^{-6} ohm-cm   (446)

               250×10^{-6} ohm-cm   (452)

               143×10^{-6} ohm-cm   (444)

Temperature coefficient of resistivity:
               -12×10^{-6} ohm-cm/°C   (452)

               31.2×10^{-6} ohm-cm/°C   (444)

Thermal EMF: +5 μv/°C   (456, 457)

MECHANICAL

Hardness: Micro : 1005 kg/mm^2 (444)
           Vickers : 950 - 1050 kg/mm^2 (444)

STRUCTURE

Rhombic, FeSi type, B 20 space group   (444)

STRUCTURE (cont.)

Cubic, T 4 space group   (445)

a = 4.620 ± 0.002 A   (444)

Cubic, B 20 type   (19)

Radius ratio: 0.92   (444)

THERMAL

Activation energy: Diffusion of Si into Cr,   (473)
10.25 kcal/mole

$CrSi_2$

CHROMIUM DISILICIDE

Formula weight:	108.13 g/mole	
Melting point:	$1500^o \pm 20^oC$	(444)
	1570^oC	(2, 24, 50)
Pycnometric density:	4.39 g/cc	(2, 25, 26)
	5.00 g/cc	(444)
	4.4 g/cc	(19)

CHEMICAL

Theoretical analysis:	51.90% silicon	
	48.10% chromium	
Synthesis:	Cr + Si, fusion	(606, 608, 609)
	Cr + Si, hot pressed	(610)
	Cr_2O_3 + SiO_2 (excess) + C	(611)
Reactivity:	Corrosion in air becomes severe, 1100-1400^oC	(227)
	51 mg/cm^2 weight gain, 1200^oC, 4 hours in air	(444)
	Reacts with boron at 1530^oC to CrB_2	(478)
	Acids, very stable	
	HF, easily soluble	
	Fused alkali, easily soluble	

ELECTRICAL

Resistivity:	250×10^{-6} ohm-cm	(444)
	260×10^{-6} ohm-cm	(446)
	1420×10^{-6} ohm-cm	(453)
	6670×10^{-6} ohm-cm	(452)

ELECTRICAL (cont.)

    Temperature coefficient of resistivity:

        $+90 \times 10^{-6}$ ohm/$^{\circ}$C   (452)

        23.8×10^{-6} ohm/$^{\circ}$C   (453)

        29.3×10^{-6} ohm/$^{\circ}$C   (453)

    Thermal EMF: 90×10^{-3} μv/$^{\circ}$C rt increases to

        140×10^{-3} μv/$^{\circ}$C at 400°C; then drops to

        80×10^{-3} μv/$^{\circ}$C at 610°C   (444)

        $+153$ μv/$^{\circ}$C   (456, 457)

    Activation energy:

        Approximately 1.3 ev   (444)

    Magnetic susceptibility:

        $+30 \times 10^{-6}$ (458)

        $+48 \times 10^{-6}$ (453)

MECHANICAL

Hardness:	Vickers 100 g	:	1150 kg/mm^2	(8, 50)
	Vickers	:	880 - 1100 kg/mm^2	(444)
	Knoop 100 g	:	1550 kg/mm^2	(8)
	Micro 50 g	:	1131 kg/mm^2	(444)
			996 - 1150 kg/mm^2	(444)

OPTICAL

    Color: gray, metallic

    Form: needles   (444)

STRUCTURE

Hexagonal, D_6^4 space group   (444)

a = 4.422 $\pm$ 0.005 A,   c = 6.531 $\pm$ 0.085 A,   c/a = 1.476

C - 40 type, hexagonal, isomorphous with VSi$_2$, NbSi$_2$, TaSi$_2$

a = 4.42 A,   c = 6.35 A   (19, 22, 606)

Radius ratio: 0.92   (444)

THERMAL

Expansion: 12.0 x 10^{-6} per ^{o}C;   25-500°C   (329)

12.9 x 10^{-6} per ^{o}C;   25-700°C   (329)

Activation energy: Diffusion of Si into Cr,   (473)
10.25 kcal/mole

Cr_2Si

DICHROMIUM MONOSILICIDE

Formula weight: 132.08 g/mole

CHEMICAL

Theoretical analysis: 21.2% silicon
                        78.8% chromium

Synthesis: 2 Cr + Si, fusion   (603, 604, 605)

$$Cr_2O_3 + SiO_2 + Al \qquad\qquad (612)$$

Reactivity: Acids, very stable

HF, easily soluble

Fused alkali, easily soluble

ELECTRICAL

Thermal EMF: -4 $\mu v/^{\circ}C$ (456, 457)

OPTICAL

Form: Prismatic   (444)

STRUCTURE

Probably orthorhombic   (19)

THERMAL

Activation energy: Diffusion of Si into Cr,
                      10.25 kcal/mole   (473)

Cr_2Si_3

DICHROMIUM TRISILICIDE

Formula weight: 188.20 g/mole

CHEMICAL

Theoretical analysis: 44.7% silicon

55.3% chromium

Synthesis: Existence proposed by Frilley   (605)

Cr_2Si_7

DICHROMIUM HEPTASILICIDE

Formula weight: 300.44 g/mole

CHEMICAL

Theoretical analysis: 65.5% silicon

34.5% chromium

Synthesis: Existence proposed by Frilley   (605)

Cr_3Si

TRICHROMIUM MONOSILICIDE

Formula weight:	184.09 g/mole	
Melting point:	1710 $\pm$ 50°C	(444)
Pycnometric density:	6.52 g/cc	(444)

CHEMICAL

Theoretical analysis: 15.22% silicon
84.78% chromium

Synthesis: $Cr_2O_3 + SiO_2 + Cu$ (612, 613)

Reactivity: Corrosion in air becomes severe
above 1050°C (227, 300)

< 1 mg/cm^2 in air after 2000 hours
at 1050°C (227)

Acids, very stable

HF, easily soluble

Fused alkali, easily soluble

ELECTRICAL

Resistivity: 45.5 x 10^{-6} ohm-cm (446, 452)

Temperature coefficient of resistivity: 65 x 10^{-6} per °C (444)

Thermal EMF: -14 μv/°C (456, 457)

MECHANICAL

Hardness:	Micro	: 1005 kg/mm^2	(444)
	Vickers	: 900 - 980 kg/mm^2	(444)
	Rockwell C	: 37	(444)

OPTICAL

Form: Small prismatic crystals (444)

STRUCTURE

Cubic

a = 4.555 $\pm$ 0.003 A   (606)

Cubic, β W type, O_h^3 space group   (444)

See Fe_3Si for details

A - 15 type, isomorphous with V_3Si (19)

Radius ratio: 0.92   (444)

THERMAL

Expansion: 9.90 x 10^{-6} per ^{o}C; 25-500°C (329)

11.1 x 10^{-6} per ^{o}C; 25-1000°C (329)

Activation energy: Diffusion of Si into Cr,

10.25 kcal/mole   (473)

Cr_3Si_2

TRICHROMIUM DISILICIDE

Formula weight:	212.15 g/mole	
Melting point:	1750°C	(2, 24)
	1560 ± 50°C	(444)
Pycnometric density:	6.52 g/cc	(2, 11, 25)
	5.6 g/cc	(444)

CHEMICAL

Theoretical analysis:	26.45% silicon	
	73.55% chromium	
Synthesis:	Cr + Si, fusion	(606, 607)
	$Cr_2O_3 + SiO_2 + C$ (SiC)	(587)
Reactivity:	Acids, very stable	
	HF, easily soluble	
	Fused alkali, easily soluble	

ELECTRICAL

Resistivity:	114×10^{-6} ohm-cm	(446)
Temperature coefficient of resistivity:	85.0	(444)

MECHANICAL

Hardness:	Micro	: 1280 kg/mm^2	(444)
	Vickers	: 1050 - 1200 kg/mm^2	(444)
	Rockwell C :	52	(444)

OPTICAL

Form:	Long, rectangular prisms	(444)

STRUCTURE

    Cubic

    Tetragonal, Cr_3B_2 type, D_{4h}^{18} space group (444)

    a = 9.16 A,   c = 4.65 A,   c/a = 0.508 (444)

    Radius ratio: 0.92 (444)

THERMAL

    Activation energy: Diffusion of Si into Cr,

                             10.25 kcal/mole (473)

PENTACHROMIUM TRISILICIDE

Formula weight: 344.23 g/mole

X-ray density: 5.9 g/cc   (614)

Pycnometric density: 5.6 g/cc   (614)

CHEMICAL

Theoretical analysis: 24.4% silicon
                       75.6% chromium

Synthesis: Cr + Si, quench from above $1500^{\circ}C$ (614)

Reactivity: Acids, very stable

              HF, easily soluble

              Fused alkali, easily soluble

ELECTRICAL

Resistivity: 667×10^{-6} ohm-cm   (452)

Temperature coefficient of thermal EMF:
                 $-2 \ \mu v/^{\circ}C$ (452)

STRUCTURE

Tetragonal
a = 9.170 $\pm$ 0.006 A,   c = 4.636 $\pm$ 0.003 A,   Z = 4

Space group D_{4h}^{12} - I4/mcm, isomorphous with Mo_5Si_3 and
                W_5Si_3 (444)

THERMAL

Activation energy: Diffusion of Si into Cr,
                      10.25 kcal/mole   (473)

IRON MONOSILICIDE

Formula weight:	83.91 g/mole	
Melting point:	1410°C, congruent	(227, 298)

CHEMICAL

Theoretical analysis:	33.4% silicon	
	66.6% iron	
Reactivity:	See $FeSi_2$	

ELECTRICAL

Resistivity:	240×10^{-6} ohm-cm	(451)
Thermal EMF:	$+9 \ \mu v/°C$	(456, 457)

STRUCTURE

Rhombic, B 20 space group	(444)
Cubic, T^4 space group	(445)
a = 4.489 A	(445)
4 metal, 4 silicon atoms per unit cell	(630)
Radius ratio: 0.93	(444)

THERMAL

Activation energy of formation:

Diffusion of Si into $\propto$ Fe: 2.90 kcal/mole (473)

Activation energy of formation:

Diffusion of Si into $\propto$ Fe: 25.17 kcal/mole (473)

Activation energy of formation:

Diffusion of Si into $\propto$ Fe: 48 kcal/mole (633)

THERMAL (cont.)

Heat of fusion:	40.96 kcal/mole	(631)
Heat of formation:	-19.2 kcal/mole	(636)

$FeSi_2$

IRON DISILICIDE

Formula weight:	111.97 g/mole	
Melting point:	1210°C, decomposes	(444)
Pycnometric density:	4.75 g/cc	(444)

CHEMICAL

Theoretical analysis: 50.1% silicon
49.9% iron

Reactivity: "Silicided" iron resists 10% aqueous solution
of H_2SO_4, HNO_3, H_3PO_4, 3% NaCl, 5% $CaCl_2$,
5% Na_2SO_4, and air at 800°C   (444)

ELECTRICAL

Resistivity: 350×10^{-6} ohm-cm   (451)

Thermal EMF: $+ 10 \ \mu v/°C$ (456, 457)

Temperature coefficient of thermal EMF:
$+ 2.12 \pm 2.0 \ \mu v/°C$ (454)

Magnetic susceptibility:
$K = 85 \times 10^{-6}$.   (458)

MECHANICAL

Hardness: Micro 50 g: 1074 kg/mm^2 (444)

STRUCTURE

Tetragonal, D_{4h}^1 space group   (444)

A rectangular parallel piped with metal atoms at apices,
silicon atoms at the center of each half of the parallel piped
cell contains 1 iron atom (1/8 x 8 comers) and 1 Si-Si
pair within.   (444)

STRUCTURE (cont.)

(In equilibrium with high Si phase)

a = 2.679 A,   c = 5.120 A,   c/a = 1.911   (444)

(In equilibrium with low Si phase)

a = 2.657 A,   c = 5.127 A   c/a = 1.908   (444)

Radius ratio: 0.93   (444)

THERMAL

Activation energy: Diffusion of Si into

α Fe: 2.90 kcal/mole   (473)
γ Fe: 25.17 kcal/mole   (473)
α Fe: 48 kcal/mole   (633)

Fe_2Si_5

DIIRON PENTASILICIDE

Suggested by Haughton & Becker (632), probably corresponds to
$FeSi_2$ phase   (444)

    Formula weight: 252.00 g/mole

CHEMICAL

    Theoretical analysis: 55.7% silicon
                            44.3% iron

    Reactivity: See $FeSi_2$

Fe_3Si

TRIIRON MONOSILICIDE

Formula weight: 195.61 g/mole

Melting point: approx. 1300°C (444)

CHEMICAL

Theoretical analysis: 14.3% silicon
85.7% iron

Reactivity: See $FeSi_2$

ELECTRICAL

Resistivity: Approx. 55×10^{-6} ohm-cm (451)

Therma EMF: -2 $\mu v/°C$ (456, 457)

STRUCTURE

Cubic, β W type, O_h^3 space group (444)
a = 5.64 A

Each metal atom is the center of an irregular tetrahedron
of 4 Si atoms; each Si is surrounded by 12 metal atoms
at the corners of an icosahedron.

Radius ratio: 0.93 (444)

THERMAL

Activation energy: Diffusion of Si into
$\propto$ Fe: 2.90 kcal/mole (473)
γ Fe: 25.17 kcal/mole (473)
$\propto$ Fe: 48 kcal/mole (633)

TRI IRON DISILICIDE

Formula weight: 223.67 g/mole

CHEMICAL

Theoretical analysis: 25.1% silicon
74.9% iron

Reactivity: See $FeSi_2$

STRUCTURE

Quite complex, not yet determined

Most likely this phase is Fe_3Si_2 (444)

      Formula weight: 363. 43 g/mole

      Melting point: 1195°C, decomposes   (444)

CHEMICAL

      Theoretical analysis: 23. 1% silicon
                            76. 9% iron

      Reactivity: See $FeSi_2$

ELECTRICAL

      Resistivity: 170 x 10^{-6} ohm-cm   (451)

STRUCTURE

      (η form)   Hexagonal, Mn_5Si_3 type, $D8_8$ space group   (444)

      Radius ratio: 0. 93   (444)

THERMAL

      Activation energy: Diffusion of Si into
                        $\propto$ Fe: 2. 90 kcal/mole   (473)
                        γ Fe: 25. 17 kcal/mole   (473)
                        $\propto$ Fe: 48 kcal/mole   (633)

HAFNIUM MONOSILICIDE

Formula weight: 206.66 g/mole

CHEMICAL

Theoretical analysis: 13.55% silicon
                         86.45% hafnium

Synthesis: Hf + Si, vac, 1125°C, 88 hours   (81)

STRUCTURE

Hexagonal, isomorphous with ZrSi   (81)
a = 6.86 A,   c = 12.60 A   (81)

Radius ratio: 0.74   (47, 444)

HfSi$_2$

HAFNIUM DISILICIDE

Formula weight: 234.72 g/mole

X-ray density: 8.02 g/cc   (82)

Pycnometric density: 7.2 g/cc   (82)

CHEMICAL

Theoretical analysis: 23.9% silicon
                        76.1% hafnium

Synthesis: HfO$_2$ + SiO$_2$ + Al (S)   (153)

                        Hf + Si, 88 hours, 1125°C in vac.   (81)

MECHANICAL

Hardness: Micro 50 g : 930 kg/mm^2 (444)
                  100 g : 865 kg/mm^2 (444)

STRUCTURE

Orthorhombic, pseudo tetragonal   (81, 82)

a = 3.69 A,   b = 3.64 A,   c = 14.46 A   (81, 82)

C - 49 Orthorhombic, ZrSi$_2$ type, D_{2h}^{17} - Cmcm space group (444)

4 molecules per unit cell

a = 3.69 $\pm$ 0.01 A,   b = 14.46 $\pm$ 0.02 A,   c = 3.64 $\pm$ 0.01 A
                                                         (153)

Radius ratio: 0.74   (47, 444)

$HgSi_x$

MERCURY SILICIDE

No compounds or interaction even at elevated temperature   (486)

IRIDIUM MONOSILICIDE

Formula weight: 221.16 g/mole

CHEMICAL

    Theoretical analysis: 12.67% silicon
                        87.33% iridium

    Existence reported   (644)

STRUCTURE

    Orthorhombic   (644)

    a = 5.558 A,   b = 3.211 A,   c = 6.273 A   (644)

    Isomorphous with IrB_2 (444)

Ir_2Si_3

DI IRIDIUM TRISILICIDE

Formula weight: 470.38 g/mole

CHEMICAL

Theoretical analysis: 17.89% silicon
82.11% iridium

Existence reported   (644)

STRUCTURE

Not yet studied   (644)

Ir_3Si_2

<u>TRI IRIDIUM DISILICIDE</u>

Formula weight: 635. 42 g/mole

<u>CHEMICAL</u>

    Theoretical analysis: 8. 83% silicon
                               91. 17% iridium

<u>STRUCTURE</u>

    Hexagonal   (644)
    a = 3.96 A,   c = 5. 12 A

    Radius ratio: 0. 86   (444)

$LaSi_2$

LANTANUM DISILICIDE

Formula weight:	195.04 g/mole	
X-ray density:	5.14 g/cc	(567)
	4.95 g/cc	(569)
Pycnometric density:	5.05 ± 0.02 g/cc	(567)

CHEMICAL

Theoretical analysis:	28.77% silicon	(567)
	71.23% lantanum	(567)
Synthesis:	La_2O_3 + excess Si, vacuum at 1500°C	(444)
Reactivity:	Air, O_2: oxidized to silicates, 1500-1600°C	(444)
	Aq. HF : decomposed	(444)
	Aq. HCl : decomposed	(444)
	K_2CO_3 fused : decomposed	(444)
	Na_2CO_3 fused : decomposed	(444)

STRUCTURE

Tetragonal, $ThSi_2$ type, space group D_{4h}^{19} (444)

4 molecules per unit cell

metal atms at 000, 0 1/2 1/4, 1/2 1/4 1/2, 1/2 0 3/4

silicon atoms at 00Z, 0Z0, Z00, 0 1/2 (1/4 + Z),

    0 1/2 (1/4 - Z), 1/2 1/2 (1/2 + Z), 1/2 1/2 (1/2 - Z)

    1/2 0 (3/4 + Z), 1/2 0 (3/4 - Z), where Z = 5/12 (444)

a = 4.272 A; c = 13.72 A; c/a = 3.21 (567, 568)

a = 4.374 A; c = 13.565 A (569)

MAGNESIUM MONOSILICIDE

Formula weight: 52.38 g/mole

CHEMICAL

Theoretical analysis: 53.57% silicon
                        46.43% magnesium

Synthesis: Thermal decomposition of Mg_2Si (481)

DIMAGNESIUM MONOSILICIDE

Formula weight: 76.70 g/mole

Melting point: 1102^oC (481, 490)
                       Decomposes to MgSi   (481)

CHEMICAL

Theoretical analysis: 36.9% silicon
                       63.1% magnesium

ELECTRICAL

Coefficient of thermal EMF: 180-240 mv $/^oC$ (against Cu)   (444)

MECHANICAL

Hardness: Micro: 457 kg/mm^2 at rt   (491)
                       320 kg/mm^2 at 300^oC (491)
                       180 kg/mm^2 at 600^oC (491)

Young's modulus: 7.74 x 10^6 psi   (444)

STRUCTURE

Face-centered cubic, CaF_2 type, O_h^5 space group   (444)
a = 6.338 A

THERMAL

Expansion: 14.8 x 10^{-6} per ^oC (444)

Heat of formation: 18.5 ± 1.5 kcal/mole   (482)

Note: For additional data see 492, 493, 494, 495.

MANGANESE MONOSILICIDE

Formula weight: 82.99 g/mole

Melting point: 1275°C   (444)

CHEMICAL

Theoretical analysis: 33.81% silicon
                        66.19% manganese

Synthesis: Mn + Si fused   (587, 605)

ELECTRICAL

Thermal EMF: + 102 μv/°C   (456, 457)

STRUCTURE

Rhombic, FeSi type, B 20 space group   (444)

Cubic, T^4 space group,   a = 4.548 A   (445)

Cubic, analogous to FeSi   (606)

Radius ratio: 0.93   (444)

MANGANESE DISILICIDE

Formula weight: 111.05 g/mole

CHEMICAL

Theoretical analysis: 50.5% silicon
49.5% manganese

Synthesis: Mn + Si, fused (587, 605)

ELECTRICAL

Thermal **EMF**: + 46 μv/$^{\circ}$C (456, 457)

Magnetic susceptibility: 35 x 10^{-6} (458)

STRUCTURE

Tetragonal

DIMANGANESE MONOSILICIDE

Formula weight: 137.92 g/mole

CHEMICAL

Theoretical analysis: 20.3% silicon
79.7% manganese

Synthesis: Mn + Si, fused   (587, 605)

TRIMANGANESE MONOSILICIDE

Formula weight: 192.85 g/mole

Melting point: 1030°C, incongruent   (444)

Pycnometric density: 6.71 g/cc - low temperature form   (628)
                       6.60 g/cc - high temperature form   (628)

CHEMICAL

Theoretical analysis: 14.52% silicon
                       85.48% manganese

ELECTRICAL

Thermal EMF: + 18 μv/°C   (456, 457)

STRUCTURE

Cubic, β W type, O_h^3 space group
a = 2.65 A   (444)

See Fe$_3$B for description of structure type

Hexagonal, 4 molecules per unit cell   (606)

Two polymorphic forms shown to exist, transforms
600 - 700°C with absorption of 530 kcal/gram atom of Mn

Radius ratio·   0.93   (444)

Mn_5Si_3

PENTAMANGANESE TRISILICIDE

Formula weight: 358.83 g/mole

Melting point: 1285°C   (444)

CHEMICAL

Theoretical analysis: 23.46% silicon
                        76.54% manganese

ELECTRICAL

Thermal EMF: + 14 μv/°C   (456, 457)

STRUCTURE

Hexagonal, $D8_8$ space group   (444)
a = 6.898 A,   c = 4.802 A,   c/a = 0.696   (444)

2 molecules per unit cell, of the 10 metal atoms, 6 form
distorted octahedra with the silicon and 4 form octahedra
about the vacant points (000) and (001/2)

Radius ratio: 0.93   (444)

MOLYBDENUM DISILICIDE

Formula weight: 152.07 g/mole

Formula volume: 24.35 cc/mole

Melting point: 2030°C ± 50   (2, 50, 227)
                       2120°C

X-ray density: 6.24 g/cc   (22, 48)
                       6.29 g/cc   (126, 167)

Pycnometric density: 5.987 g/cc   (126)
                       5.9 - 6.3 g/cc   (444)

CHEMICAL

Theoretical analysis: 36.9% silicon
                       63.1% molybdenum

Synthesis: Direct fusion of elements   (603, 615, 28, 11)

                  Sinter Mo + Si   (27, 50)

                  $MoO_3 + SiO_2 + Al + S$ (48, 616)

                  $Mo + CuSi_x$ (617)

Reactivity: Carbon reduces mp to 1870°C   (19)

                  Corrosion in air becomes severe
                  above 1700°C   (227)

                  0.4% weight loss in air; 3 hr at
                  1400°C   (227)

                  No attack by O_2 to 1100°C   (227)

                  $<3.1 \times 10^{-6}$ gm/cm^2/hr in air,
                  100 - 1565°C   (227, 282)

                  0.3 mg/cm^2 weight gain; 1200°C,
                  4 hrs in air   (444)

                  9 mg/cm^2 weight gain; 1500°C,
                  4 hrs in air   (444)

CHEMICAL (cont.)

Reactivity: (cont.)

6 mg/cm^2 weight gain; 1200°C, 5 hrs in air   (444)

8 mg/cm^2 weight gain; 1200°C, 40 hrs in air   (444)

24 mg/cm^2 weight gain; 1200°C, 100 hrs in air   (444)

Reacts with boron at 1750°C to MoB   (478)

Heating with graphite yields Mo$_4$CSi$_3$ phase of variable composition   (444)

Fused metals: stable to all which do not form stable silicides   (444)

Pb, Sn, Na, no reaction to 1000°C

Ag, Hg, practically no reaction

Fe, Cu, Cr, Pt, react

Oxides - reduces all those less stable than SiO$_2$ (CuO, PbO, Fe$_2$O$_3$, ZnO, Cr$_2$O$_3$, MnO)   (444)

O$_2$, air: below 650°C to MoO$_2$ + SiO$_2$ 700°C and higher, MoO$_3$ + SiO$_2$ (621)

SO$_2$, CO$_2$, NO$_2$ at 1000°C, good resistance

Mineral acids, aqua regia, HF: insoluble

Hf + HNO$_3$: soluble

Fused alkali: soluble

F$_2$, reacts at rt

CHEMICAL (cont.)

Reactivity: (cont.)

· Cl$_2$, reacts at elevated temperature

Br$_2$ + H$_2$, very slight reaction at red heat

O$_2$, not attacked at red heat

ELECTRICAL

Resistivity: 21.5 x 10^{-6} ohm-cm at 22°C   (49)

18.9 x 10^{-6} ohm-cm at -80°C   (49)

75-80 x 10^{-6} ohm-cm at 1600°C   (49)

Superconductive < 1.3°K   (459)

Temperature coefficient of thermal EMF:

1.41 ± 0.32 μv/$^{\circ}$C   (454)

Thermal EMF: MoSi$_2$ - Pt:

5.13 + 1/2 (3.761 - 2 t^2) - 1/3 (15.651 - 6 t^3)
from 60-600°C   (455)

Magnetic susceptibility:

-29.5 x 10^{-6} per mole   (453)

Thermionic work function:

5 - 6 ev   (444)

MECHANICAL

Strength: Bending (MOR): 18,570 psi; rt; ρ = 5.57 g/cc   (170)

50,700 psi; rt, sintered   (227, 239)

36-57,000 psi; rt, hot
pressed   (227, 239)

MECHANICAL (cont.)

Strength: Bending (MOR): (cont.)

67,250 psi, 980°C, sintered (227, 283)

86,000 psi, 1090°C, sintered (227)

72,000 psi, 1090°C, hot pressed (227, 283)

55,000 psi, 1200°C, hot pressed (227)

30,000 psi, 980°C, hot pressed (623, 624)

15,000 psi, 1040°C, hot pressed (623, 624)

8,500 psi, 1100°C, hot pressed (623, 624)

Tensile : 40,000 psi, 980°C (623, 624, 227 239, 283)

42,160 psi, 1090°C (623, 624, 227)

42,800 psi, 1200°C (623, 624, 227)

41,070 psi, 1300°C (623, 624, 227)

Stress - rupture:

30,000 psi, 980°C (227, 283)

13,500 psi, 1090°C (227)

8,500 psi, 1200°C (227)

Short time strength:

67,000 psi, 980°C (624)

86,000 psi, 1100°C (624)

72,000 psi, 1200°C (624)

MECHANICAL (cont.)

    Strength: (cont.)

        Creep:

°C	Stress psi	Hrs to Failure	% Per Hr Creep
870	35,000	107	0.0024
980	20,000	224	0.0028
1040	12,000	110	0.073
1095	10,000	85	0.18

        Elongation: $<$0.5%, 27-1320°C   (623)

Hardness: Vickers 100 g : 1290 kg/mm^2 (8)

          Knoop 100 g : 1257 kg/mm^2 (2, 27)

          Vickers   : 1320 - 1550

          Micro 50 g   : 1200 kg/mm^2 (444)

               100 g : 1290 kg/mm^2 (444)

          Rockwell   : 87   (2, 29)

                       57 - 87   (227)

Young's modulus: 42.415 x 10^6 psi, rt; ρ= 5.57 g/cc (170)

                    55.04 x 10^6 psi, rt; ρ= 5.966 g/cc (167)

                    39.3   x 10^6 psi, rt;   (444)

Shear modulus: 23.63 x 10^6 psi, rt; ρ = 5.966 g/cc (167)

Bulk modulus: 27.37 x 10^6 psi, rt; ρ = 5.966 g/cc (167)

Poisson's ratio: 0.165 psi, rt; ρ = 5.966 g/cc   (167)

Speed of sound: 26,180 fps, rt; ρ= 5.966 g/cc   (167)

Poisson	E_{fw} Young's	E_{fc}	Shear	Bulk	Density	(126)
.166	55.33	55.19	23.73	27.70	5.987	
.158	53.14	54.45	22.95	25.96	5.874	
.172	55.68	55.22	23.76	28.01	5.974	

MECHANICAL (cont.)

Poisson	E_{fw} Young's	E_{fc}	Shear	Bulk	Density	(cont.)
.167	56.36	56.07	24.15	28.06	6.041	
.169	55.22	55.09	23.61	27.85	5.966	
.165	55.68	55.17	23.90	27.82	5.991	
.165	55.04	55.17	23.63	27.37	5.966	

OPTICAL

Form: Tetrahedral prisms with tetrahedral
             pyramids on faces

STRUCTURE

Tetragonal, D - 11   (22, 48, 167)

a = 3.20 A,   c = 7.86 A,   c/a = 2.46   (22, 48)

  3.20 A,   7.85 A   (167)

  3.197 A,   7.871 A,   2.463   (444)

D_{4h}^{17} Tetragonal space group   (444)

Two molecules per unit cell; the Si atoms form double close
packed layers, alternating with Mo layers in the direction of
the Z axis.   (444)

Radius ratio: 0.86   (47)
                 0.84   (444)

Isomorphous with WSi_2

THERMAL

Conductivity: 0.075 CGS   (2, 19)

                 0.074 CGS at 425°C   (227)

THERMAL (cont.)

Conductivity: (cont.)

0.057 CGS at 650°C

0.046 CGS at 875°C

0.041 CGS at 1100°C

0.129 CGS at 150°C   (623)

0.093 CGS at 540°C   (623)

k = (0.629 - 9.30 x 10^{-4} T + 6.64 x 10^{-7} T^2)
.245 CGS   where T is °C to 700°C   (227)

Expansion: 7.79 x 10^{-6} per °C; 25-500°C   (329)

8.51 x 10^{-6} per °C; 25-1000°C   (329)

8.41 x 10^{-6} per °C;   0-1000°C   (7)

8.56 x 10^{-6} per °C;   0-1400°C   (7)

9.00 x 10^{-6} per °C; 25-1500°C   (227, 239)

9.18 x 10^{-6} per °C; 25-1500°C   (227, 282)

Specific heat: 10 - 14 cal/gm/°C; 425-1000°C   (171)

Heat of formation: ΔH^{o}_{298} = -27.2 kcal/mole   (444)
-31   kcal/mole   (108)
-47.9 kcal/mole   (472)

Activation energy: Diffusion of Si into Mo, 9.47 kcal/mole
(473)

Heat capacity: 0.092   (444)

MoSi$_3$

MOLYBDENUM TRISILICIDE

Formula weight: 180.13 g/cc

CHEMICAL

Theoretical analysis: 46.7% silicon
53.3% molybdenum

Synthesis: Direct fusion of elements   (615)

Existence in doubt

TRIMOLYBDENUM MONOSILICIDE

Formula weight: 315.91 g/mole

Formula volume: 35.1 cc/mole

Melting point: 2290°C
                       2050 ± 50°C   (2, 27, 50)

X-ray density: 8.97 g/cc   (2, 11, 28)

Pycnometric density: 8.4 ± 0.3 g/cc   (28)

CHEMICAL

Theoretical analysis: 8.9% silicon
                         91.1% molybdenum

Synthesis: Mo + Si, sinter   (27, 50)

                  Mo + Si, fusion   (11, 28)

Reactivity: Corrosion in air becomes severe
                  above 1700°C   (227)

ELECTRICAL

Resistivity: Superconductive <1.3°K   (459)

MECHANICAL

Hardness: Vickers 10 g :   891 kg/mm^2 (8)
                  100 g : 1310 kg/mm^2 (8)
          Vickers   : 1320 - 1550 kg/mm^2 (444)
          Knoop 100 g : 1312 kg/mm^2 (2, 27)
          Micro 100 g : 1310 kg/mm^2 (444)

STRUCTURE

Cubic, (A - 15), βW type, O$_h^3$ space group   (28)
a = 4.890 ± 0.002 A   (28)

See Fe$_3$Si for details of structure type

STRUCTURE (cont.)

    Radius ratio: 0.86   (47)
                      0.84   (444)

THERMAL

    Expansion: 5.69×10^{-6} per $^{\circ}C$; 25-500°C   (329)

                        6.98×10^{-6} per $^{\circ}C$; 25-1000°C   (329)

    Heat of formation: ΔH^{O}_{298} = -21 kcal/mole   (444)

                                   -24 kcal/mole   (593)

    Entropy: 24.5 eu   (622

    Activation energy: Diffusion of Si into Mo:
                              9.47 kcal/mole   (473)

Mo_3Si_2

TRIMOLYBDENUM DISILICIDE

Formula weight:	343.97 g/mole	
Melting point:	2100°C	(2, 27, 50)
Pycnometric density:	7.40 g/cc	(2, 27)

CHEMICAL

Theoretical analysis: 16.35% silicon
83.65% molybdenum

Synthesis:	Mo + Si, sinter	(27, 50)
	Mo + Si fusion	(11, 28)
Reactivity:	Corrosion in air becomes severe above 1700°C	(227)
	Stable in argon to mp	(227)

MECHANICAL

Hardness: Knoop 100 g : 1168 kg/mm^2 (2, 27)

STRUCTURE

Tetragonal, Cr_3B_2 type, D_{4h}^{18} space group   (444)

a = 9.64A,   c = 4.93A,   c/a = 0.517   (444)

Radius ratio: 0.86   (47)

0.84   (444)

THERMAL

Activation energy: Diffusion of Si into Mo,
9.47 kcal/mole   (473)

Mo_5Si_3

PENTAMOLYBDENUM TRISILICIDE

Formula weight: 553.93 g/mole

Formula volume: 67.9 cc/mole

Melting point: 2150 ± 50°C   (27)

X-ray density: 8.14 g/cc   (619)
                       7.94 g/cc   (108)

Pycnometric density: 7.4 g/cc   (444)

CHEMICAL

Theoretical analysis: 15.2% silicon
                        84.8% molybdenum

MECHANICAL

Hardness: Micro 100 g : 1170 kg/mm^2 (444)
           Vickers   : 1200-1320 kg/mm^2 (444)

STRUCTURE

Hexagonal, Mn_5Si_3 type, D8$_8$ space group   (444)

a = 9.62 A,   c = 4.90 A   (pure)   (618, 619, 620)

Stabilized with traces of carbon
a = 7.27 A,   c = 4.992 A   c/a = .687   (619)

Radius ratio: 0.86   (47)
                0.84   (444)

THERMAL

Heat of formation: ΔH^O_{298} = -66 kcal/mole   (593)

Activation energy: Diffusion of Si into Mo,
                      9.47 kcal/mole   (473)

NIOBIUM MONOSILICIDE

Formula weight: 120.97 g/mole

Melting point: 2180°C
                    <1900°C   (227, 298)

CHEMICAL

Theoretical analysis: 23.2% silicon
                        76.8% niobium

Reactivity: Reacts with N_2 to form nitrides   (227, 272)

                Oxidized at 800-1100°C   (227)

                Corrosion in air becomes severe,
                800-1100°C   (227)

STRUCTURE

Radius ratio: 0.82   (47)
                0.80   (444)

THERMAL

Activation energy: Diffusion of Si into Nb,
                        11.72 kcal/mole   (473)

NIOBIUM DISILICIDE

Formula weight:	148.04 g/mole	

Melting point:
	1950°C	(12, 19, 50, 590)
	>2000°C	(22)
	1950-2000°C	(227, 2, 298)
	1930°C	(596)
	2150°C	(444)

Pycnometric density:
	5.45 g/cc	(590)
	5.29 g/cc	(50)

CHEMICAL

Theoretical analysis: 37.7% silicon
62.3% niobium

Synthesis:	Nb + Si	(22)

Reactivity: Reacts with N$_2$ to form nitrides (108, 227, 272)

Stable in air at red heat (227)

Corrosion in air becomes severe, 800-1100°C (227)

54 mg/cm^2 weight loss; 1200°C, 4 hrs in air (444)

Heating with graphite yields ternary phase (444)

ELECTRICAL

Resistivity: 6.3 x 10^{-6} ohm-cm (449)

24.5 x 10^{-6} ohm-cm (444)

Superconductive <1.2°K (459)

MAGNETIC

Susceptibility: -30 x 10^{-6} per mole (453)

MECHANICAL

Hardness: Knoop 100 g : 1050 kg/mm^2 (8, 50)
            Micro 50 g : 1082 kg/mm^2 (444)
            Vickers   : 600-700 kg/mm^2 (590)

STRUCTURE

C - 40 Hexagonal, Cr_3B_2 type, D_6^4 space group   (22)
a = 4.785 A,   c = 6.576 A,   c/a = 1.374   (22)

Radius ratio: 0.82   (47)
                       0.80   (444)

THERMAL

Heat of formation: ΔH^o_{298} = -30 kcal/mole   (593)

Activation energy: Diffusion of Si into Nb,
                           11.72 kcal/mole   (473)

Nb$_2$Si

DINIOBIUM MONOSILICIDE

Probably Nb$_5$Si$_3$ (444)

Formula weight: 213.88 g/mole

Pycnometric density: (α) 7.75 g/cc (595)
(β) 7.34 g/cc (595)

CHEMICAL

Theoretical analysis: 13.1% silicon
86.9% niobium

Synthesis: Nb + Si (595)

STRUCTURE

2 forms reported: α & β

THERMAL

Heat of formation: ΔH^O_{298} = -21 kcal/mole (444)

Activation energy: Diffusion of Si into Nb,
11.72 kcal/mole (473)

Nb_3Si_2

TRINIOBIUM DISILICIDE

Probably Nb_5Si_3 (444)

Formula weight: 334.85 g/mole

CHEMICAL

Theoretical analysis: 16.75% silicon
83.25% niobium

ELECTRICAL

Resistivity: Superconductive $< 1.2^{\circ}K$ (459)

THERMAL

Activation energy: See $NbSi_2$

Nb_4Si

TETRANIOBIUM MONOSILICIDE

Formula weight: 399.70 g/mole

Melting point: Approx. 1950°C, peritectic   (596)
                            2580°C, congruent   (597)

Pycnometric density: 8.01 g/cc

CHEMICAL

Theoretical analysis: 7.02% silicon
                        92.98% niobium

MECHANICAL

Hardness: Micro 100 g : 630-820 kg/mm^2 (444)
           Vickers   : 475-550 kg/mm^2 (590)

STRUCTURE

Hexagonal
a = 3.59 A,   c = 4.46 A,   c/a = 1.24   (597)

Isomorphous with Ta_4Si and Zr_4Si (596)

Radius ratio: 0.82   (47)
                     0.80   (444)

THERMAL

Activation energy: See $NbSi_2$

Nb_5Si_3

PENTANIOBIUM TRISILICIDE

Formula weight: 547.74 g/mole

Melting point: 2405°C   (12, 227)
                       2480°C, congruent   (596)

Pycnometric density: (α) 6.56 g/cc   (444)
                       (β) 6.26 g/cc   (590)
                       (β) 7.34 g/cc   (595)

CHEMICAL

Theoretical analysis: 15.34% silicon
                        84.66% niobium

Reactivity: Not stable in presence of carbon, converts
                to $D8_8$ phase.   (444)

MECHANICAL

Hardness: Vickers 100 g : 400-600 kg/mm^2 (590)
           Micro 100 g   : 700 kg/mm^2 (444)

STRUCTURE

α form: Tetragonal, Cr_3B_2 type, D_{4h}^{18} space group   (599)
a = 6.557 A,   c = 11.860 A,   c/a = 1.815   (599)

β form: Tetragonal, Cr_3B_2 type, D_{2h}^{11} space group   (592)
a = 9.998 A,   c = 5.067 A,   c/a = 0.51   (592)

α form transforms to β form at 1900-2100°C   (596)

Radius ratio: 0.82   (47)
                     0.80   (444)

THERMAL

Heat of formation: ΔH_{298}^O = -63 kcal/mole   (593)
Activation energy: See $NbSi_2$

NEODYMIUM DISILICIDE

Formula weight: 200.39 g/mole

X-ray density: 5.84 g/cc   (567)

CHEMICAL

Theoretical analysis: 28.01% silicon   (567)
                        71.99% neodymium   (567)

Synthesis: Nd$_2$O$_3$ + excess Si, vacuum, 1500°C   (444)

Reactivity: Air, O$_2$; oxidized to silicates 1500-1600°C   (444)

                Aq HF   :   decomposed   (444)

                Aq HCl   :   decomposed   (444)

                K$_2$CO$_3$ fused :   decomposed   (444)

                Na$_2$CO$_3$ fused :   decomposed   (444)

STRUCTURE

Tetragonal,   ThSi$_2$ type, space group D$_{4h}^{19}$ (444)

See LaSi$_2$

a = 4.103 A,   c = 13.53 A,   c/a = 3.30   (567, 568)

NICKEL MONOSILICIDE

Formula weight:	86.75 g/mole	
Melting point:	1000°C, congruent	(444)
X-ray density:	5.86 g/cc	(637)

CHEMICAL

Theoretical analysis:	32.35% silicon	
	67.65% nickel	
Synthesis:	Ni + Si, fusion	(637)

ELECTRICAL

Thermal EMF:	$+ 8 \ \mu v/°C$	(456, 457)

MECHANICAL

Hardness: Micro:	400 kg/mm^2 at 20°C	(639)
	300 kg/mm^2 at 500°C	(639)

STRUCTURE

Rhombic, FeSi type, B 20 space group	(444)
Cubic, T^4 space group	(445)
Rhombic, distorted NiAs structure	(637)
Ni atoms at x = 0.184, y = 0.006, z = 0.000	(637)
Si atoms at x = 0.080, y = 0.330, z = 0.500	(637)
a = 5.62 A, b = 5.18 A, c = 3.34 A	(637)
Radius ratio: 0.95	(444)

THERMAL

Activation energy:	See NiSi$_2$	
Heat of formation:	$\Delta H = -20.5$ kcal/mole	(636)

NICKEL DISILICIDE

Formula weight:	114.81 g/mole
Melting point:	Decomposes 1025°C (444)

CHEMICAL

Theoretical analysis: 48.88% silicon
51.12% nickel

ELECTRICAL

Thermal EMF: + 7 μv /°C   (456, 457)

Temperature coefficient of thermal EMF:
- 1.25 $\pm$ 0.0 μv/°C   (454)

MECHANICAL

Hardness: Micro 50 g : 1019 kg/mm^2 (444)

STRUCTURE

Cubic, Fluorite type, O_h^5 space group   (444)
a = 5.395 $\pm$ 0.005 A

Radius ratio: 0.95   (444)

THERMAL

Activation energy: Diffusion of Si into Ni,
24.95 kcal/mole   (473)

DINICKEL MONOSILICIDE

Formula weight: 145.44 g/mole

Melting point: 1290°C, congruent   (444)

CHEMICAL

Theoretical analysis: 19.29% silicon
                        80.71% nickel

MECHANICAL

Hardness: Micro: 450 kg/mm^2 at 20°C   (639)
                   320 kg/mm^2 at 500°C   (639)

STRUCTURE

Hexagonal, NiAs type, D_{6h}^4 space group   (444)
a = 3.797 A,   c = 4.898 A,   c/a = 1.290

Radius ratio: 0.95   (444)

THERMAL

Activation energy: See $NiSi_2$

Heat of formation: ΔH = -33.6 kcal/mole   (636)

Ni_2Si_3

DINICKEL TRISILICIDE

Formula weight: 201.56 g/mole

CHEMICAL

Theoretical analysis: 41.9% silicon
58.1% nickel

ELECTRICAL

Thermal EMF: + 9 $\mu v/^{\circ}C$ (456, 457)

THERMAL

Activation energy: See $NiSi_2$

TRINICKEL MONOSILICIDE

Formula weight: 204.13 g/mole

Melting point: 1150°C, decomposes   (444)

CHEMICAL

Theoretical analysis: 13.5% silicon
                         86.5% nickel

Synthesis: Anodic solution of 93 Si, 7 Ni alloys   (638)

This phase is stable to melting point.

ELECTRICAL

Thermal EMF: - 2 μv/°C   (456, 457)

MECHANICAL

Hardness; Micro: 400 kg/mm^2

STRUCTURE

Cubic, Cu$_3$Au type, O$_h^1$ space group   (444)
a = 3.50 A

See Fe$_3$Si for description of structure type

Radius ratio: 0.95   (444)

THERMAL

Activation energy: See NiSi$_2$

Heat of formation: ΔH = -35.5 kcal/mole   (636)

Ni_3Si_2

TRINICKEL DISILICIDE

| Formula weight: | 232.19 g/mole | |
| Melting point: | 830°C, decomposes | (444) |

CHEMICAL

| Theoretical analysis: | 24.1% silicon |
| | 75.9% nickel |

STRUCTURE

Not determined   (444)

THERMAL

Heat of formation: $\Delta H = -32.1$ kcal/mole   (640)

Ni_5Si_2

PENTANICKEL DISILICIDE

Formula weight: 349.57 g/mole

CHEMICAL

Theoretical analysis: 16.05% silicon
                        83.95% nickel

ELECTRICAL

Thermal **EMF**:   - 2 $\mu v/^oC$ (456, 457)

THERMAL

Activation energy: See $NiSi_2$

<u>NEPTUNIUM DISILICIDE</u>

Formula weight: 293.12 g/mole

<u>CHEMICAL</u>

Theoretical analysis: 19.15% silicon
80.85% neptunium

Synthesis: NpF$_3$ + Si at 1500°C in vacuo   (140)

Reactivity: H$_2$O, insoluble   (140)

6 M HCl, reacts violently   (140)

<u>STRUCTURE</u>

Tetragonal, ThSi$_2$ type, D$_{4h}^{19}$ space group   (332)

Isomorphous with ThSi$_2$ and $\propto$ USi$_2$ (332)

a = 3.97 A,   c = 13.70 A   (332)

OSMIUM MONOSILICIDE

Formula weight: 218.26 g/mole

CHEMICAL

Theoretical analysis: 12.83% silicon
87.17% osmium

STRUCTURE

Cubic   (644)
a = 4.729 A   (644)

OsSi$_2$

<u>OSMIUM DISILICIDE</u>

Formula weight: 246.32 g/mole

<u>CHEMICAL</u>

Theoretical analysis: 22.75% silicon
77.25% osmium

<u>STRUCTURE</u>

Isomorphous with RhSi$_2$ (444)

Radius ratio: 0.87   (444)

Os_2Si_3

DIOSMIUM TRISILICIDE

Formula weight: 464.58 g/mole

CHEMICAL

Theoretical analysis: 18.1% silicon
                        81.9% osmium

STRUCTURE

Tetragonal   (644)
a = 5.57 A,   c = 4.47 A   (644)

Radius ratio: 0.87   (444)

PALLADIUM MONOSILICIDE

Formula weight: 134.76 g/mole

Melting point: 900°C, congruent   (444)

CHEMICAL

Theoretical analysis: 20.82% silicon
                       79.18% palladium

Synthesis: Alloy Pd + 60% or more Si,
                leach with caustic   (444)

STRUCTURE

Rhombic, MnP type, D_{2h}^{16} space group   (444)
a = 6.121 A,   b = 5.588 A,   c = 3.374 A

Hexagonal
a = 6.52 A,   c = 3.42 A   (644)

Radius ratio: 0.86   (444)

Pd_2Si

DIPALLADIUM MONOSILICIDE

Formula weight: 241.46 g/mole

Melting point: 1400°C, peritectic (?)   (444)

CHEMICAL

Theoretical analysis: 11.62% silicon
                        88.38% palladium

STRUCTURE

Not yet studied

PRASEODYMIUM DISILICIDE

Formula weight: 197.04 g/mole

X-ray density: 5.64 g/cc   (567)

Pycnometric density: 5.46 $\pm$ 0.02 g/cc   (567)

CHEMICAL

Theoretical analysis: 28.43% silicon   (567)
                                   71.57% praseodymium   (567)

Synthesis: Pr$_2$O$_3$ + excess Si, vacuum, 1500°C   (444)

Reactivity: Air, O$_2$; oxidized to silicates, 1500-1600°C (444)

        Aq HF   :   decomposed   (444)

        Aq HCl   :   decomposed   (444)

        K$_2$CO$_3$ fused :   decomposed   (444)

        Na$_2$CO$_3$ fused :   decomposed   (444)

STRUCTURE

Tetragonal,   ThSi$_2$ type, space group D$_{4h}^{19}$ (444)

See LaSi$_2$

a = 4.140 A,   c = 13.64 A,   c/a = 3.30   (567, 568)

PLATINUM MONOSILICIDE

Formula weight: 223.29 g/mole

Melting point: 1229°C, congruent   (647)
                        1100°C   (648)

CHEMICAL

Theoretical analysis: 12.57% silicon

                            87.43% platinum

Synthesis: Pt + Si, violent reaction   (649)

STRUCTURE

Rhombic, MnP type, B-31   (444)
a = 5.92 A,   b = 5.58 A,   c = 3.59 A

Radius ratio: 0.84   (444)

Pt_2Si

DIPLATINUM MONOSILICIDE

Formula weight: 418.52 g/mole

Melting point: 1099°C, congruent   (647)

CHEMICAL

Theoretical analysis: 6.70% silicon
                         93.30% platinum

STRUCTURE

Tetragonal   (644)

a = 2.77 A,   c = 2.95 A   (644)

Isomorphous with Pt_3B_2 (444)

Radius ratio: 0.84   (444)

PLUTONIUM DISILICIDE

Formula weight: 295.32 g/mole

Pycnometric density: (β) 9.18 g/cc   (652)
                          9.12 g/cc   (653)

CHEMICAL

Theoretical analysis: 19.0% silicon
                       81.0% plutonium

Synthesis: PuF$_3$ + Si, 1300°C, 5 x 10^{-5} mm of Hg   (652)

Reactivity: Si, heating with excess failed to form
             PuSi$_3$ - analogous to USi$_3$ (652)

             Air, 700°C, PuO$_2$ + SiO$_2$ (444)

STRUCTURE

Tetragonal, ThSi$_2$ type, D$_{4h}^{19}$ space group   (332)
a = 3.98 A,   c = 13.58 A   (332)

Isomorphous with ThSi$_2$ and USi$_2$ (332)

(β form ?) Hexagonal, AlB$_2$ type, P 6/mmm   (652)
a = 3.884 A,   c = 4.082 A   (652)

THERMAL

Heat of formation: ΔH = -211 kcal/mole   (653)

RHENIUM MONOSILICIDE

Formula weight: 214.37 g/mole

Melting point: >1700°C   (444)

Vapor pressure: $\log P_{Si}$ = 7.444 - 25.800 T   (444)

CHEMICAL

Theoretical analysis: 13.02% silicon
                       86.98% rhenium

Synthesis: Re + Si, vacuum

STRUCTURE

Rhombic, FeSi type, B 20 space group   (444)

Cubic, T^4 space group   (445)

Isomorphous with FeSi, CoSi, NiSi, and MnSi

Cubic
a = 4.775 A,   Z = 4,   P 23 space group   (629)

Radius ratio: 0.85   (444)

THERMAL

Heat of formation: ΔH^O_{298} = -10.2 kcal/mole   (593)

Free energy of formation: ΔF^O = -26,580 - 0.94 T cal/mole
                                           below 1500°K   (444)

RHENIUM DISILICIDE

Formula weight:	242.43 g/mole	
Melting point:	>1700°C	(444)
Vapor pressure:	log P$_{Si}$ = 7.518 - 25.610 T	(444)

CHEMICAL

Theoretical analysis: 23.1% silicon
                         76.9% rhenium

Synthesis: Re + Si in vacuum

STRUCTURE

Tetragonal, MoSi$_2$ type, D$_{4h}^{17}$ space group   (444)

Radius ratio: 0.85   (444)

THERMAL

Heat of formation: ΔH^o_{298} = -16.6 kcal/mole   (593)

Free energy of formation: ΔF^o = -55,030 + 0.74 T cal/mole
                           below 1500°K   (444)

TRIRHENIUM MONOSILICIDE

Formula weight: 586,99 g/mole

Melting point: >1700°C   (444)

Vapor pressure: log P$_{Si}$ = 5.953 - 24.04 T

CHEMICAL

Theoretical analysis: 4.78% silicon
                        95.22% rhenium

Synthesis: Re + Si in vacuum

STRUCTURE

Probably cubic, β W type   (444)

See Fe$_3$Si for description of structure type

Radius ratio: 0.85   (444)

THERMAL

Heat of formation: ΔH^{o}_{298} = -12.6 kcal/mole   (593)

Free energy of formation: ΔF^{o} = 21,250 - 5.49 T cal/mole
                                   below 1500°K   (444)

RHODIUM MONOSILICIDE

Formula weight: 130.97 g/mole

CHEMICAL

Theoretical analysis: 20.4% silicon
79.6% rhodium

STRUCTURE

Cubic

a = 4.675 $\pm$ 0.01A,   Z = 4   (646)

C_{2h}^5 - P_{23} space group   (646)

Radius ratio: 0.87   (444)

RHODIUM DISILICIDE

May be identical with Rh$_2$Si$_3$ phase.   (444)

     Formula weight: 159.03 g/mole

CHEMICAL

     Theoretical analysis: 35.3% silicon
                           64.7% rhodium

STRUCTURE

     Isomorphous with OsSi$_2$ (444)

     Radius ratio: 0.87   (444)

Rh_2Si_3

DIRHODIUM TRISILICIDE

May be identical with phase designated $RhSi_2$ (444)

Formula weight: 290.00 g/mole

CHEMICAL

Theoretical analysis: 29.0% silicon
71.0% rhodium

Rh_3Si_2

TRIRHODIUM DISILICIDE

Has been identified   (444)

    Formula weight: 364.85 g/mole

CHEMICAL

    Theoretical analysis: 15.35% silicon
                              84.65% rhodium

RUTHENIUM MONOSILICIDE

Formula weight: 129.76 g/mole

CHEMICAL

Theoretical analysis: 21.62% silicon
                        79.38% ruthenium

Synthesis: Ru + Si react, caustic leach, HF-HNO$_3$ leach (643)

STRUCTURE

∝ form: body centered cubic, CsCl type   (644)
a = 2.909 A   (645)

β form: face-centered cubic   (644)
a = 4.70 A   (444)

Radius ratio: 0.87   (444)

Ru_2Si_3

DIRUTHENIUM TRISILICIDE

Formula weight: 287.58 g/mole

CHEMICAL

Theoretical analysis: 29.3% silicon
72.7% ruthenium

STRUCTURE

Tetragonal   (444)
a = 5.52 A,   c = 3.42 A   (444)

Radius ratio: 0.87   (444)

Ru_3Si_2

TRIRUTHENIUM DISILICIDE

Has been detected   (444)

      Formula weight: 361.22 g/mole

CHEMICAL

      Theoretical analysis: 15.5% silicon
                                84.5% ruthenium

SCANDIUM DISILICIDE

Formula weight: 101.22 g/mole

CHEMICAL

Theoretical analysis: 55.56% silicon
                         44.44% scandium

Synthesis: Sc_2O_3 + excess Si, vacuum, $1500^{\circ}C$ (444)

Reactivity: Air, O_2; oxidized to silicates, $1500\text{-}1600^{\circ}C$ (444)

                 Aq HF   :   decomposed   (444)

                 Aq HCl   :   decomposed   (444)

                 K_2CO_3 :   decomposed   (444)

                 Na_2CO_3 :   decomposed   (444)

STRUCTURE

Not yet established   (444)

SmSi$_2$

SAMARIUM DISILICIDE

Formula weight: 206.55 g/mole

X-ray density: 6.26 g/cc   (567)

Pycnometric density: 6.04 $\pm$ 0.02 g/cc   (567)

CHEMICAL

Theoretical analysis: 27.17% silicon   (567)
         72.83% samarium   (567)

Synthesis: Sm$_2$O$_3$ + excess Si, vacuum, 1500°C (444)

Reactivity: Air, O$_2$; oxidized to silicates, 1500-1600°C (444)

      Aq HF   : decomposed   (444)

      Aq HCl   : decomposed   (444)

      K$_2$CO$_3$ fused : decomposed   (444)

      Na$_2$CO$_3$ fused : decomposed   (444)

STRUCTURE

Tetragonal, resembles ThSi$_2$ (444)

Not completely understood   (444)

a = 4.041 A, c = 13.33 A, c/a = 3.39   (567, 568)

STRONTIUM MONOSILICIDE

Formula weight: 115.69 g/mole

CHEMICAL

Theoretical analysis: 24.25% silicon   (444)
                         75.75% strontium

Synthesis: See Ref. 485

STRUCTURE

Not yet established   (444)

THERMAL

Heat of formation: 112.8 kcal/mole   (444)

STRONTIUM DISILICIDE

Formula weight: 143.75 g/mole

CHEMICAL

Theoretical analysis: 32.04% silicon   (444)
                         67.96% strontium

Synthesis: See Ref. 485

STRUCTURE

Not yet established   (444)

THERMAL

Heat of formation: 147.4 kcal/mole   (444)

TANTALUM MONOSILICIDE

Formula weight: 208.94 g/mole

CHEMICAL

    Theoretical analysis: 13.4% silicon
                            86.6% tantalum

ELECTRICAL

    Resistivity: Superconductive $4.25^{\circ}K$ (460)

                       Superconductive $4.25-4.40^{\circ}K$ (333)

THERMAL

    Activation energy: See $TaSi_2$

TANTALUM DISILICIDE

Formula weight: 237.00 g/mole

Formula volume: 25.9 cc/mole

Melting point: 2400°C
                         2200°C   (2, 12, 23, 227, 444)
                         >2050°C   (22)

Vapor pressure: P_{Si} = 80 x 10^{-7} atm at 2200°K   (601)

X-ray density: 9.14 g/cc   (2, 23, 444)

Pycnometric density: 8.83 g/cc   (444)

CHEMICAL

Theoretical analysis: 16.65% silicon
                         83.35% tantalum

Synthesis: Ta + Si   (11, 22)

                Ta$_2$O$_5$ + SiO$_2$ + Al(S)   (571, 572, 573)

Reactivity: Corrosion in air becomes severe,
                1100-1700°C   (227)

                Maximum oxidation resistance at
                75% silicon   (227)

                Unstable in N$_2$ at 1350°C   (227, 235)

                Good resistance to oxidation in air
                at 1500°C   (227)

                Heating with graphite yields ternary phase   (444)

                51 mg/cm^2 weight loss; 1200°C, 4 hrs
                in air   (444)

                Aqueous acids, insoluble

                HF, soluble

                Fused alkali, soluble

ELECTRICAL

Resistivity: 8.5 x 10^{-6} ohm-cm   (449)

               38.0 x 10^{-6} ohm-cm   (453)

               Superconductive $< 1.2^{o}$K   (459)

               Superconductive $> 4.2^{o}$K   (444)

Temperature coefficient of resistivity:
3. 32   (453)

Temperature coefficient of thermal EMF:
-6. 42 $\pm$ 1. 0 μv/^{o}C   (454)

MAGNETIC

Susceptibility: -33 x 10^{-6} per mole   (453)

MECHANICAL

Hardness: Vickers 100 g : 1560 kg/mm^2 (8)
             Vickers 40 kg : 1000-1200 kg/mm^2 (444)
             Knoop 100 g   : 1000-1200 kg/mm^2 (2, 23)
             Micro 50 g   : 1407 kg/mm^2 (444)
             Micro 100 g   : 1510-1610 kg/mm^2 (444)

OPTICAL

Color - blue-gray, metallic   (19)

Form - Tetrahedral prisms with pyramids developed
       on faces.   (19)

STRUCTURE

Hexagonal, CrSi$_2$ type, D$_6^4$ space group   (444)
a = 4.771 A,   c = 6.551 A,   c/a = 1.373   (444)

STRUCTURE (cont.)

Radius ratio: 0.82   (47)

                        0.80   (444)

THERMAL

Specific heat: 8-9 cal/gm/$^{\circ}$C ;   425-1550°C   (171)

C_p = 15.54 + 4.74 x 10^{-3}T cal/mole/$^{\circ}$K ;

484-1125°K   (128)

Heat content: $H_T - H_{293}$ = 15.54 T + 2.37 x 10^{-3}T^2 - 4,844

cal/mole ; 484-1125°K   (128)

Heat of formation: ΔH°_{298} = -26.2 kcal/mole   (593)

ΔH°_{298} = -23.2 kcal/mole   (601)

ΔH°_{298} = -27.9 kcal/mole   (472)

Activation energy: Formation from diffusion of

Si into Ta, 6.04 kcal/mole   (473)

DITANTALUM MONOSILICIDE

Formula weight:	389.82 g/mole	
Formula volume:	28.4 cc/mole	
Melting point:	Decomposes	(2, 23)
	Approx. 2460°C, incongruent	(444)
Vapor pressure:	P_{Si} = 7.57 x 10^{-7} atm at 2200°K	(601)
	= 42.8 x 10^{-7} atm at 2300°K	(601)
	= 112 x 10^{-7} atm at 2400°K	(601)
X-ray density:	13.54 g/cc	(2, 23)
Pycnometric density:	12.4 g/cc	(444)

CHEMICAL

Theoretical analysis: 7.2% silicon
92.8% tantalum

Synthesis:	Ta + Si	(11)
Reactivity:	Unstable in N_2 at 1350°C	(227, 235)
	Maximum oxidation resistance at 75% silicon	(227)

MECHANICAL

Hardness: Knoop 100 g	: 1200-1500 kg/mm^2	(2, 23)
Vickers 40 kg	: 1200-1500 kg/mm^2	(444)

STRUCTURE

Tetragonal, C 16 space group	(444)
Similar to Ta_2B, Mo_2B, Fe_2B, Mn_2B, Co_2B, Ni_2B	(444)
a = 6.142 A, c = 5.029 A, c/a = 0.818	(444)
Radius ratio: 0.82	(47)
0.80	(444)

THERMAL

Heat of formation: ΔH^O_{298} = -30.7 kcal/mole   (593)

ΔH^O_{298} = -29.3 kcal/mole   (601)

Activation energy: See TaSi$_2$

Ta_4Si

TETRATANTALUM MONOSILICIDE

Formula weight: 751.58 g/mole

CHEMICAL

Theoretical analysis: 3.72% silicon
96.28% tantalum

STRUCTURE

Hexagonal, DO_{16} space group   (444)

Radius ratio: 0.82   (47)
0.80   (444)

THERMAL

Activation evergy: See $TaSi_2$

PENTATANTALUM MONOSILICIDE

Probably Ta$_9$Si$_2$ (444)

Formula weight: 932.46 g/mole

CHEMICAL

Theoretical analysis: 3.01% silicon
96.99% tantalum

Synthesis: Brewer (11) obtained a TaSi$_{0.2}$
phase by sintering Ta + Si

THERMAL

Heat of formation: ΔH^o_{298} = -32.2 kcal/mole   (444)

Activation energy: See TaSi$_2$

PENTATANTALUM DISILICIDE

Probably Ta_2Si (444)

Formula weight: 960.52 g/mole

CHEMICAL

Theoretical analysis: 5.73% silicon
94.27% tantalum

Synthesis: Brewer (11) obtained a $TaSi_{0.4}$
phase by sintering Ta + Si

Ta_5Si_3

PENTATANTALUM TRISILICIDE

Formula weight: 988.58 g/mole

Formula volume: 75.7 cc/mole

Melting point: ~2500°C   (2, 12, 23, 227)

                ~2500°C, congruent   (444)

X-ray density: 13.06 g/cc   (2, 23, 444)

Pycnometric density: 11.6 g/cc   (444)

CHEMICAL

Theoretical analysis: 8.51% silicon

                       91.49% tantalum

Synthesis: Ta + Si   (11)

Reactivity: Corrosion in air becomes severe,

              1100-1400°C   (227)

              Maximum oxidation resistance at

              75% silicon   (227)

              Unstable in N_2 at 1350°C   (227, 235)

MECHANICAL

Hardness: Knoop 100 g : 1200-1500 kg/mm^2 (2, 23)

            Vickers 40 kg : 1200-1500 kg/mm^2 (444)

STRUCTURE

Hexagonal, Mn_5Si_3 type, $D8_8$ space group   (444)

Similar to Ti_5Si_3, Ti_5Ge_3, Ti_5Sn_3 (444)

a = 7.459 A,   c = 5.215 A,   c/a = 0.699   (444)

Radius ratio: 0.82   (47)

                    0.80   (444)

THERMAL

Heat of formation: ΔH^O_{298} = -86.7 kcal/mole   (593)

ΔH^O_{298} = -80.1 kcal/mole   (601)

ΔH^O_{298} = -74.7 kcal/mole   (602)

Activation energy: See $TaSi_2$

$$Ta_9Si_2$$

$$(Ta_{4.5}Si)$$

NONATANTALUM DISILICIDE

Formula weight: 1684.04 g/mole

Formula volume: 131.0 cc/mole

Melting point: 2500°C   (444)

                  Approx. 2510°C   (227, 2)

Vapor pressure: P_{Si} = 2.62×10^{-7} atm at 2200°K   (601)

                        = 7.14×10^{-7} atm at 2300°K   (601)

                        = 37.5×10^{-7} atm at 2400°K   (601)

                        = 78.4×10^{-7} atm at 2500°K   (601)

X-ray density: 12.86 g/cc   (444)

Pycnometric density: 12.7 g/cc   (444)

CHEMICAL

Theoretical analysis: 3.33% silicon

                       96.67% tantalum

Synthesis: Ta + Si   (11)

Reactivity: Unstable in N_2 at 1350°C   (227, 235)

              Oxidizes in O_2, 1100-1400°C   (227)

              Oxidation resistance increases with silicon

              concentration, maximum at 75% silicon   (227)

MECHANICAL

Hardness: Vickers 40 kg : 1000-1200 kg/mm^2 (444)

STRUCTURE

Hexagonal   (444)

a = 6.093 A,   c = 4.909 A,   c/a = 1.612   (444)

Ta_9Si_2

$(Ta_{4.5}Si)$

(Cont.)

THERMAL

Heat of formation: $\Delta H = -32.2$ kcal/mole   (593)

                      $\Delta H = -34.4$ kcal/mole   (601)

Activation energy: See $TaSi_2$

THORIUM MONOSILICIDE

Formula weight: 260.18 g/mole

Melting point: < 1700°C   (212)

X-ray density: 9.03 g/cc   (444)

CHEMICAL

Theoretical analysis: 10.75% silicon
                       89.25% thorium

Synthesis: ThO_2 + Si   (654)

Reactivity: Air, very easily oxidized   (444)

STRUCTURE

Rhombic, USi type, D_{4h}^{16} space group   (444)
a = 5.89 A,   b = 7.88 A,   c = 4.15 A

Isomorphous with USi and ZrSi

Cell contains 4 ThSi molecules

Radius ratio: 0.65   (444)

THORIUM DISILICIDE

Formula weight: 288.24 g/mole

Melting point:	>1670°C	(213, 214)
	>1700°C	(444)
X-ray density:	7.79 g/cc	(215)
Pycnometric density:	7.63 g/cc	(215)
	7.96 g/cc at 16°C	(214)

CHEMICAL

Theoretical analysis: 19.45% silicon
                                  80.55% thorium

Synthesis:
   a. ThO_2 + Si, electric arc furnace   (214)

   b.   Th + Si + Al, vacuum, 1000°C (571, 572, 573)

   c. K_2SiF_6 + K_2ThF_6 + Al   (213, 214)

Reactivity:
73 mg/cm^2 weight gain, 4 hrs, 1200°C in air   (444)

HNO_3, no reaction   (444)

H_2SO_4 (conc), no reaction   (444)

Halogen acids, dissolves   (444)

H_2SO_4 dilute, dissolves   (444)

HCl gas, reacts vigorously   (444)

10% acqueous alkali, no reaction   (444)

Fused alkali, oxidizes   (444)

Air, easily oxidized   (444)

ELECTRICAL

Resistivity: Becomes superconductive below 3.16°K   (216)

MECHANICAL

Hardness: Micro : 1120 kg/mm^2 (444)

OPTICAL

Form - quadratic plates   (214)

STRUCTURE

Body-centered tetragonal   (215)
a = 4.126 A,   c = 14.346 A   (215)

Tetragonal   (217)

Tetragonal, D$_{4h}^{19}$ space group   (444)

4 molecules per unit cell - Prisms of Th atoms with Si
chains parallel X & Y axis passing through Th prisms
at various heights. Chains are so spaced that Si - Si
distance between chains is the same as that within the
chains giving in effect a Si atom skeleton.   (444)

Radius ratio: 0.65   (47, 444)

THERMAL

Heat of formation: ΔH = -39.9 kcal/mole   (472)

$$Th_2Si_3$$

DITHORIUM TRISILICIDE

Previously thought to be high temperature form of $ThSi_2$ (444)

    Formula weight: 548.42 g/mole

    Melting point: > 1700°C   (444)

    X-ray density: 8.23 g/cc   (444)

CHEMICAL

    Theoretical analysis: 15.35% silicon
                                 84.65% thorium

    Synthesis: ThO_2 + Si   (654)

    Reactivity: Air, easily oxidized   (444)

STRUCTURE

    Hexagonal, USi_2 type, D_{6h}^1 (C 6/mmm) space group   (444)
    a = 3.985 ± 0.002 A,   c = 4.220 ± 0.002 A

    Radius ratio: 0.65   (444)

Th_3Si_2

TRITHORIUM DISILICIDE

Formula weight: 752.48 g/mole

X-ray density: 9.80 g/cc   (444)

CHEMICAL

Theoretical analysis: 7.45% silicon
                       92.55% thorium

Synthesis: ThO_2 + Si   (654)

Reactivity: Air, easily oxidized   (444)

STRUCTURE

Tetragonal, D_{4h}^5 (P 4/mbm) space group   (444)

Isomorphous with U_3Si_2
a = 7.835 $\pm$ 0.003 A,   c = 4.154 $\pm$ 0.005 A   (444)

Radius ratio: 0.65   (444)

TITANIUM MONOSILICIDE

Formula weight: 75.96 g/mole

Melting point: 1950°C
                       d. 1760°C   (2, 15)
                       < 1675°C   (227, 298)
                       1760°C incongruent   (576)

X-ray density: 4.34 g/cc   (578)

Pycnometric density: 4.21 ± 0.05 g/cc   (578)

CHEMICAL

Theoretical analysis: 36.9% silicon
                       63.1% titanium

Synthesis: a) TiH + Si   (570)
                   b) SiO_2 + K_2TiF_6 + S + Al   (571, 572, 573)

                   For complete study of Ti - Si system
                   see Hansen   (576)

Reactivity: Reacts with N_2 to form nitrides   (227, 272)

                   Slow oxidation in O_2, 800-1100°C   (227)

                   Dissolves considerable Mo, converting to
                   Ti_5Si_3 (444)

ELECTRICAL

Resistivity: 39.3×10^{-6} ohm-cm   (453)

                   Temperature coefficient 4.13   (453)

MAGNETIC

Susceptibility: $+59 \times 10^{-6}$ per gram atom (453)

MECHANICAL

Hardness: Knoop 100g: 1039 kg/mm^2 (15)

Micro 100g: 986 kg/mm^2 (576)

STRUCTURE

Orthorhombic C_{2v}^1 space group   (578)

a = 3.61 A; b = 4.960 A; c = 6.470 A   (578)

Radius ratio: 0.80   (47)
0.79   (444)

THERMAL

Heat of formation: ΔH_{298}^o = -31.0 kcal/mole   (584)

ΔH_{298}^o = -39.2 $\pm$ 3.0 kcal/mole   (471)

ΔH_{298}^o = -31.0 $\pm$ 3.0 kcal/mole   (108)

ΔH_{298}^o = -31.0 $\pm$ 3.0 kcal/mole   (472)

Activation energy: See TiSi$_2$

TITANIUM DISILICIDE

Formula weight:	104.02 g/mole	
Melting point:	1540°C	(2, 15, 576)
	approx. 1500°C	(227, 298)
X-ray density:	4.40 g/cc	(19)
	3.85 g/cc	(82)
Pycnometric density:	4.39 g/cc	(14)
	4.13 g/cc	(2, 17)
	3.90 g/cc	(82)

CHEMICAL

Theoretical analysis:	53.96% silicon	
	46.04% titanium	
Synthesis:	a) Ti + Si	(444)
	b) 10 K$_2$SiF$_6$ + TiO$_2$, fuse, electrolyze	(574)
	c) TiH + Si	(570)
Reactivity:	Reacts with N$_2$ to form nitrides	(227, 272)
	Corrosion in air becomes severe, 800-1100°C	(227)
	3 mg/cm^2 weight gain, 4 hrs., 1200°C in air	(444)
	Reacts with boron at 1450 to TiB$_2$	(478)
	Converted by N$_2$ to TiN	(108)
	HClO$_4$: insoluble	(444)
	H$_2$SO$_4$: insoluble	(444)
	HCl : insoluble	(444)
	HNO$_3$ (1:1) with H$_2$O: insoluble	(444)
	KOH (50%): insoluble	(444)

CHEMICAL (cont.)

Reactivity: (cont.)

KHSO$_4$ (fused 200 - 300°C): insoluble   (444)

Na$_2$B$_4$O$_7$ fused: decomposed   (444)

NaOH fused: decomposed   (444)

KOH fused: decomposed   (444)

HF (1:1) with H$_2$0: decomposed   (444)

Oxidation: 0.3 mg/cm^2, 4 hrs. in air,
1200°C   (444)

ELECTRICAL

Resistivity: 123 x 10^{-6} ohm-cm   (449)

16.7 x 10^{-6} ohm-cm   (453)

18 x 10^{-6} ohm-cm at rt.   (450)

135 x 10^{-6} ohm-cm at 1200°C max.   (450)

100 x 10^{-6} ohm-cm at 1400°C min.   (450)

150 x 10^{-6} ohm-cm at 1500°C   (450)

Temperature coefficient of resistivity: 4.63   (453)

Temperature coefficient of thermal EMF; + 3.67 $\pm$
1.56 μv. /^{o}C   (454)

Magnetic susceptibility: +136 x 10^{-6} per gram atom   (453)

MECHANICAL

Hardness: Knoop 100g: 618 kg/mm^2 (2, 15, 576)

MECHANICAL (cont.)

Knoop 100g: (cont.)

: 870 kg/mm^2		(50)
: 840 - 390 kg/mm^2		(153)
Micro 50g: 892 kg/mm^2		(444)
: 824 kg/mm^2		(444)
Micro 100g: 1039 kg/mm^2		(576)

STRUCTURE

Orthorhombic - pseudo tetragonal	(82, 153)
a = 3.62 A; b = 13.76 A; c = 3.60 A	(82, 153)
Suggest two modifications to explain wide deviation	(153)
Orthorhombic, D_{2h}^{24} Space group	(444)
Rhombic, slightly deformed	(577)
a = 8.236 A; b = 4.773 A; c = 8.523 A	(577)
Z = 8	
Orthorhombic, C-54	(19)
a = 8.24 A; b = 4.77 A; c = 8.52 A	(19)
Radius ratio: 0.80	(47)
0.79	(444)

THERMAL

Heat of Formation: $\Delta H_{298}^o = -32.2$ kcal/mole (584)

$\Delta H_{298}^o = -42.9 \pm 4.5$ kcal/mole (471)

THERMAL (cont.)

Heat of Formation: (cont.)

$$\Delta H^O_{298} = -32.2 \pm 4.5 \text{ kcal/mole} \qquad (108)$$

$$\Delta H^O_{298} = -32.2 \pm 4.5 \text{ kcal/mole} \qquad (472)$$

Activation energy: diffusion of Si into Ti, 5.69 kcal/mole(473)

<u>DITITANIUM MONOSILICIDE</u>

Formula weight: 123.86 g/mole

<u>CHEMICAL</u>

Theoretical analysis: 22.6% silicon
77.4% titanium

Synthesis: $Si + TiCl_4$ (575)

Ti_5Si_3

PENTATITANIUM TRISILICIDE

Formula weight: 323.68 g/mole

Melting point: 2120°C   (2, 576, 15, 227, 298)
                    2150°C   (13)

Pycnometric density: 4.32 g/cc   (2, 16)

CHEMICAL

Theoretical analysis: 26.0% silicon
                       74.0% titanium

Synthesis: a.   TiH + Si   (570)

                b.   Ti + Si   (444)

Reactivity: Reacts with N_2 to form nitrides   (227, 272)

                0.4% weight change, 3 hrs in air
                at 1100°C   (227)

                Corrosion in air becomes severe,
                800-1100°C   (227)

ELECTRICAL

Resistivity: 350×10^{-6} ohm-cm   (453)

                Temperature coefficient: 0.80   (453)

                Superconductive below 1.2°K   (459)

Magnetic susceptibility: $+164 \times 10^{-6}$ per gram atom (453)

MECHANICAL

Hardness: Vickers 100 g : 986 kg/mm^2 (8)
           Knoop 100 g   : 986 kg/mm^2 (15, 576)

STRUCTURE

Hexagonal, Mn_5Si_3 type, $D8_8$ space group   (444)

Rhombic, D_{2h}^{24} - Fddd   (16)

a = 7.465 ± 0.002 A; c = 5.162 ± 0.002A   (16)
c/a = 0.692   (16)

Isomorphous with Ti_5Ge_3 and Ti_5Sn_3 (16)

Radius ratio 0.79   (444)
            0.80   (47)

THERMAL

Heat of formation: ΔH^o_{298} = -138.6 kcal/mole   (108, 584)

ΔH^o_{298} = -147 ± 12 kcal/mole   (471)

ΔH^o_{298} = -139 kcal/mole   (472)

Activation energy: See $TiSi_2$

URANIUM MONOSILICIDE

Formula weight:	266.13 g/mole	
Melting point:	1570°C	(227)
	Forms peritectically	(444)
X-ray density:	10.40 g/cc	(444)

CHEMICAL

Theoretical analysis: 10.54% silicon
                        89 46% uranium

Reactivity:	Poor resistance to N_2 at 500°C	(227, 309)
	Oxidation rate is linear	(227, 309)
	Very poor oxidation resistance at 300°C	(227, 277)
	Disintegrates in air at 400°C	(227, 277)

MECHANICAL

Hardness: Knoop 25 g :	645 kg/mm^2	(227, 309)
100 g :	745 kg/mm^2	(227, 309)

STRUCTURE

Rhombic, FeB type, D_{2h}^{16} - Penm   (444)
a = 5.65 $\pm$ 0.01,   b = 7.65 $\pm$ 0.01,   c = 3.90 $\pm$ 0.01 A

Radius ratio: 0.77   (444)

THERMAL

Expansion:	0-1000°C, See Ref. 309	(227)
	See Ref. 233	(227)

URANIUM DISILICIDE

Formula weight:	294.19 g/mole	
Melting point:	1700°C, congruent	(227, 261, 444)
X-ray density:	(α) 8.98 g/cc	(332)
	(β) 9.25 g/cc	(332)

CHEMICAL

Theoretical analysis: 19.05% silicon
80.95% uranium

Synthesis:	$U_3O_8 + SiO_2 + Al + S$	(655)
Reactivity:	Suitable for short term use in N_2 at 500°C	(227, 309)
	Oxidation is parabolic	(227, 309)
	Fairly stable in air at 400°C	(227, 277)
	Burns in air at 800°C	(227)

MECHANICAL

Hardness: Knoop	25 g : 745 kg/mm^2	(227, 309)
	100 g : 700 kg/mm^2	(227, 309)

STRUCTURE

(α form)	Tetragonal, $ThSi_2$ type, D_{2h}^{19}	(444)
(β form)	Hexagonal, AlB_2 type, D_{4h}^1	(444)

USi_2 type structure uranium atoms at corners of cube, Si and U atoms statistically at face centers: Unit cell contains 1.07 U, 2.14 Si atoms. A defect Cu_3Au type structure.   (444)

α form   Body-centered tetragonal, D_{4h}^{10} -14/ama
a = 3.97 $\pm$ 0.03,   c = 13.71 $\pm$ 0.08   (332)

STRUCTURE (cont.)

β form   Hexagonal, AlB$_2$ type, D$_{6h}^1$ -C6/mmm

a = 3.85 $\pm$ 0.01,   c = 4.06 $\pm$ 0.01 A   (332)

Face-centered cubic

a = 4.053   (657)

Radius ratio: 0.77   (444)

THERMAL

Expansion: From 0-1000°C   See Ref. 309   (227)

See Ref. 233   (227)

57 x 10^{-6} per $^{\circ}$C parallel a axis   (656)

-26 x 10^{-6} per $^{\circ}$C parallel c axis   (656)

16 x 10^{-6} per $^{\circ}$C parallel U - Si bond   (656)

URANIUM TRISILICIDE

Formula weight:	322.25 g/mole	
Melting point:	Approx. 1500°C	(227, 195)
	Forms peritectically	(444)

CHEMICAL

Theoretical analysis:	26.1% silicon	
	73.9% uranium	
Reactivity:	Moderate resistance to N$_2$ at 500°C	(227, 309)
	Oxidation is parabolic	(227, 309)
	Stable in air at 400°C	(227, 277)

MECHANICAL

Hardness: Knoop 25 g :	485 kg/mm^2	(227, 309)
100 g :	445 kg/mm^2	(227, 309)

STRUCTURE

Tetragonal		
c/a = 4.03		(444)
Radius ratio:	0.77	(444)

THERMAL

Expansion:	13.4 x 10^{-6} per ^{o}C; 20-250°C	(227, 277)
	14.9 x 10^{-6} per ^{o}C; 20-600°C	(227, 277)
	16.3 x 10^{-6} per ^{o}C; 20-950°C	(227, 277)
	See Ref. 233	(227)

U_2Si_3

DIURANIUM TRISILICIDE

Formula weight: 560.32 g/mole

Melting point: 1600°C   (227)

CHEMICAL

Theoretical analysis: 15.0% silicon
                        85.0% uranium

TRIURANIUM MONOSILICIDE

Formula weight: 742.27 g/mole

Pycnometric density: 14.5 g/cc   (501)

CHEMICAL

Theoretical analysis: 3.78% silicon
                         96.22% uranium

ELECTRICAL

Resistivity: 58 x 10^{-6} ohm-cm at 0°C   (501)

                 63 x 10^{-6} ohm-cm at 100°C   (501)

                 67 x 10^{-6} ohm-cm at 200°C   (501)

                 72 x 10^{-6} ohm-cm at 400°C   (501)

                 77 x 10^{-6} ohm-cm at 600°C   (501)

STRUCTURE

Tetragonal, D_{4h}^{18} - I4/mcm   (444)

Tetragonal unit cell with U atoms at apices, Si atoms
at face centers.

a = 6.017 ± 0.002,   c = 8.679 ± 0.003 A

Radius ratio: 0.77   (444)

THERMAL

Specific heat: Cp = 3.16 x 10^{-6}T + 0.0412 cal/gm/°C
                  50 - 430°C   (501)

U_3Si_2

TRIURANIUM DISILICIDE

Formula weight:	770.33 g/mole	
Melting point:	Approx. 1650°C	(227, 277)
	1665°C, congruent	(444)
X-ray density:	12.20 g/cc	(444)
Pycnometric density:	11.45 g/cc	(136)

CHEMICAL

Theoretical analysis: 7.28% silicon
92.72% uranium

Reactivity:		
	Slight reaction with H_2O at 100°C	(136)
	Very poor resistance to N_2 at 500°C	(227, 309)
	Oxidation is parabolic	(227, 309)
	May be used for short term at 300°C in air	(227, 277)
	Disintegrates in air at 400°C	(227, 277)

ELECTRICAL

Resistivity: 150×10^{-6} ohm-cm; $\rho = 11.2$ g/cc   (136)

MECHANICAL

Strength: Bending (MOR) : 12,500 psi
11,300 psi

Hardness: Knoop 100 g   : 796 kg/mm^2 (227, 309)
Knoop 100 g   : 593 kg/mm^2 (522-728)   (38)

Young's modulus: 11.3×10^6 psi

Shear modulus: 4.8×10^6 psi

Poisson's ratio: 0.17

Creep: 800°C   (136)

STRUCTURE

Tetragonal, D_{4h}^5 -P 4/mbn   (444)

6 uranium atoms, 4 silicon atoms per unit cell. Si atom
pairs parallel 001 plane, uranium atoms in layers alternating (444)
a = 7.315 A,   c = 3.895 A

Radius ratio: 0.77   (444)

THERMAL

Conductivity: 0.06 CGS at 800^oC (136)

Expansion: 13.54×10^{-6} per ^oC; 25-200^oC (136)

14.03×10^{-6} per ^oC; 25-400^oC (136)

14.43×10^{-6} per ^oC; 25-600^oC (136)

14.57×10^{-6} per ^oC; 25-800^oC (136)

14.85×10^{-6} per ^oC; 25-1000^oC (136)

14.99×10^{-6} per ^oC; 25-1200^oC (136)

15.65×10^{-6} per ^oC; 20-250^oC (227, 277)

15.20×10^{-6} per ^oC; 20-600^oC (227, 277)

14.58×10^{-6} per ^oC; 20-950^oC (227, 277)

See Ref. 233   (227)

U_5Si_3

PENTAURANIUM TRISILICIDE

Reported by Katz   (140)

Actually is U_3Si_2 (332)

    Formula weight: 1274.53 g/mole

    Melting point: 1665°C   (227)

CHEMICAL

    Theoretical analysis: 6.60% silicon
                                     93.40% uranium

$U_{10}Si_3$

DECAURANIUM TRISILICIDE

Reported by Katz (140)

Actually is U_3Si (332)

VSi_2

VANADIUM DISILICIDE

Formula weight: 107.07 g/mole

Melting point:

1750°C	(14)
1654°C	(589, 11)
1650-1700°C	(12)
1655°C	(227, 298)
1660-1750°C	(227)
1660°C	(589)
Approx. 1670°C	(590)

Pycnometric density: 4.71 g/cc   (14)

CHEMICAL

Theoretical analysis: 52.4% silicon
                      47.6% vanadium

Synthesis:

V_2Si + Si (excess)   (587, 588)

V_2Si (22)

Cu Silicide + V_2O_5 (586)

V_2O_3 + Si + Al   (586)

Reactivity: 4.9 mg/cm^2 weight gain,
4 hours, 1200°C in air   (444)

Aqueous acids, insoluble

Aqueous alkali, insoluble

HF, soluble

Fused alkali, soluble

ELECTRICAL

Resistivity: 9.5×10^{-6} ohm-cm   (449)

13.3×10^{-6} ohm-cm   (453)

Superconductive < 1.2°K   (459)

MAGNETIC

    Susceptibility + 168 x 10^{-6} per mole   (453)

MECHANICAL

    Hardness: Knoop 100 g : 1090 kg/mm^2 (2, 19)

                Micro   : 890 - 960   (890)

STRUCTURE

    Hexagonal, CrSi$_2$ type, D_6^4 space group   (22)

    a = 4.562 A,   c = 6.359 A   (22)

    Radius ratio: 0.87   (444)

V_2Si

DIVANADIUM MONOSILICIDE

Formula weight:	129.96 g/mole	
Melting point:	Higher than VSi_2	(586)
Pycnometric density:	5.48 g/cc	(14)

CHEMICAL

Theoretical analysis:	21.6% silicon	
	78.4% vanadium	
Synthesis:	V_2O_3 + Si	(586, 588)
	V_2O_5 + Si + C	(586, 588)

THERMAL

Heat of Formation:	$\Delta H = -35.0$ kcal/mole	(594)

TRIVANADIUM MONOSILICIDE

Formula weight:	180.91 g/cc	
Melting point:	Approx. 1730°C	(590)
Pycnometric density:	5.67 g/cc	(591)

CHEMICAL

Theoretical analysis: 15.5% silicon
                          84.5% vanadium

ELECTRICAL

Resistivity: Superconductive below 17°K   (459)

MECHANICAL

Hardness: Micro: 1430 - 1560 kg/mm^2 (590)

STRUCTURE

Cubic   (14)

Cubic, β W type, O_h^3 space group   (591)
a = 4.712 ± 0.003 A   Z = 2   (591)

See Fe_3Si for description of structure type.

Radius radio: 0.87   (444)

THERMAL

Heat of formation: ΔH_{298} = -36.9 kcal/mole   (593)

V_5Si_3

PENTAVANADIUM TRISILICIDE

Formula weight: 338.93 g/mole

Melting point: Approx. 2150°C   (590)

Pycnometric density: 4.80 g/cc   (590)

CHEMICAL

Theoretical analysis: 24.8% silicon
                                 75.2% vanadium

Synthesis: 5 V + 3 Si in absence of C   (590)

MECHANICAL

Hardness: Micro: 1350 - 1510 kg/mm^2 (590)

STRUCTURE

Tetragonal, Ni3P type, D_{2d}^{11} (590, 592)

a = 9.410 A,   c = 4.74,   c/a = 0.504   (590, 592)

Isomorphous to Ta_5Si_3, β Nb_5Si_3, Cr_5Si_3, Mo_5Si_3,

         and W_5Si_3

Radius ratio: 0.87   (444)

TUNGSTEN MONOSILICIDE

Formula weight: 211.98 g/mole

Melting point: 2150°C

CHEMICAL

Theoretical analysis: 13.21% silicon

86.79% tungsten

TUNGSTEN DISILICIDE

Formula weight: 240.04 g/mole

Melting point:	$2150^{\circ}C$	(30, 449)
	$2180^{\circ}C$	(2, 14, 227)
	$2165^{\circ}C$	(444)

Pycnometric density:	9.3 g/cc	(14)
	9.25 g/cc	(444)

CHEMICAL

Theoretical analysis: 23.35% silicon
76.65% tungsten

Synthesis:	W + Si	(11, 449, 30)
	$WO_3 + SiO_2 + Al$	(573)
	$W + CuSi_x$	(617)

Reactivity:	Corrosion in air becomes severe above $1950^{\circ}C$	(2)
	Good to $1930^{\circ}C$ in air	(227)
	17 mg/cm^2 weight loss; $1200^{\circ}C$, 4 hrs in air	(444)
	23 mg/cm^2 weight loss; $1500^{\circ}C$, 4 hrs in air	(444)
	Mineral acids, insoluble	
	HF, soluble	
	Fused alkali, decomposed	

ELECTRICAL

Resistivity:	54.9×10^{-6} ohm-cm	(444)
	33.4×10^{-6} ohm-cm	(27)
	38.2×10^{-6} ohm-cm	(453)

ELECTRICAL (cont.)

Resistivity: (cont.)

Superconductive	$<1.2\,^{\circ}K$	(459)
	$<1.9\,^{\circ}K$	(444)

Temperature coefficient of thermal **EMF**:

$+9.60 \pm 0.31\ \mu v/^{\circ}C$ (454)

Magnetic susceptibility:

-75×10^{-6} per mole   (453)

Thermionic work function:

5 - 6 ev   (444)

MECHANICAL

Hardness: Vickers 100 g :	$1090\ kg/mm^2$	(8)
10 μ :	$1632\ kg/mm^2$	(8)
Knoop 100 g :	$1090\ kg/mm^2$	(14)
Micro 50 g :	$1260\ kg/mm^2$	(444)

OPTICAL

Color: blue-gray, metallic

Form: "hexahedral" prisms   (444)

STRUCTURE

Tetragonal   (14)

Tetragonal, $MoSi_2$ type, D_{4h}^{17} space group   (28, 627)

$a = 3.212 \pm 0.005$ A, $c = 7.880 \pm 0.005$,   c/a = 2.454

Radius ratio:	0.86	(47)
	0.84	(444)

THERMAL

Expansion: 7.79×10^{-6} per °C;   25-500°C   (329)

8.31×10^{-6} per °C;   25-1000°C   (329)

8.21×10^{-6} per °C;   0-1000°C   (7)

8.81×10^{-6} per °C;   0-1400°C   (7)

Heat capacity: 8 cal/gm/°C, constant from 425-1450°C   (171)

$$Cp = 17.04 + 2.12 \times 10^{-3} \, T \text{ cal/mole/°K}$$
$$460 - 1068°K \quad (128)$$

Heat content: $H_T - H_{293} = 17.04 + 1.06 \times 10^{-3} \, T^2 - 5.175$
$$\text{cal/mole, } 460 - 1068°K \quad (128)$$

Heat of formation: $\Delta H^o_{298} = -22.4 \text{ kcal/mole}$ (593)

Activation energy of formation:

By diffusion of Si into W : 5.78 kcal/mole   (473)

W_2Si_3

DITUNGSTEN TRISILICIDE

Formula weight: 452.02 g/mole

CHEMICAL

Theoretical analysis: 18.62% silicon
81.38% tungsten

Synthesis: $WO_3 + Si$ (625)

Existence not confirmed

TRITUNGSTEN MONOSILICIDE

Formula weight: 579.82 g/mole

CHEMICAL

Theoretical analysis: 4.85% silicon
                         95.15% tungsten

STRUCTURE

Probably cubic, β W type   (444)

See Fe$_3$Si for description of structure

Radius ratio: 0.86   (47)
                     0.84   (444)

$$W_3Si_2$$

TRITUNGSTEN DISILICIDE

Formula W_5Si_3 should be applied to this compound   (626)

      Formula weight: 607.88 g/mole

      Melting point: 2340°C   (2, 29)
                                    2350°C   (30, 449, 227, 2)
                                    2320°C   (444)

      Pycnometric density: 12.21 g/cc   (444)

CHEMICAL

      Theoretical analysis: 9.23% silicon
                                      90.77% tungsten

      Synthesis: W + Si   (11, 449, 30)

      Reactivity: Oxidizes in O_2, $1400-1700^\circ$C   (227)

ELECTRICAL

      Resistivity: Superconductive at 2.84°K   (459)

MECHANICAL

      Hardness: Knoop 100 g : 770 kg/mm^2 (2, 30)
                 Rockwell A : 91   (2, 29)

STRUCTURE

      Tetragonal, Cr_3B_2 type, D_{4h}^{18} space group   (444)
      a = 9.54 A,   c = 4.93 A,   c/a = 0.517

      Radius ratio: 0.86   (47)
                     0.84   (444)

THERMAL

      Heat of formation: $\Delta H = -30$ kcal/mole   (593)

PENTATUNGSTEN TRISILICIDE

Schoenberg suggests this as high temperature form of W_3Si_2,
stabilized by traces of carbon.   **(444)**

Formula weight: 1003.78 g/mole

CHEMICAL

Theoretical analysis: 8.4% silicon
                         91.6% tungsten

Synthesis: W + Si, sinter in C tube furnace   **(626)**

STRUCTURE

Face-centered tetragonal, Cr_3B_2 type, D_{4h}^{18} space group   **(626)**
a = 9.645 A,   c = 4.97 A

Radius ratio: 0.86   **(47)**
                         0.84   **(444)**

THERMAL

Heat of formation: ΔH_{298}^O = -46.5 kcal/mole   **(444)**

YTTRIUM DISILICIDE

Formula weight: 147.04 g/mole

Pycnometric density: 4.35 $\pm$ 0.05 g/cc   (567)

CHEMICAL

Theoretical analysis: 38.69% silicon   (567)
              61.31% yttrium   (567)

Synthesis: Y$_2$O$_3$ + excess Si, vacuum, 1500°C   (444)

Reactivity: Air, O$_2$; oxidized to silicates, 1500-1600°C (444)

  Aq HF   : decomposed   (444)

  Aq HCl   : decomposed   (444)

  K$_2$CO$_3$ fused : decomposed   (444)

  Na$_2$CO$_3$ fused : decomposed   (444)

STRUCTURE

Tetragonal,   resembles ThSi$_2$ (444)

Not completely understood   (444)

Radius ratio: 0.65   (444)

$YbSi_2$

YTTERBIUM DISILICIDE

Formula weight: 145.04 g/mole

CHEMICAL

Theoretical analysis: 31.80% silicon
                        68.20% ytterbium

Synthesis: Yb_2O_3 + excess Si, vacuum, 1500°C   (444)

Reactivity: Aq HF   : decomposed   (444)

              Aq HCl   : decomposed   (444)

              Air, O_2 : oxidized at 1500-1600°C   (444)

              K_2CO_3 fused : decomposed   (444)

              Na_2CO_3 fused : decomposed   (444)

STRUCTURE

Not yet established   (444)

ZINC SILICIDE

According to Hansen (486) no binary compounds form. Solubility of

Si in Zn is   0.06% at 600°C

                  0.15% at 650°C

                  0.57% at 730°C

                  0.92-1.62 at 850°C   (444)

ZIRCONIUM MONOSILICIDE

Formula weight: 119.28 g/mole

Formula volume: 21.46 cc/mole

Melting point: 1950°C
                  d. 2095°C   (2, 18, 227, 298, 581)
                  ~ 2100°C, peritectic   (582)
                      2150°C   (444)

X-ray density: 5.56 g/cc   (444)

CHEMICAL

Theoretical analysis: 23.52% silicon
                       76.48% zirconium

Synthesis: Zr + Si, sintered

Reactivity: Reacts with N_2 to form nitrides   (227, 272)

                Oxidizes in O_2, 1100°C - 1400°C   (227)

ELECTRICAL

Resistivity: 49.4×10^{-6} ohm-cm   (453)

                Superconductive below 1.2°K   (459)

MAGNETIC

Susceptibility: -63×10^{-6} per mole (453)

MECHANICAL

Hardness: Knoop 100 g : 1030 kg/mm^2 (14)
           Micro 50 g : 1020 - 1180 kg/mm^2 (444)

STRUCTURE

Hexagonal   (14, 210, 581, 582)

STRUCTURE (cont.)

$a = 7.01\,A,\quad c = 12.77\,A$ (210)

$\quad\quad 7.005\,A,\quad c = 12.772\,A,\quad c/a = 1.823$ (581)

Orthorhombic, FeB type

$a = 6.698\,A,\quad b = 3.778\,A,\quad c = 5.291\,A$

Radius ratio: 0.75   (47)

0.73   (444)

THERMAL

Heat of formation: $\Delta H^{o}_{298} = $ -37 kcal/mole   (584)

-58 $\pm$ 10 kcal/mole   (108)

<u>ZIRCONIUM DISILICIDE</u>

See Reference 211

Formula weight: 147.34 g/mole

Formula volume: 30.1 cc/mole

Melting point: d. 1520°C to ZrSi + liquid   (2, 18, 581)
                  d. 1525°C   (2, 227, 298)
                      1520°C, peritectic   (582)
                      1680 - 1700°C   (444)

X-ray density: 4.90 g/cc   (82)

Pycnometric density: 4.88 g/cc   (2, 19, 20, 21, 211)
                      5.2 g/cc   (82)

<u>CHEMICAL</u>

Theoretical analysis: 38.09% silicon
                       61.91% zirconium

Synthesis: Fusion of ZrO$_2$ + Si   (579)

                  K$_2$ZrF$_6$ + excess Si   (579)

                  Zr + Si   (580)

Reactivity: Insoluble in HClO$_4$; conc. H$_2$SO$_4$;
                  HCl; HNO$_3$ (82)

                  Reacts with N$_2$ to form nitrides   (227, 272)

                  Oxidizes in O$_2$, 800-1100°C   (227)

                  12 mg/cm^2 weight gain; 4 hours,
                  1200°C in air   (444)

                  Reacts with boron at 1650 to ZrB$_2$ (478)

                  Less stable than ZrN   (108)

                  KOH, 50%, insoluble

CHEMICAL (cont.)

    Reactivity: (cont.)

        NaOH, 50%, insoluble

        $KHSO_4$, fused, soluble

        $Na_2B_4O_7$, fused, soluble

        KOH, fused, soluble

        NaOH, fused, soluble

        HF, dilute or concentrated, soluble

ELECTRICAL

    Resistivity: 161×10^{-6} ohm-cm   (449)

                106.2×10^{-6} ohm-cm   (453)

    Temperature coefficient:

        3.52×10^{-6} ohm/$^{\circ}$C   (453)

    Temperature coefficient of thermal EMF:

        $+ 15.95 \pm 2.0$ μv/$^{\circ}$C   (454)

MAGNETIC

    Susceptibility: $- 96 \times 10^{-6}$ per mole   (453)

MECHANICAL

    Hardness: Knoop 100 g : 1030 kg/mm^2 (2, 19)

           Micro 50 g : 1063 kg/mm^2 (444)

                       830 - 980 kg/mm^2 (444)

OPTICAL

    Color: gray

OPTICAL (cont.)

Form: Rhombic pillars (prisms) (572)

STRUCTURE

Orthorhombic, pseudo tetragonal (21, 82)

a = 3.72 A,   b = 14.69 A,   c = 3.66 A (82)
    3.72 A,   14.61 A,   3.67 A (21)

Rhombic, D$_{4h}^{17}$ space group (444)

4 molecules per unit cell, Si atoms form chains parallel
to X + Z axis (444)

Radius ratio: 0.75 (47)
                        0.73 (444)

THERMAL

Heat of formation: ΔH^o_{298} = -38 kcal/mole (444)
                                          -30.5 kcal/mole (108)

DIZIRCONIUM MONOSILICIDE

Formula weight:	210.50 g/mole	
Formula volume:	35.1 cc/mole	
Melting point:	d. 2100°C	(2, 227, 298)
	d. 2110°C, peritectic	(18, 581)
	~ 2200°C, peritectic	(582)
	2220°C	(444)
X-ray density:	5.99 g/cc	(444)

CHEMICAL

Theoretical analysis:	13.33% silicon	
	86.67% zirconium	
Synthesis:	Zr + Si, sintered	
Reactivity:	Oxidizes in O_2, 800-1100°C	(227)
	Reacts with N_2 to form nitrides	(227, 272)

ELECTRICAL

Resistivity:	Superconductive below 1.2°K	(459)

MECHANICAL

Hardness: Micro 50 g : 1180 - 1280 kg/mm^2 (444)

STRUCTURE

Hexagonal, C16 space group	(444)
Tetragonal, $CuAl_2$ type, isomorphous with Ta_2Si	(583)
a = 6.568 A, c = 5.361 A, c/a = 0.816	(444)
Radius ratio: 0.75	(47)
0.73	(444)

THERMAL

Heat of formation: $\Delta H = -74 \pm 10$ kcal/mole (108)

$$Zr_3Si_2$$

TRIZIRCONIUM DISILICIDE

Formula weight: 329.78 g/mole

Melting point: d. 2200°C   (227, 298)
                        2210°C, peritectic   (18, 581)

CHEMICAL

Theoretical analysis: 17.02% silicon
                         82.98% zirconium

Reactivity: Oxidizes in O_2, 800-1100°C   (227)

                      Reacts with N_2, to form nitrides (227, 272)

THERMAL

Heat of formation: ΔH^o_{298} = -92 kcal/mole   (444)

TETRAZIRCONIUM MONOSILICIDE

Melting point: 1630°C, peritectic   (18, 581)

CHEMICAL

Theoretical analysis: 7.14% silicon
                        92.86% zirconium

THERMAL

Heat of formation: ΔH^{o}_{298} = -52 kcal/mole   (444)

Zr_4Si_3

TETRAZIRCONIUM TRISILICIDE

Formula weight:	449.06 g/mole

Melting point: d. 2220°C   (227, 298)
                  2210°C, peritectic   (18, 581)

CHEMICAL

Theoretical analysis: 18.74% silicon
                         81.26% zirconium

Reactivity: Oxidizes in O_2, 800-1100°C   (227)

               Reacts with N_2 to form nitrides (227, 272)

Zr_5Si_3

PENTAZIRCONIUM TRISILICIDE

Formula weight:	540.28 g/mole	
Melting point:	2250°C, congruent	(582)
X-ray density:	5.90 g/cc	(444)

CHEMICAL

Theoretical analysis:	15.58% silicon	
	84.42% zirconium	
Synthesis:	Sinter Zr + Si	
Reactivity:	Reacts with N_2 to form nitrides	(227, 272)

MECHANICAL

Hardness: Micro 50 g : 1280 - 1390 kg/mm² (444)

STRUCTURE

Hexagonal, Mn_5Si_3 type, $D8_8$ space group		(583)
a = 7.870 A, c = 5.547 A, c/a = 0.7048		(444)
Isomorphous with Ti_5Si_3 and Ta_5Si_3		(583)
Radius ratio:	0.75	(47)
	0.73	(444)

THERMAL

Heat capacity:	$Cp = 43.99 + 2.16 \times 10^{-2} T$ cal/mole/°K; 573 - 1113°K	(128)
Heat content:	$H_T - H_{293} = 43.99 T + 1.08 \times 10^{-2}T^2 - 14,079$ cal/mole; 573 - 1113°K	(128)
Heat of formation:	$\Delta H^o_{298} = -138$ kcal/mole	(444)
	-72 ± 10 kcal/mole	(108)

Zr_6Si_5

HEXAZIRCONIUM PENTASILICIDE

Formula weight: 687.62 g/mole

Melting point: $2250^{\circ}C$ (227, 298)
                       $2225^{\circ}C$, congruent   (18, 581)

CHEMICAL

Theoretical analysis: 20.40% silicon
                       79.60% zirconium

Reactivity: React with N_2 to form nitrides   (227, 272)

                       Oxidizes in O_2, 800-1100$^{\circ}C$ (227)

APPENDIX A

BIBLIOGRAPHY - NUMERICAL

BIBLIOGRAPHY
(Numerical)

1. Sidgwick, N. V., "The Chemical Elements and Their Compounds," Oxford, London (1951).

2. Campbell, I. E., "High Temperature Technology," Wiley, New York (1956).

3. Lange, N. A., "Handbook of Chemistry," Handbook Pub. Co., Sandusky, O. (1946).

4. Materials and Methods, 35 [1], 98 (1952).

5. Warde, J. M., "Refractories for Nuclear Energy," Technical Bulletin #94, Refractories Institute, Pittsburgh.

6. Johnson, J. R., "Ceramic Fuel Materials for Nuclear Reactors," Jour. Metals, 662 (1956).

7. Lynch, J. F., Slyh, J. A., Duckworth, W. H., "Molybdenum Disilicide Coatings for Graphite," WADC-TR-53-457 (1954).

8. Mott, B. W., "Micro Indentation Hardness Testing," Butterworth's, London (1956).

9. Gangler, J. J., Robards, C. F., McNutt, J. E., "Physical Properties at High Temperatures of Seven Hot Pressed Ceramics," NACA-TN-1911 (1949).

10. "Ceramic Materials for High Temperatures," Alfred University Progress Report No. 237, Vol. XXI No. 3 (1956).

11. Dauben, C. H., Searcy, A. W., Templeton, D. H., Brewer, L., "High Melting Silicides," Jour. Amer. Ceram. Soc., 33, 291 (1950).

12. Benesovsky, F., "Sintered High Temperature Corrosion Resistant Materials," Plansee Proc., 1955, 154-172

13. Arbiter, W., "New High Temperature Intermetallic Compounds," WADC-TR-53-190 (1953).

14. Lambertson, W. A., Belliotti, J. V., "A Literature Survey of Silicides For The Protection of Silicon Carbide," Carborundum.

15. Kessler, H., Hansen, M., McPherson, D., Trans. Amer. Soc. Metals, 44, 518 (1952).

16. Duwez, P., Pietrobansky, P., Jour. Metals, 3, 772 (1951).

17. Wallbaum, H., Laves, F., Z. Krist., 101, 78 (1939).

18. Lundin, C., Hansen, M., McPherson, D., Trans. Amer. Soc. Metals, 45, 901 (1953).

19. Schwartzkopf, P., Kieffer, R., "Refractory Hard Metals," MacMillan, New York (1953).

20. Seyforth, H., Z. Krist., 67, 295 (1928).

21. Naray-Szabo, S. V., Z. Krist., 97, 223 (1937).

22. Wallbaum, H., Z. Metallkunde, 33, 378 (1941).

23. Kieffer, R., Nowotny, H., Benesovsky, F., Schachner, H., Z. Metallkunde, 44, 242 (1953).

24. Kurnakov, N., Compt. rend. acad. sci. USSR., 34, 110 (1942).

25. Honigschmid, O., "Karbid u. Silizid," W. Knapp, Halle/Saale (1914).

26. Boren, B., Mineral. Geol., 11A [10], 28 (1933).

27. Kieffer, R., Cerwenka, E., Z. Metallkunde, 43, 101 (1952).

28. Templeton, D. H., Dauben, C. H., Acta Cryst., 3, 261 (1950).

29. Fansteel Data Sheet.

30. Kieffer, R., Benesovsky, F., Gallistl, E., Z. Metallkunde, 43, 284 (1952).

31. Wilhelm, H. A., Armstrong, P.E., Carlson, O. N., "Zirconium-Germanium Alloy Systems," Trans. Amer. Soc. Metals, Reprint No. 48 (1955).

32. Whittemore, O. J., Jr., "Special Refractories for Use Above 1700°C," Ind. Eng. Chem., 47, 2510 (1955).

33. Norton, F. H., "Refractories," McGraw-Hill, New York (1949).

34. Mong, L. E., National Bureau of Standards Report No. 2427.

35. Kieffer, R., Kölbl, F., Powd. Met. Bull., 4, 4-17 (1949).

36. Parche, M. C., "Facts About Fused Alumina," Carborundum (1954).

37. Ryshkewitch, E., "Compressive Strength of Porous Sintered Alumina and Zirconia," Jour. Amer. Ceram. Soc., 36, 65 (1953).

38. Carborundum Company Data.

39. Cronin, L. J., "Refractory Cermets," Amer. Ceram. Soc. Bull., 30[7], 234 (1951).

40. Blumenthal, W. B., "Chemical Behavior of Zirconium," Van Nostrand, New York (1958).

41. Moch, M., Jour. Amer. Ceram. Soc., 77, 304 (1955).

42. Moers, K., Z. anorg. u. allgem. Chem., 198, 262 (1931).

43. Eberhart, J. L., Materials and Methods, (August 1954).

44. Glaser, F. W., Jour. Metals, 5; Trans. AIME., 197, 1119 (1953).

45. Same as Ref. 10.

46. Schwartzkopf, P., Sindeband, J. J., Electrochemical Society Meeting, Cleveland (1950)

47. Evans, R. C., "Crystal Chemistry," University Press, Cambridge (1948).

48. Zachariasen, W. H., Z. physik. Chem., 128, 39 (1927).

49. Glaser, F W., Jour. Appl. Phys., 22, 103 (1951).

50. Cerwenka, E., Thesis, Tech. Hochschule, Graz (1951).

51. Goldschmidt, H. J., Jour. Iron and Steel Inst., 160, 345 (1948).

52. Agte, C., Moers, K., Z. anorg. u. allgem. Chem., 198, 233 (1931).

53. Friederich, E., Settig, L., Z. anorg. u. allgem. Chem , 144, 169 (1925)

54. Agte, C., Älterthum, H., "Investigations of Systems of High Melting Carbides and Contributions to the Problem of Melting Carbon," Z. techn. Physik. 11, 182 (1930).

55. Nowotny, H., Kieffer, R., Metallforschung, 2, 257 (1947).

56. Norton, J. T., Mowry, A. L., Trans. AIME, 185, 133 (1949).

57.   Ellinger, F. H., Trans. Amer. Soc. Metals, 31, 89 (1943).

58.   Burgers, W. G., Basart, J. C. M., Z. anorg. u. allgem. Chem.,
      216, 209 (1934).

59.   McKenna, P. M., Ind. Eng. Chem., 28, 767 (1936).

60.   Duwez, P., O'Dell, F., Jour. Electrochem. Soc., 97, 299 (1950).

61.   Rundle, R. E., Baensiger, N. C., Wilson, A. S., McDonald, R. A.,
      Jour. Amer. Chem. Soc., 70, 99 (1948).

62.   Litz, L. M., Garrett, A.B., Croxton, F. C., Jour. Amer. Chem Soc.,
      70, 1718 (1948).

63.   Mallett, M. W., Gerds, A. F., Nelson, H. R., Jour. Electrochem. Soc.,
      99, 197 (1952).

64.   Nowotny, H., Kieffer, R., Benesovsky, F., Laube, E., Monatsh.,
      88, 336 (1957).

65.   Kieffer, R., Benesovsky, F., Honak, E. R., Z. anorg. u. allgem. Chem.,
      268, 191 (1952).

66.   Sindeband, J. J., Trans AIME , 185, 198 (Feb. 1949).

67.   Andrieux, J. L., Thesis, University of Paris (1929).

68.   Kieffer, R., Benesovsky, F., "High Melting Hard Metals," (1953).

69.   Andrieux, J. L., Ann. chim., 12,[10], 423 (1929).

70.   Andrieux, J. L., Compt. rend., 189, 1279 (1929).

71.   Honak, E., Thesis, Tech. Hochschule, Graz (1951).

72.   Post, B., Glaser, F. W., Moskowitz, D., "Transition Metal Borides,"
      Acta Met., 2, 20 (1954).

73.   Moeller, T., "Inorganic Chemistry," Wiley, New York (1952).

74.   Becker, K., "Hochschmelzende Hartstoffe und ihre technische Anwendung,"
      Verlag, Berlin (1933).

75.   de Boer, J. H., Fast, J. D., Z. anorg. u. allgem. Chem., 187, 177
      (1930).

76. Curtis, C. E., Doney, L. M., Johnson, J. R., "Some Properties of Hafnium Oxide, Hafnium Silicate, Calcium Hafnate and Hafnium Carbide," Jour. Amer. Ceram. Soc., 37[10], 458 (1954).

77. Becker, K., Ebert, F., Z. Physik, 31, 368 (1925).

78. Thielke, N. R., "Application of Crystal Chemistry to the Search for New Refractories," Jour. Amer. Ceram. Soc., 33[10], 304 (1950).

79. Humphrey, G. L., "Heats of Formation of Hafnium Oxide and Hafnium Nitride," Jour. Amer. Chem. Soc., 75, 2806 (1953).

80. Same as Reference 76.

81. Post, B., Glaser, F, W., Moskowitz, D., "Hafnium Silicides," J. Chem. Phys., 22, 1264 (1954).

82. Cotter, P. G., Kohn, J, A., Potter, R. A., "Physical and X-Ray Study of the Disilicides of Titanium, Zirconium and Hafnium," Jour. Amer. Ceram. Soc., 39[1], 11 (1956).

83. Brewer, L., Jour. Amer. Ceram. Soc., 33[10], 273 (1950).

84. NACA-TN-1918 (1949).

85. Long, R. A., Metal Progress, 68[9], 983 (1952).

86. Steinitz, R., Trans, AIME, Jour. Metals, 4 [9], 983 (1952).

87. Brewer, L., Jour. Amer. Ceram. Soc., 34, 173 (1951).

88. Weber, W. P., Quirk, J. F., Lemmon, A. W., Filbert, R. B., "Properties of Beryllium Oxide and Carbides of Beryllium, Molybdenum, Niobium, Tantalum, and Titanium," BMI - 1165 (1957).

89. Norton, J T., Blumenthal, H,, Sindeband, S. J., Trans. AIME. 185, 749 (1949).

90. Kiessling, R., Acta Chem. Scand , 1, 893 (1947).

91. Accountius, O. E., Stoop, R, F,, Konrad, H, E., Greenhouse, H, M., McBride, C,, "Study of the Systems TiC - SiC - B$_4$C and TiC - U C - ZrC," WADC-TR-53-287.

92. Boach, J. D., "Effect of Chromium on the Oxidation Resistance of Titanium Carbide," Jour. Electrochem. Soc., 98, 160 (1951).

93. Cadoff, J., Nielson, J, P,, "Titanium Carbon Phase Diagram," Jour. Metals, 5, 248 (1953).

94. Gangler, J. J., "Some Physical Properties of Eight Refractory Oxides and Carbides," Jour. Amer. Ceram. Soc., 33, 367 (1950).

95. Glaser, F. W., Ivanick, W., "Sintered TiC," Jour. Metals, 4, 387 (1952).

96. Kuo, K., Hagg, G., "A New Molybdenum Carbide," Nature, 170, 245 (1952).

97. Mallett, M. W., Sheipline, V. M., "Carbides," AEC-TIS Reactor Handbook, 3, Sec. 1, Chap. 17 (1955).

98. Quirk, J. F., "Beryllium Carbide," AEC-TIS Reactor Handbook, 3, Sec. 1, Chap. 15 (1955).

99. Wilson, R. E., Jones, G. A., Tinklepaugh, J. R., "Oxidation of Hot Pressed Titanium Carbide," Amer. Ceram. Soc. Bull., 30, 103 (1957).

100. Chiotti, P., "Experimental Refractory Bodies of High Melting Nitrides, Carbides, and UO$_2$," Jour. Amer. Ceram. Soc., 35, 123 (1952).

101. Davey, F. K., Alaball, E. R., Lorey, G. E., "Titanium Nitride Cermets," WADC-TR-55-155.

102. Hower, L. D., Londeree, J. W., Welty, H. F., "Bodies Derived From Mixtures of TiO - TiN - NiO," Jour. Amer. Ceram. Soc., 34, 309 (1951).

103. Humphrey, G. L., "Heats of Combustion and Formation of TiN and TiC," Jour. Amer. Chem. Soc., 76, 978 (1954).

104. Anon, "Refractory Materials for Use in High Temperature Areas of Aircraft," WADC-TR-54-467.

105. Finlay, G. R., "Refractories for 4000°F and Higher," Chemistry in Canada, 4, 41 (1952).

106. Fairchild Airplane Co., Ceramics and Refractories Division, NEPA-1262.

107. Ryshkewitch, E., "Properties and Physical Constants of Highly Refractory Materials," WADC-TR-50-633 (1950).

108. Brewer, L., Krikorian, O., "Reactions of Refractory Silicides with Nitrogen and Carbon," Jour. Electrochem. Soc., 38, 103 (1956).

109.   Long, R. A., "Fabrication and Properties of Hot Pressed $MoSi_2$," NACA-RM-E-50 F 2 W (1950).

110.   Stavrolakis, J. A., Barr, H. N., Rice, H. H., "An Investigation of Boride Cermets," Amer. Ceram. Soc. Bull., 35, 47 (1956).

111.   Crandall, W. B., Lawrence, W. G., "Fundamental Properties of Metal Ceramic Mixtures at High Temperatures," N. P. 1779, Report No. 22, (September 1950).

112.   Same as Reference 9.

113.   Glaser, F. W., "Progress Report on Cermets," Metal Progress, 67 [4] 77 (1955).

114.   Kingery, W. D., Economos, G., Homonich, J., Berg, M., "Metal Ceramic Composition Suitable for Service at Elevated Temperature," NEPA - 1446 (1950).

115.   Koenig, J. H., Snyder, N. H., "Ceramics," Ind. Eng. Chem., 43, 2008 (1951).

116.   Nelson, J. A., Willmore, T. A., Bennett, D. G., "Metal Bonded TiB_2," WADC-TR-52-111.

117.   Norton, F. H., Kingery, W. D., "Measurement of Thermal Conductivity of Refractory Materials," NYO - 601 (1952).

118.   Rice, W. H., Earhart, W. H., Thielke, N. R., "Refractory Materials for Use in High Temperature Areas of Aircraft," N. P. - 4970 (June 1953).

119.   Wheelock, N. R., Liedeholm, C. A., "Reproducibility of the Mechanical Properties of Ceramic Materials Part II. Thermal Shock Resistance of High Purity Al_2O_3 Specimens," CWR - 481 (August 1957).

120.   Shevlin, T. S., Hauck, C. A., "Fundamental Study and Equipment for Sintering and Testing of Cermet Bodies," Jour. Amer. Ceram. Soc., 38, 450 (1955).

121.   Shevlin, T. S., Hauck, C. A., "Alumina Base Cermets," WADC-TR-54-173 Part I.

122.   Stavrolakis, J. A., Barr, H. N., Rice, H. H., "Investigation of Boride Cermets" Amer. Ceram. Soc. Bull., 35, 47 (1956).

123.   Swartz, E. L., Crandall, W. B., "Fundamental Properties of Metal Ceramic Mixtures at High Temperature," N. P. 5796 (January 1955).

124. Norton Co., Technical Bulletin No. 515.

125. Kingery, W. D., Francl, J., Coble, R. L., Vasilos, T., "Thermal Conductivity: Data for Several Pure Oxide Materials Corrected to Zero Porosity," Jour. Amer. Ceram. Soc., 37, 107 (1954).

126. Lang, S. M., "Properties of High Temperature Ceramics and Cermets - Elasticity and Density at Room Temperature," N.B.S. Monograph No. 6, (March 1, 1960).

127. Same as Reference 105.

128. Margrave, J., University of Wisconsin, Private Communication.

129. Whittemore, O. J. Jr., Jour. Can. Ceram. Soc., 28, 43 (1959).

130. Same as Reference 94.

131. Renaux, L., Thesis, University of Paris (1900).

132. Hasselman, D. P. H., Carborundum Research Notebook 5793.

133. Vig, G., Carborundum Research Notebook 5843.

134. Post, B., Moskowitz, D., Glaser, F. W., "Borides of Rare Earth and Related Metals," Plansee Proc., 173-186 (1955).

135. Cadoff, I., Nielson, J. P., Miller, E., "Properties of Arc Melted versus Powder Metallurgy TiC," Plansee Proc., 50-55 (1955).

136. Taylor, K. M., McMurtry, C. H., "Synthesis and Fabrication of Refractory Uranium Compounds," Summary Report, Contract AT(40-1)-2558 (February 1961).

137. Wilson, W. B., Jour. Amer. Ceram. Soc., 43, 77 (1960).

138. Kempter, C. P., McGuire, J. C., Nadler, M. R., "Uranium Mononitride," Anal. Chem., 31 [1], 156-7 (1959).

139. Saller, H. A., Rough, F. A., "Compilation of United States and United Kingdom Uranium and Thorium Constitutional Diagrams," BMI - 1000 (June 1955).

140. Katz, J. J., Rabinowitch, E., "The Chemistry of Uranium" Part I, McGraw-Hill, New York (1951).

141. Same as Reference 100.

142. Tripler, A. B., Snyder, J. M., Duckworth, W. H., "Further Studies of Sintered Refractory Uranium Compounds," BMI - 1313 (January, 1959).

143. Pollard, F. H., Woodward, P., Trans. Faraday Soc., 46, 190-199 (1950).

144. Coffman, J. A., Kibler, G. M., Riethof, T. R., Watts, A. A., "Carbonization of Plastics and Refractory Materials Research," WADD-TR-60-646 Part I (February, 1961).

145. Naylor, B. F., "High Temperature Heat Contents of TiC and TiN," Jour. Amer. Chem. Soc., 68, 370-1 (1946).

146. Taylor, R. E., Nakata, M. M., "Study of Thermal Properties of Refractories," Second Quarterly Report, AF33(657)-7136 (January, 1962).

147. Green, L., "Observations on the High Temperature Elastic and Anelastic Properties of Polycrystalline Graphites," Ford-Aeronutronics Division, Report U-659, (June, 1959).

148. Larrabee, R. D., "Spectral Emissivity of Tungsten," Opt. Soc. Amer., 49, [6], 619 (1959).

149. Shaffer, P. T. B., Carborundum Research Notebook 5625.

150. Taylor, K. M., Lenie, C., Jour. Electrochem. Soc., 107 [4], 308 (1960).

151. Ott, H., Z. Physik, 22, 201 (1924).

152. Stackelberg, M. v., Spiess, K. F., Z. physik. Chem., A-175, 140 (1935).

153. Kohn, J. A., Cotter, P. G., Potter, R. A., Amer. Mineral., 41, 355 (1956).

154. A.S.T.M. X-Ray Powder Data File.

155. Myers, D. K., "Aluminum Nitride Literature Search," Carborundum (May 3, 1957).

156. Carborundum Data Sheet.

157. Coors Porcelain Co. Data Sheet No. 0001.

158. Kingery, W. D., "Property Measurements at High Temperature," Wiley, New York (1959).

159. Taylor, K. M., "Boron Nitride, A New Material," Materials and Methods, (January, 1956).

160. Popper P , "Special Ceramics, " Heywood, London (1960)

161. Kubaschewski O , Evans. E L , "Metallurgical Thermochemistry, " Pergamon, London (1958)

162. Pease, R S , Acta Cryst , 5, 356 (1952)

163. Taylor, K M.,Ind Eng. Chem , 47, 2506 (1955).

164. Larach, S. , Schrader, R. E. , Phys. Rev., 102, 582 (1956).

165. Larach, S. Schrader, R. E. Phys. Rev., 104, 68 (1956).

166. Neshpor, V S , Samsonov, G. V., Fiz. metal. metalloved , 4 [1], 181-182 (1957).

167. Same as Reference 126.

168. Grisaffe, S. J., "Thermal Expansion of Hafnium Carbide, " Jour. Amer. Ceram. Soc., 43 [9] 494 (1960).

169. Nowotny H., Laube, E. "Thermal Expansion of High Melting Phases, " Planseeber , 9, [1/2] 54-59 (1960)

170. Same as Reference 149.

171. Booker, J , Paine, R. M., Stonehouse, A. J., "Investigation of Refractory Metal Beryllides and Silicides as Very High Temperature Materials, " Progress Report, Contract AF33(616)-6540.

172. Kieffer, R., Kölbl, F., Powd. Met. Bull , 4, 4 (1949).

173. Foster, L. S., Forbes, L. W., Briar, L. B.,Moody, L. S.,Smith, W. H., Jour. Amer. Ceram. Soc., 33, 27 (1950).

174. Same as Reference 46.

175. Köster, W., Rauscher, W., Z. Metallkunde, 39, 111 (1948).

176. Elliot, D. E., Kempter, C. P., Jour. Phys. Chem., 62, 630 (1958).

177. Becker, K., Ebert, F., Z. techn. Phys. 11, 216 (1930).

178. Same as Reference 53.

179. Becker, K., Z. Physik, 51, 481 (1928).

180. Styri, H., Metals and Alloys, 3, 273 (1932).

181. Becker, K., Z. Physik, 34, 185 (1933).

182. Lehman, G. W., WADC-TR-60-581 (July, 1960).

183. Shaffer, P. T. B., Hasselman, D. P. H., Chaberski, A. Z., "Factors Affecting Thermal Shock Resistance of Polyphase Ceramic Bodies," WADD-TR-60-749 Parts I and II,(February,1961) and (July,1962).

184. Baratta, F., Amer. Rocket Soc. Jour., 32 [1], 83 (1962).

185. Data converted to common set of units for convenience.

186. Blum, S. L., Pappis, J., "Pyrographite," Electronic Prog., 17 (May-June, 1960).

187. Wachtman, J. B. Jr., Tefft, W. E., Lam, D. G. Jr., "Youngs Modulus of Single Crystal Corundum," from "Mechanical Properties of Engineering Ceramics" by Kriegel, W. W. and Palmour, H. III, Interscience Publishers (1961).

188. Wachtman, J. B. Jr., Tefft, W. E., Lam, D. G. Jr., Stinchfield, R. P., "Elastic Constants of Single Crystal Corundum at Room Temperature," Jour. Res. N.B.S., 64A, 213-228 (1960).

189. Stehsel, M. L., Hale, R. M., Waller, C. E., "Modulus of Rupture Measurements on Beryllium Oxide at Elevated Temperatures," from "Mechanical Properties of Engineering Ceramics" by Kriegel, W. W. and Palmour, H. III, Interscience Publishers (1961).

190. Carniglia, S. C., "Some Thermal and Mechanical Properties of Dense Beryllium Oxide," from "Mechanical Properties of Engineering Ceramics" by Kriegel, W. W. and Palmour, H. III, Interscience Publishers (1961).

191. Crandall, W. B., Chung, D. H., Gray, T. J., "Mechanical Properties of Ultra-Fine Grain Hot-Pressed Alumina," from "Mechanical Properties of Engineering Ceramics," by Kriegel, W. W. and Palmour, H. III, Interscience Publishers (1961).

192. Lowrie, R., "Research on Physical and Chemical Principles Affecting High Temperature Materials for Rocket Nozzles," Semi Annual Report, Contract DA-30-069-ORD-2787 (December 31, 1961).

193. de Klerk, J., Bolef, D. I., Bull Amer. Phys. Soc., 6, 76 (1961).

194. Spinner, S., Jour. Res. N.B.S., 65C, 89 (1961).

195. Snyder, J. M., Engle, G. B., Loch, L. D., "Properties of Some Refractory Fuel Compounds." USAEC-TID-7350, Part I, 141-3 (1957).

196. Ehrlich, P., Angew. Chem., 59, 163 (1947).

197. Ehrlich, P., Z. anorg. Chem., (1) 259 (1949).

198. Blumenthal, H., Jour. Amer. Chem. Soc., 74, 2942 (1952).

199. Schissel, P. O., Trulson, O. C., "Mass Spectrometric Study of the Vaporization of the Titanium Boron System." ARPA-DA-30-069-ORD-2787, Report C-11 (February 5, 1962).

200. Pechman, A., "Ceramics for High Temperature Applications," Ceramics, II, 19 (March, 1954).

201. Weiss, G., Ann. chim., 1, 446 (1946).

202. Glaser, F. W., "Contribution to the Metal-Carbon-Boron Systems," Jour. Metals, 4, 391 (1952).

203. Lowrie, R., Schomaker V., Crist, R., "Research in Physical and Chemical Principles Affecting High Temperature Materials for Rocket Nozzles," Progress Report, DA-30-069-ORD-2787 (June 30, 1960); Appendix: Null, M. R., Lozier, W. W., "Spectral Emissivities of TiC, NbC, and TiB_2."

204. Lowrie, et al., ibid., (December 31, 1960); Appendix: Pike, J. N., "Some Physical Measurements on Hot-Pressed HfC."

205. Lowrie, et al., ibid., (June 30, 1961).

206. Lowrie, et al., ibid., (March 31, 1962).

207. Leitnaker, J. M., Bowman, M. G., Gilles, P. W., "High Temperature Evaporation and Thermodynamic Properties of ZrB_2," Jour. Chem. Phys., 36 [2], 350 (1962).

208. Wood, W. D., Deem, H. W., Lucks, C. F., "Emittance of Ceramics and Graphite," DMIC Memo 148, (March 28, 1962).

209. Taylor, R. E., Nakata, M. M., "Study of Thermal Properties of Refractories," AF33(657)-7136, (April, 1962).

210. Lundin, L. E., "The System Zirconium-Silicon," Symposium, Amer. Soc. Metals, Cleveland (1953).

211. Miller, G. L., "The Metallurgy of the Rarer Metals," Vol. 2, "Zirconium," Academic Press, New York (1954).

212. Jacobson, E. L., Freeman, R. D., Tharp, A. G., Searcy, A. W., "Preparation,Identification and Chemical Properties of Thorium Silicides," Jour. Amer. Chem. Soc., 78 [19], 4850 (1956).

213. Honigschmid, O., Compt. rend., 142, 157, 280 (1906).

214. Honigschmid, O., Monatsh., 27, 205 (1906).

215. Brauer, G., Mitius, A., Z. anorg. u. allgem. Chem., 249, 325 (1942).

216. Hardy, G. F. Phys. Rev., 89, 884 (1953).

217. Frevel, L. K., et al., Ind. Eng. Chem; Anal. Edit., 18, 83 (1946).

218. Wedekind, E., Chem. Ztg., 29, 1032 (1905).

219. Knudsen, F. P., Jour. Amer. Ceram. Soc., 45[2], 94 (1962).

220. Kibler, G. M., Lyon, T. F., de Santis, V. T., "Carbonization of Plastics and Refractory Materials Research," AF33(616)-6841, Quarterly Report (March 31, 1962).

221. Smagina, E. I., Kutsev, V. S. Ormont, B. F., Zhur. Fiz. Khim., 34, 2329 (1960).

222. Hagg, G., Z. phys. Chem., 11, 433 (1930).

223. Burgers, W. G., Jacobs,   , Z. Krist., 94, 299 (1936).

224. Fast, J. D., Z. anorg. u. allgem. Chem., 241, 42 (1939).

225. Jaeger, et al., Proc. Acad. Sci. Amsterdam, 39, 442 (1936).

226. Campbell, I. E., Proc. Electrochem. Soc., 93, 284 (1948).

227. Bradshaw, W. G. Mathews, C. O., "Properties of Refractory Materials: Collected Data and References," LMSD-2466 (June 24, 1958).

228. Adenstedt, H. K., "Physical, Thermal, and Electrical Properties of Hafnium and High Purity Zirconium," Trans. ASM, 44, 949 (1952).

229. Alliegro, R., Coffin, L., and Tinkelpaugh, J., "Metal and Self-Bonded Silicon Carbide," WADC-TR-54-38 (January, 1954).

230. Andrews, M. R. "Reactions of Gases with Incandescent Tantalum," Jour. Amer. Chem. Soc., 54 1845-54 (1932).

231. Baker T. W., Spindler, W. E., and Wilkinson, D., "The Coefficient of Thermal Expansion of ZrN," AERE-M/M-143 (1957).

232. Baur, J. P., Bridges, D. W., Fassel, W. M., Jr., Jour. Electrochem. Soc. 103, 266-271 (1956).

233. Beckman, C., and Kiessling. R., Nature, 178, 1341 (December 15, 1956).

234. Bidwell, C. C., "Electrical and Thermal Properties of Iron Oxide," Phys Rev., 10, 756 (1917).

235. Brewer L., and Krikorian, O. "Reactions of Refractory Silicides with Carbon and with Nitrogen," Jour. Electrochem Soc 103 38-50 (1950), UCRL-2544 (April 29 1954)

236. Burdick, M. D., Moreland. R. E., Geller, R. F., "Strength and Creep Characteristics of Ceramic Bodies at Elevated Temperatures," NACA-TN-1561 (April, 1949).

237. Calhoun. H. "Effect of Temperature and Additions on the Creep Properties of Alumina," NP-5539 (January, 1955).

238. Ceramic Age, "Ceramic Protection for Jet Aircraft," p. 22 (September, 1956)

239. "Refractory Molybdenum Silicides," Climax Molybdenum Co., Bulletin cdb-6 (January, 1956).

240. "Arc-Cast Molybdenum," Climax Molybdenum Co., Data Sheet (1953).

241. Couch, D. Shapiro H. Brenner, A., "Research on Protection of Molybdenum from Oxidation at Elevated Temperatures," NBS Report-4670 (June 30, 1956)

242. Currie, L. M., Hamister, V. C., and MacPherson, H. G., "The Production and Properties of Graphite for Reactors," National Carbon Co., presented at Geneva Conf., (1955).

243. Curtis, C. E. "Development of Zirconia Resistant to Thermal Shocks," Jour. Amer. Ceram. Soc., 30, 180-196 (1947).

244. Dushman, S., "Vacuum Technique," Wiley, New York (1949).

245. Edwards, R. K., "Studies of Materials at High Temperatures," AECD-3394, UCRL-1639 (March 12, 1952).

246. Fieldhouse, I. B., Hedge, J. C., Lang, J. I., Waterman, T. E., "Thermal Properties of High Temperature Materials," WADC-TR-57-487 (July. 1957).

247. Frangos, T. F., "Silicon Nitride Refractory," Materials, 47 [(1], 115-117 (January, 1958).

248. Wells, A. F., "Structural Inorganic Chemistry," Oxford, London (1962).

249. Geller, R. F. and Yavorsky, P. J., "Effect of Some Oxide Additions on the Thermal Length Changes of Zirconia," Jour. Res. Nat'l. Bur. Stands., 35 [1], 87-110 (1945).

250. Gleiser, M., Larsen, W. L., Speiser, R., and Spretnak, J. W., "The Properties of Oxidation-Resistant Scales Formed on Molybdenum Base Alloys," presented in "Basic Effects of Environment on the Strength, Scaling, and Embrittlement of Metals at High Temperatures," ASTM, Cincinnati Meeting, Publ. #171 (February 2, 1955).

251. Goodwin, T. C., Jr., and Ayton, M. W., "Thermal Properties of Certain Metals," WADC-TR-56-423 (August, 1956).

252. Grant, N. J., "Choice of High Temperature Alloys; Influence of Fabrication History," Metal Progress.

253. Green, L., "The Erosion of Graphite by High Temperature Helium Jets," NAA-SR-77 (May 24, 1950).

254. Green, L., "The Behavior of Graphite Under Alternating Stress," NAA-SR-report (declassified).

255. Greiner, E. S., and Ellis, W. C., "Thermal and Electrical Properties of Titanium," Metals Technology, 15 (September, 1948).

256. Gulbransen, E. A., and Andrew, K. F., "Reactions of Columbium and Tantalum with Oxygen, Nitrogen, and Hydrogen," Jour. Metals, 188, 586-589 (March, 1950).

257. Hamjian, H., and Lidman, W., "Influence of Structure on Properties of Sintered Chromium Carbide," NACA-TN-2731 (June, 1952).

258. Hampel, C. A., "Rare Metals Handbook," Reinhold, New York (1954).

259. Handwerk, J. H. and Noland, R. A., "Oxide Fuel Elements for High Temperature," Chem. Eng. Prog., 53 [2], 60F-62F (February, 1950).

260. Harman, C. G. and Mixer, W. G., Jr., "A Review of Silicon Carbide," BMI-748 (June, 1952).

261. Hiester, N. K., Ferguson, F. A., Fishman, N., "High Temperature Technology," <u>Chem. Eng.</u>, 237 (March, 1957).

262. Hodge, J. C., "Measurement of the Thermal Conductivity of Urania," AECU-3380.

263. Hoffman, C., "Investigation of Chromium-Alumina Metal-Ceramic Body for Possible Gas Turbine Application," NACA-RM-E53007 (November, 1953).

264. Hoffman, G. A., "Fibered Materials for Flight Structures," Rand Corp. RN-1868 (February, 1957).

265. Holland, L., "Vacuum Deposition of Thin Film," Wiley, New York (1957).

266. Hove, J. E., "Some Physical Properties of Graphite as Affected by High Temperature Irradiation"; presented at American Nuclear Society Meeting, New York, Fall, 1957.

267. Inouye, J., "Scaling of Columbium in Air," ONRL-1565 (February 24, 1953).

268. Institution of Metallurgists, "Behavior of Metals at Elevated Temperatures," Philosophical Library, New York (1957).

269. Kebler, R., "Optical Properties of Synthetic Sapphire," Linde Industrial Sapphire Bulletin.

270. Klopp, W. D., Sims, C. T., and Jaffee, R. I., "High Temperature Oxidation and Contamination of Niobium," BMI-1170 (1957).

271. Knowles, P. R., "The Niobium-Hydrogen System," IGR-RIC-190 (March, 1957).

272. Krikorian, O. H., "High Temperature Studies, I: Reactions of Refractory Silicides with Carbon and with Nitrogen," UCRL-2888 (April, 1955).

273. Kubaschewskii, O., and Catterall, J. A., "Thermochemical Data of Alloys," Pergamon Press, New York (1956).

274. Kubaschewskii, O. and Hopkins, B. E. "Oxidation of Metals and Alloys," Academic Press, New York (1953).

275. Lang, S. M., Knudsen F. P., Fillmore, C. L., and Roth, R. S., "High Temperature Reactions of Uranium Dioxide with Various Metal Oxides," NBS Circular-568 (February, 1956).

276. Lambertson W. A. and Handwerk, J H., "The Fabrication and Physical Properties of Urania Bodies," ANL-5053 (February, 1956).

277. Loch, L. D., Engle, L. B., Snyder, M. J., Duckworth, W. H., "Survey of Refractory Uranium Compounds," BMI-1124 (August 7, 1956).

278. Livey, D. T. and Murray, P., "The Stability of Beryllia and Magnesia in Different Atmospheres at High Temperature," Jour. Nuclear Energy, 2, 202-212 (1956).

279. Mallett, M. W., Belle, J., Cleland, B. B., "Kinetics of the Zr-N and Zr-Sn-N System," BMI-829 (May 26, 1953); Declassified 1955.

280. Mash, D. R., "Nuclear Reactor Metallurgy," ASM Seminar No. 8, Monterey, California (May 3, 1957).

281. Materials and Methods (January, 1950).

282. "Try Molybdenum Disilicide," Materials and Methods (January, 1956).

283. Meissner, W. and Franz, H., Z. Physik, 65, 39 (1930).

284. Michaelson, H., "High Temperature Materials for Vacuum Service," Materials and Methods (December, 1953).

285. Miller, A. R., "A Thermodynamic Study of Metals and Ceramics," Temescal Corporation (August 23, 1957).

286. Miller, G. L., "Columbium and Its Uses," Materials and Methods, 131-135 (May, 1957).

287. Mitchell, L., "Ceramics," Ind. Eng. Chem., 48, 1702-1709 (September, 1956); 47, 1956-1961 (September, 1955).

288. Monack, A. J., "Electrical Insulating Materials," Materials and Methods, 158-159 (April, 1957).

289. Neeley, J. J., and Teeter, C. E., Jr., "Thermal Conductivity and Heat Capacity of Beryllium Carbide," Jour. Amer. Ceram. Soc. 33, 363 (1950).

290. "Report on Tantalum Carbide," NEPA Div., Fairchild Engine and Airplane Corporation, AECU-104 (April 20, 1948).

291. Norton, F. H., Kingery, W. D., Economos, G., and Humenik,, M.,"Study of Metal Ceramic Interactions at High Temperature," NYO-3144 (February 1, 1953).

292. "Beryllium," Nuclear Engineering Data Sheet No. 4 (January, 1958).

293. "Beryllium, Physical, and Mechanical Properties," Nuclear Power Data Sheet No. 5 (September, 1957).

294. Oak Ridge National Lab. Quarterly Progress Reports ORNL 1952 (February, 1955) and ORNL 1945 (June. 1955).

295. Paprocki, S. J. and Stacy, J. T., "Investigation of Some Niobium-Base Alloys," BMI-1143 (October 31, 1956).

296. Phalniker, C. A., Evans, E. B., and Baldwin, W. M., Jr., "High Temperature Scaling of Cobalt-Chromium Alloys," Jour. Electrochem. Soc. 103, 429-438 (1956).

297. Porter H B . "Rocket Refractories," NAVORD Report 4893, NOTS 1191 (August 26, 1955).

298. Powell, C. F. Campbell, I. E., and Gonser, B. W., "Vapor Plating," Wiley, New York (1955).

299. Powell, R. W., "The Thermal and Electrical Conductivities of Beryllium," Philosophical Mag , 44 [7], 645 (June, 1953).

300. Reed, E. R., "Report on Tungsten Carbide," AECU-110, NEPA-465 (March 31, 1948).

301. Rossinni. F. D., Cowie, P. A., Ellison, F. O., and Browne, C. N., "Properties of Titanium Compounds and Related Substances," ONR Report ACR-17 (October. 1946).

302. Ryshkewitch . E., "Properties and Physical Constants Data of High Refractory Materials," AF-TR-6330 (August, 1950).

303. Saller, H. A. Stacy, J. and Porembka, P., "Initial Investigation of Niobium Base Alloys," BMI-1003 (May, 1955).

304. Seibel R. D: "Survey and Bibliography on the Determination of Thermal Conductivity of Metals at Elevated Temperatures," WAL 821/9 (August 15, 1954).

305. Semchyshen. M., and Barr, R., "Arc-Cast Molybdenum Base Alloys," Fourth Annual Report, Climax Molybdenum Co. (1953).

306. Sherwood, E. M., "Less Common Metals," Ind. Eng. Chem. 48, 1735-1741 (September, 1956), 47, 2044-2050 (September, 1955).

307. Sims, C. T., Klopp, W. D., Jaffee, R. I., "Studies of the Oxidation Resistance and Contamination of Binary Niobium Alloys," BMI-1169 (February 19, 1957)

- 654 -

308. Smithells, C. J., "Metals Reference Book," Vols. 1 and 2, Interscience Publishers, New York (1955).

309. Snyder, M. J. and Duckworth, W. H., "Properties of Some Refractory Uranium Compounds," BMI-1223 (September 9, 1957).

310. Stavrolakis, J. and Norton, F., "Measurements of the Torsion Properties of Alumina and Zirconia at Elevated Temperatures," Jour. Amer. Ceram. Soc. (September, 1950).

311. Stull, D. R. and Sinke, G. C., "Thermodynamic Properties of the Elements," American Chemical Society, Washington, D. C. (November, 1956).

312. Technical Information Services, "General Properties of Materials," Vol. 3, Sect. 1 of the Reactor Handbook, USAEC (February, 1955).

313. Trice, J. B., Neeley, J. J., and Teeter, C. E., Jr., "The Thermal Conductivity of a Hot-Pressed Beryllium Carbide Cylinder," NEPA-818 (July 3, 1948).

314. Udy, M. C., "Chromium," Vols. I and II, Reinhold, New York (1956).

315. Udy, M. C., Shaw, H. L., and Boulger, F. W., "The Properties of Beryllium," AECD-3382 (July 15, 1949).

316. Webb, W. T., Norton, J. T., and Wagner, C., "Oxidation of Tungsten," Jour. Electrochem. Soc. 103, 107-111 (1936).

317. "Development of Niobium Base Alloys," Westinghouse Electric Corp., Dept. No. A-2173 (April, 1956).

318. Westphal, R. C., "Thermal Conductivity of Fuel-Element Materials," AECD-3864 (January 23, 1954).

319. White, D. W. and Burke, J. E., "The Metal Beryllium," American Society for Metals, Cleveland (1955).

320. "Development, Properties, and Investigation of a Cermet Containing 28% Alumina and 72% Chromium," WADC-TD-53-17 (1952).

321. Wygant, J., "Elastic and Flow Properties of Dense Pure Oxide Refractories," Jour. Amer. Ceram. Soc. (December, 1951).

322. Yosim, S. J. and Milne, T. A., "Basic Chemistry of High Temperature Inorganic Systems," Semi-Annual Progress Report, NAA-SR-2124 (December 15, 1957).

323. Ziegler, W. T., and Young, R. A., Oxford Conference on Low Temperature Physics (1951).

324. Grain, C. F., Campbell, W. J., "Thermal Expansion and Phase Inversion of Six Refractory Oxides," Bureau of Mines, RI-5982 (1962),

325. Dolloff, R. T., "Phase Equilibrium Relations of Selected Metal Carbides at High Temperature," Progress Report, AF33(616)-6286 (December 31, 1959).

326. Taylor, R. E., "Thermal Conductivity of Zirconium Carbide at High Temperature," Jour. Amer. Ceram. Soc., 45 [7], 353 (1962).

327. Wachtman, J. B., Jr., Scuderi, T. G., Cleek, G. W., "Linear Thermal Expansion of Aluminum Oxide and Thorium Oxide from 100 - 1000°K," Jour. Amer. Ceram. Soc., 45 [7], 319 (1962).

328. Plummer, W. A., Campbell, D. E., Comstock. A. A., "Method of Measurement of Thermal Diffusivity to 1000°C," Jour. Amer. Ceram. Soc., 45 [7], 310 (1962).

329. Krikorian, O, H., "Thermal Expansion of High Temperature Materials," UCRL-6132 (September, 1960).

330. Post, B., Glaser, F. W., Jour. Chem. Phys., 20, 1050 (1952).

331. Ogden, H. R., Jaffee, R. I., Jour. Metals, 3, 335 (1951).

332. Zachariasen, W. H., Acta Cryst. 2, 94 (1949).

333. Meissner, W., Franz, H., Westerhoff, H., Z. Physik. 75, 521 (1932).

334. Campbell, I. E., Powell, C. F., Nowicki, D. H., Gonser, B. W., Jour. Electrochem. Soc., 96, 318 (1949),

335. Kiessling, R., Acta Chem. Scand., 3, 90 (1949).

336. Morgan, F. H., Jour. Appl. Phys., 22, 108 (1951).

337. Goldwater, D. L., Haddad, R. E., Jour. Appl. Phys., 22, 70 (1951)

338. Glaser, F. W., Post, B., Jour. Metals, 4, 631 (1952),

339. Andrieux, L., Rev. met., 45, 49 (1948).

340. Hulm, J. K., Matthias, B. T., Phys. Rev., 82, 273 (1951).

341. Andersson, L. H., Kiessling, R., Acta Chem. Scand., 4, 160, 209 (1950).

342. Kiessling, R., Acta Chem. Scand., 3, 603 (1949).

343. Kiessling, R., Acta Chem. Scand., 3, 595 (1949).

344. Frueh, A. J., Jr., Acta Cryst., 4, 66 (1951).

345. Moissan, H., Compt. rend., 119, 185 (1894).

346. Moissan, H., Ann. chim. phys., 8, 565 (1896).

347. Tucker, S. A., Moody, A. R., Jour. Chem. Soc., 81, 14 (1902).

348. Wedekind, E., Fetzer, K., Ber. Dtsch. Chem. Ges., 40, 297 (1907).

349. Binet du Jassoneix, A., Compt. rend., 143, 897, 1149 (1906).

350. Cole, N. W., Edmonds, W. H., U. S. Patent 2,088,838 (1937).

351. Kiessling, R., Liu, Y. H., Jour. Metals, 3, 639 (1951).

352. Binet du Jassoneix, A., Compt. rend., 143, 169 (1906).

353. Wedekind, E., Jochem. O., Ber. Dtsch. Chem. Ges., 46, 1205 (1913).

354. Tucker, S. A., Moody, A. R., Proc. Chem. Soc., 17, 129 (1901).

355. Andrieux, L., Weiss, G., Thesis, U. of Grenoble (1946).

356. Andrieux, L., Weiss, G., Bull. soc. chim. France, 15, 598 (1948).

357. Bertaut, F., Blum, P., Acta Cryst., 4, 72 (1951).

358. Naray-Szabo, S. V., Tobias, C. W., Jour. Amer. Chem. Soc., 71, 1882 (1949).

359. Steinitz, R., Powd. Met. Bull., 6, 54 (1951).

360. Steinitz, R., Jour. Metals, 4, 148 (1952).

361. Agte, C., Thesis, Tech. Hochsch. Berlin (1931).

362. Calvert, E. D., Kirk, M. M., Beall, R. A., Bureau of Mines, Report of Investigation, No. RI 5951 (1962).

363. Wedekind, E., Horst, C., Ber. Dtsch. Chem. Ges., 46, 1203 (1913).

364. Weiss, G., Blum, P., Ann. soc. chim. France, 14, 1077 (1947).

365. Moissan, H., Compt. rend., 123, 15 (1896).

366. Zalkin, A., Templeton, D. H., Jour. Chem. Phys. 18, 391 (1950).

367. Allard, G., Compt. rend., 189, 108 (1929).

368. Stackelberg, M. V., Neumann, F., Z. physik. Chem., (B) 19, 314 (1932).

369. Bertaut, F., Blum, P., Compt. rend., 234, 2621 (1952).

370. Lafferty, J. M., Phys. Rev., 79, 1012 (1950).

371. Andrieux, L., Blum, P., Compt. rend., 229, 210 (1949).

372. Bertaut, F., Blum, P., Compt. rend., 229, 666 (1949).

373. Daane, A. H., Baenziger, N. C., USAEC Report, 1SC-53 (1949).

374. Ruff, O., Jellinek, E., Z. anorg. u. allgem. Chem., 97, 315 (1916).

375. Roth, W., Wolf, U., Fritz, O., Z. Elektrochem., 46, 42 (1940).

376. Stackelberg, M. V., Schnorrenberg, E., Z. physik. Chem., B-27, 37 (1934).

377. Tiede, E., Birnbräuer, E., Z. anorg. u. allgem. Chem., 87, 167 (1914).

378. Zhdanov, G. S., Sevast'yanov, N. G., Compt. rend. acad. sci. URSS , 32, 432 (1941).

379. Clark, H. K., Hoard, J. L., Jour. Amer. Chem. Soc., 65, 2115 (1943).

380. Lebeau, P., Compt. rend., 121, 496 (1895).

381. Lebeau, P., Ann. Chim. Phys., 16 [7] , 476-9 (1899).

382. Messerknecht, C., Biltz, W., Z. anorg. u. allgem. Chem., 148, 153 (1925).

383. Schmidt, J. M., Bull. Soc. Chim., 43 [4] , 49 (1928).

384. Fichter, F., Brunner, E., Z. anorg. u. allgem. Chem., 93, 91 (1915).

385. Moissan, H., Compt. rend., 118, 501 (1894).

386. Hinnüber, J., Z. VDI, 92, 111 (1950). (See Ref. 19.)

387. Bridgman, P. W., Proc. Amer. Acad. Sci., 66, 255 (1932).

388. Mott, W. R., Trans. Amer. Electrochem. Soc., 35, 255 (1919).

389. Roth, W. A., Becker, G., Z. physik. Chem., 159, 1 (1932).

390. Kelley, K. K., U. S. Bur. Mines Bull. 407 (1937).

391. Kelley, K. K., "Specific Heats at Low Temperatures of Ti and TiC," Ind. Eng. Chem., 36, 865 (1944).

392. Humphrey, G. L., Jour. Amer. Chem. Soc., 73, 2261 (1951).

393. Brewer, L., Bromley, L. A., Gilles, P. W., Lofgren, N. L., "The Chemistry and Metallurgy of Miscellaneous Materials - Thermodynamics," McGraw-Hill, New York (1950).

394. Klemm, W., Schuth, W., Z. anorg. u. allgem. Chem., 201, 24 (1931).

395. Haddad, R. E., Goldwater, D. L., Morgan, F. H., Jour. Appl. Phys., 20, 1130 (1949).

396. Ruff, O., Wallstein, R., Z. anorg. u. allgem. Chem., 128, 96 (1923).

397. Prescott, C. H., Jour. Amer. Chem. Soc., 48, 2534 (1926).

398. Haddad, R. E., Goldwater, D. L., Morgan, F. H., Jour. Appl. Phys., 20, 886 (1949).

399. King, E. G., Jour. Amer. Chem. Soc., 71, 316 (1949).

400. Elliott, R. P., Metal Progress, 246 (October 1960).

401. Joly, A., Compt. rend., 82, 1905 (1876).

402. Joly, A., Ann. Sci. Ecole Norm. 6, 148 (1877).

403. Zalabak, C. F., NASA-TN-D-761 (March 1961).

404. Kelley, K. K., Jour. Amer. Chem. Soc., 62, 818 (1940).

405. Thorne, P. S. L., Roberts, E. L., "Inorganic Chemistry" 6th Edition, pp. 657-660 (1954).

406. Thorpe, J. F., Whiteley, M. A., "Thorpe's Dictionary of Applied Chemistry," Volume 1, 4th Edition, pp. 284-5 (1943).

407. Serpek, O., "Method of Producing Aluminum Nitride," U. S. Patent 888,044 (May 19, 1908).

408. Serpek, O., "Process of Producing Aluminum Nitride," U. S. Patent 987,408 (March 21, 1911).

409. Serpek, O., "Process for the Manufacture of Aluminum Nitride," U. S. Patent 996,032 (June 20, 1911).

410. Miner, C. F., "Process of Producing Nitrides of Aluminum and Magnesium from their Minerals," U. S. Patent 1,803,720 (May 5, 1931).

411. Johnson, A. F., "Refractory and Method of Making," U. S. Patent 2,480,473 (August 30, 1949).

412. Johnson, A. F., "Refractory," U. S. Patent 2,480,475 (August 30, 1949).

413. See Reference 153, "Synthesis of Aluminum Nitride Monocrystals."

414. Urbain, E., "AlN from Direct Nitrogenation of Silico-Aluminum Compound," French Patent 677,330 (October 16, 1928); Chem. Abs., 24, 3091-1.

415. Newmann, B., Kröger, C, Haebler, H., "The Heat of Formation of Nitrides (Li, Al, Be, and Mg)" Z. anorg. u. allgem. Chem., 204, 81-96 (1932), Chem. Abs., 26, 3433-1.

416. Plotnikov, V. A., Kalita, P T., "Formation of Aluminum Nitride from Aluminum and Ammines," Jour. Gen. Chem. (USSR), 3, 872-3 (1933); Chem Abs., 28, 3019-8.

417. Sato, S., "Heat of Formation and Specific Heat of Aluminum Nitride," Bull. Inst. Phys.-Chem Res. (Tokyo), 14, 862-71 (1935); Chem. Abs., 30, 2093-9

418. Haffitte, P., Elchardus, E., Grandadum, P., "Nitride Formation by Mg and Al Metals," Rev. Ind. Minerals, 375, 861-8 (1936); Chem. Abs., 30, 8115-2

419. Meijering, J. L., Druyvesteyn, M. J., "Hardening of Metals by Internal Oxidation," Phillips Res. Reports, 2, 260-80 (1947); Chem. Abs., 1168 a

420. Tsumura, Y., "Aluminum Nitride," Japan Patent 179,269 (June 6, 1949); Chem. Abs., 9425 g (1951).

421. Stanton, R., "Solid-Liquid Reaction Processes," U. S. Patent 2,615,906 and 907 (October 28, 1952); Chem. Abs., 1996 c (1953).

422. Born, K., Koch, W., "Effect of Aluminum on Properties of Low Carbon Steels," Stahl. u Eisen. 72, 1268-77 (1952); Chem Abs., 3778 g (1953).

423. Urbain, M., "Determination of Aluminum and Its Nitride in low-Carbon Steel," Rev Met., 50, 617-23 (1953); Chem. Abs., 6311 g (1954)

424. Bardgett, W. E., Gemmill, M. G., "Causes of Variable Creep Strength in Basic Open-hearth Carbon Steel," J. Iron. Steel Inst. (London), 179, 211-219 (1955); Chem. Abs., 14620 c (1955).

425. Fuks, M. Y., Aronson, E. V., "Rontgenographic Investigation of Nitrided Layer of Carbon and Alloy Steels," Zhur. Tekh. Fiz., 24, 1448-54 (1954); Chem. Abs., 13854 h (1955).

426. Kasatkin, B. S., Kakhovskii, N. I., "Bessemer Steel and Peculiarities in Its Welding," Automat. Svarka, 7 [5], 24-37 (1954); Chem. Abs., 13062 b (1955).

427. Pearson, J., Ende, J. C., "Thermodynamics of Metal Nitrides and Nitrogen in Iron and Steel," Jour. Iron Steel Inst. (London), 175, 52-8 (1953); Chem. Abs., 8762 c (1955).

428. Yajima, E., Furusawa, K., "Brittleness of Steel at High Temperature," Bull. Nagoya Inst. Technol., 5, 260-6 (1953); Chem. Abs., 8767 d (1955).

429. Flament, P., "Behavior of Aluminum Nitride with Respect to Chemical Treatments," Compt. Rend. 27th Cong. Intern. Chim. Belge., 20, 372-7 (1955); Chem. Abs., 10586-7 (1956).

430. Janeff, W., "Preparation of Metal Nitrides in a Glow Discharge Tube and Some of Their Properties," Z. Physik, 142, 619-36 (1955); Chem. Abs., 11870 c (1956).

431. Long, G., Foster, L. M., Jour. Amer. Ceram. Soc., 42[2] 53 (1959).

432. Pallmer, P. G., "Thermal Expansion of Plutonium Carbides," Contract AT(45-1)-1350 (February 1962).

433. Pease, R. S., Nature, 165, 722 (1950).

434. Wentorf, R. F., "The Cubic Form of Boron Nitride," Jour. Chem. Phys., 26, 956 (1957).

435. Mulford, R., Olson, W., Reported in Chem. Eng. News, p. 56 (July 30, 1962).

436. van Arkel, A. E., Physica, 4, 286 (1924).

437. Neumann, B., Kroger, C., Kunz, H., Z. anorg. u. allgem. Chem., 218, 379 (1934).

438. Sato, S., Sci. Pap. Inst. Phys. Chem. Res. Tokyo, 34, 888 (1938).

439. Shomate, C. H., Jour. Amer. Chem. Soc., 68, 310 (1946).

440. Shukow, J., Jour. Russ. Phys. Chem., 42, 40 (1910).

441. Clausing, P., Z. anorg. u. allgem Chem., 208, 401 (1932).

442. Foster, L. S., U. S. AEC (declassified) AECD-2942 (1950).

443. Foster, L. S., Met. Prog., 62[2], 160 (1952).

444. Samsonov, G. V., "Silicides and Their Uses In Engineering," Akad. Nauk Ukrain. SSR (1959). Translation FTD-TT-61-409, Wright Patterson AFB (January 29, 1962).

445. Bokiy, G. B., "Introduction to Crystal Chemistry," Moscow (1954).

446. Kurnakov, N. S., Izv. Sekt. Fiz.-Khim. Analiza, IONKh AN SSSR, 16, 77 (1951).

447. Chubb, W., Dickerson, R. F., "Properties of Uranium Carbides," Jour. Amer. Ceram. Soc., 41[9], 564 (1962).

448. Rough, F. A., Dickerson, R. F. "Uranium Carbide - Fuel of the Future," Nucleonics, 18[3], 74-77 (1960).

449. Gallistl, E., Thesis, Tech. Hoch. Graz. (1951).

450. Glaser, F., Moskowitz, D., Powd. Met. Bull., 6[6], 178 (1953).

451. Glaser, F., Jour. Metals, 1, 475 (1949).

452. Guseva, L. N., Ovechkim, B. I., DAN SSSR, 112, 681 (1957).

453. Robins, D., Phil. Mag., 3, 313 (1958).

454. Sirota, N. N., Samsonov, G. V., Strel'nikova, N. S., "Physics and Physico-chemical Analysis," Moscow (1957).

455. Arvin, M., Jour. Appl. Phys., 24, 498 (1953).

456. Nikitin, Y. N., ZhTF, 28, 23 (1958).

457. Nikitin, Y. N., ZhTF, 28, 26 (1958).

458. Foëx, G., Helv Phys. Acta., 26, 199 (1954).

459. Same as Ref. 216.

460. Dorfman, Y. G., Kikoin, I. K., Fiz. metallov., GTTI , 405 (1934).

461. Corak - cited from Kieffer and Schwartzkoff, "Hartstoffe und Hartmetalle," Wien (1953).

462. Kiessling, R., Fortschr. Chem. Forschung., 3, 41 (1954).

463. Meissner, W., Erg. des Exakt. Naturwiss., XI (1932).

464. Matthias, B., et al., Phys. Rev., 93, 1415 (1954).

465. Meissner, W., Franz. H., Z. Phys., 63, 558 (1930).

466. Meissner, W., Z. Ges. Kälteind., 39, 104 (1932).

467. Meissner., W., Franz, H., Westerhoff, H., Ann. Phys,., 17, 593 (1933).

468. McLennan, J., Allen, J., Wilhelm, J., Trans. Roy. Soc. Canada, 25, 13 (1931).

469. Horn, F., Ziegler, W., Jour. Amer. Electrochem. Soc., 69, 2762 (1947).

470. Cook, D., Zemansky, M., Boorse, H., Phys. Rev., 79, 7021 (1951).

471. Golutvin, Y. M., Zh. F. Kh., 30, 2251 (1956).

472. Robins, D., Jenkins, J., Acta metall., 3, 598 (1955).

473. Samsonov, G. V., Solonnikova, L. A., Fiz. Metal. Metallov, 5, 565 (1957).

474. Samsonov, G. F., Latysheva, V. P., Fiz. Metal. Metallov., 2, 303 (1956).

475. Gruzin, P. L., Polikarpov, Y. A., Shumilov, A. M., Zav. lab., 21, 417 (1955).

476. Gulbransen, E., Andrews, K., Jour. Metals, 187, 746 (1949).

477. Zeytts, F., Fiz. metallov., GITTL, 205 (1947).

478. Schwartzkopf, P., Glaser, F., Z. Metallkunde, 44, 353 (1953).

479. Eckerlin, P., Wölfel, E., Z. anorg. u. allgem. Chem., 280, 3215 (1955).

480. Tamaru, S., Z. anorg. u. allgem. Chem., 62, 81 (1909).

481. Wöhler, L., Schliegphake, O., Z. anorg. u. allgem. Chem., 11, 1951 (1926).

482. Kubaschewski, O., Villa, H., Z. Electrochem., 53, 32 (1949).

483. Hellner, E., Z. anorg. Chem., 261, 266 (1950).

484. Hellner, E., Angew. Chem., 62, 125 (1950).

485. Wöhler, L. Schuff, K., Z. anorg. Chem., 209, 33 (1932).

486. Hansen, M., "Structures of Binary Alloys," Metallurgizdat (1941).

487. Jette, E., Gibert, E., Jour. Chem. Phys., 1, 753 (1933).

488. Loman, H., Jour. Inst. Metals, 49, 369 (1932).

489. Masing, G., Dahl, O., Wiss. Veröff. Siemens-Konz., 8, 255 (1929).

490. Vogel, R., Z. anorg. u. allgem. Chem., 61, 46 (1909).

491. Parthé, E., Powd. Met. Bull., 8[1/2], (1957).

492. Sirota, N. N., Chizhevskaya, S. N., "Collection of Physics & Physico-chemical Analyses," Moscow, (1957).

493. Witsett, T., Iowa State College, 31, 541 (1957).

494. Ageyev, N. V., Bull. Acad. Sci. USSR, Div. Chem. Sci., 1, 31 (1952).

495. Whitsett, T., Iowa State College, Jour. of Sci., 31, 541 (1957);[Same as Ref. 493.]

496. Epelbaum, V. A., Ormont, B. F., Zav. Lab., 14, 104 (1948).

497. Epelbaum, V. A., Brager, A. K., Acta physicochim., USSR, 13, 595, 600 (1940).

498. Epelbaum, V. A., Ormont, B. F., Jour. Phys. Chem., USSR, 20, 459 (1946).

499. Epelbaum, V. A., Ormont, B. F., Jour. Phys. Chem., USSR, 21, 3 (1947).

500. Hahn, H., Z. anorg. u. allgem. Chem., 258, 58 (1949).

501. Cape, J. A., Taylor, R. E., "Thermal Properties of Refractory Materials," WADD-TR-60-581 Part II, (June 1962).

502. Leitnaker, J. M., "Thermodynamic Properties of Refractory Borides," Los Alamos, LA-2402 (April 13, 1960).

503. Brown, A. R. G., "Silicon Carbide - A Review," Royal Aircraft Est., England, Tech. Note MET/PHYS. 325 (August 1960), ASTIA No. AD. 249685.

504. Van Arkel, A. E., de Boer, J. H., Z. anorg. u. allgem Chem., 148, 345 (1925).

505. Becker, K., Z. Physik, 32, 489 (1937).

506. Todd, S. S., Jour. Amer. Chem. Soc., 72, 2914 (1950).

507. Coughlin, J. P., King, E. G., Jour. Amer. Chem. Soc., 72, 2262 (1950).

508. Pollard, F. H., Fowles, G. W. A., Jour. Chem. Soc., 2444 (1952).

509. Dawihl, W., Rix, W., Z. anorg. u. allgem. Chem., 244, 191 (1940).

510. Slade, R. E., Higson, G. I., Jour. Chem. Soc., 115, 215 (1919).

511. King, E. G., Jour. Amer. Chem. Soc., 71, 316 (1949).

512. Brauer, G., Z. Electrochem., 46, 397 (1940).

513. Muthman, W., Weiss, L., Riedelbauch, R., Liebigs Ann., 355, 92 (1907).

514. Ascherman, G., Friederich, E., Justi, E., Kramer, J., Physik, Z., 42, 349 (1941).

515. Umanski, J. S., Jour. Phys. Chem. USSR , 14, 332 (1940).

516. Friederich, E., Settig, L., Z. anorg. u. allgem. Chem., 143, 293 (1925).

517. Armstrong, C. T., Jour. Amer. Chem. Soc., 71, 3583 (1949).

518. Rundle, R. E., Acta Cryst., 1, 180 (1949).

519. Eriksson, S., Jernkont. Ann., 118, 530 (1934).

520. Blix, R., Z. physik. Chem., 3 B, 229 (1929).

521. Maier, C. G., U. S. Bur. Mines Bull. 436 (1942).

522. Hägg, G., Z. physik Chem., 7 B, 339 (1930).

523. Zachariasen, W. H., Acta Cryst., 2, 388 (1949).

524. Neumann, B., Kröger, C., Haebler, H., Z. anorg. u. allgem. Chim., 207, 145 (1932).

525. Pollock, B. D., "Vaporization and Thermodynamic Stability of ZrC at High Temperature," North American Aviation Report NAA-SR-5439, (January 1, 1961).

526. Popper, P., Ruddlesden, S. N., "Preparation Properties and Structure of Silicon Nitride," Trans. Brit. Ceram. Soc., 60, 603 (1961).

527. Swanson, H. E., Tatke, E., "Standard X-Ray Diffraction Powder Patterns," NBS Circ. 539 Vol. 1 (1953).

528. Rough, F. A., Chubb, W., "Progress in Development of Uranium Carbide-Type Fuel Elements," BMI-1488 (December 27, 1960).

529. Anthony, F. M., Pearl, H. A., "Investigations of Feasibility of Utilizing Available Heat Resistant Materials for Hypersonic Leading Edge Applications," WADC-TR-59-744 (1960).

530. Olson, O. H., Morris, J. C., "Determination of Emissivity and Reflectivity of Aircraft Structural Materials," WADC-TR-56-222 (1956).

531. Taylor, R. E., "High Temperature Thermal Conductivity Apparatus," ASD-TR-62-348 (April 1962).

532. Taylor, R. E., "Thermal Conductivity and Thermal Expansion of BeO at Elevated Temperatures," NAA-SR-4905 (July 1960).

533. McQuarrie, M., "Thermal Conductivity: V, High Temperature Methods and Results for Alumina, Magnesia and Beryllia from 100-1800°C," Jour. Amer. Ceram. Soc., 37, 84-88 (1954).

534. Adams, M., "Thermal Conductivity: III, Prolate Spheroidal Envelop Method Data for Al_2O_3, BeO, MgO, ThO_2, and ZrO_2," Jour. Amer. Ceram. Soc., 37, 74-79 (1954).

535. Taylor, R. E., "Thermal Conductivity of TiC at High Temperatures," Jour. Amer. Ceram. Soc., 44[10], 525 (1961).

536. Norton, F. H., Kingery, W. D., "Measurement of Thermal Conductivity of Refractory Materials," NYO-599 (1951).

537. McClelland, J. D., Zehms, E. H., "Thermal Conductivity of MgO from 1030 - 1880°C," Jour. Amer. Ceram. Soc., 43, 54 (1960).

538. Rasor, N. S., McClelland, J. D., "Thermal Conductivity of Graphite, Mo and Ta to Their Destruction Temperature," Int. Jour. Phys. Chem. Solids, 15, 17 (1960).

539. Allen, R. D., Glasier, L. F., Jordan, P. L., "Spectral Emissivity, Total Emissivity, and Thermal Conductivity of Mo, Ta, and W above 2300°K," Jour. Appl. Phys., 31[8], 1382 (1960).

540. Fieldhouse, J. B., et al., "Measurements of Thermal Properties," WADC-TR-55-495 Part I (September 1956).

541. Tyle, R. P., "Preliminary Measurements on The Thermal and Electrical Conductivities of Mo, Nb, Ta, and W," Jour. Less Common Metals, 3, 13 (1961).

542. Lucks, C. F., Deem, H. W., "Thermal Conductivities, Heat Capacities, and Linear Expansion of Five Materials, "WADC-TR-55-496 (August 1956).

543. Osborne, R. H., "Thermal Conductivities of W and Mo at Incandescent Temperatures," Opt. Soc. Amer. Jour., 31, 428 (1941).

544. Mikol, E. P., "Thermal Conductivity of Mo Over The Temperature Range 1000 - 2100°F," ORNL-1131 Tech. Report No. 2 (1952).

545. Barrett, T., Winter, R. M., "Thermal Conductivity of Wires & Rods," Ann. Physik, 77 (1925).

546. Nadler, M. R., Fitzsimmons, Jour. Amer. Ceram. Soc., 38, 214 (1955).

547. Hoffman, A., Naturwiss, 21, 676 (1933).

548. Megaw, H. D., Proc. Phys. Soc., (London), 58, 133 (1946).

549. Shirane, G. Phys. Rev., 84, 854 (1951).

550. Taylor, R. E., "Thermal Conductivity of BeO," Jour. Amer. Ceram. Soc., 45[2], 74-78 (1962).

551. Curtis, C. E., Thomas, E. A., "Zircon and Zirconium Oxide Refractories," TAM (1948).

552. Minutes of Chalk River Conference, TID - 7514 (1956).

553. Skinner, B. J., Amer. Mineral., 42, 39 (1957).

554. Mordike, B. L., Fitzgerald, Jour. Less Common Met., 1, 132 (1959).

555. Beals, R. J., Lauchner, J. H., "Automatic Recording Dilatometer," Amer. Ceram. Soc. Bull., 37 [11], 486 (1958).

556. Brixner, L. H., "Preparation and Structure of Strontium and Barium Tantalates," Jour. Amer. Chem. Soc., 80, 3214 (1958).

557. Jones, W. H., Milford, F. J., Fawcett, S. L., Batt. Tech. Rev., 11 [9], 2-11 (1962).

558. Rogers, A. F., Kerr, P. F., "Optical Mineralogy," McGraw, New York (1942).

559. Hofman, H. O., "Metallurgy of Lead," McGraw Hill, New York (1918).

560. Kovalskii, A. E., Petrova, L. A., Trud. Sovesk. Mikrov., Acad. Nauk SSSR , 170-186 (1950); AEC-Tr-3834.

561. Vogel, R., Z. anorg. u. allgem. Chem., 84, 323 (1914).

562. Ulik, F., Ber. Wien. Acad., (II) 52, 115 (1865).

563. Sandler, E., Thesis, Tech. Hoch., München (1911).

564. Dodero, M., Compt. rend., 199, 566 (1934).

565. Sterba, J., Compt. rend., 135, 172 (1902).

566. Dodero, M., Bull. Soc. Chim. France, 17, 545 (1960).

567. Brauer, G., Haag, H., Z. anorg. u. allgem. Chem., 267, 198 (1952).

568. Brauer, G., Haag, H., Naturwiss., 37, 210 (1950).

569. Bertaut, F., Blum, P., Acta Cryst., 3, 319 (1950).

570. Alexander, P., Metals & Alloys, 9, 179 (1938).

571. Honigschmid, O., Compt. rend., 143, 224 (1906).

572. Honigschmid, O., Monatsh., 27, 1067 (1906).

573. Honigschmid, O., Monatsh., 28, 1017 (1907).

574. Dodero, M., Thesis, U. of Grenoble.

575. Levy, L., Compt. rend., 121, 1148 (1895).

576. Hansen, M., Kessler, H., McPherson, D., Amer. Soc. Metals, Preprint No. 4 (1951); Published as Ref. 15.

577. Schroth, H., Thesis, U. Graz (1952).

578. Ageyev, N. V., Samsonov, V. P., DANSSSR , 112, 853 (1957).

579. Wedekind, E., Ber., 35, 3929 (1902).

580. Wedekind, E., Chem. Ind. Koll., 7, 249 (1900).

581. Lundin, C., McPherson, D., Hansen, M., Amer. Soc. Metals, Preprint No. 41 (1952); See also Ref. 18.

582. Kieffer, R. Benesovsky, F., Maschenschalk, R., Z. Metallkunde, 45, 493 (1954).

583. Schachner, H., Nowotny, H., Maschenschalk, R., Monatsh.. 84, 677 (1953).

584. Searcy, A., Jour. Amer. Ceram. Soc., 40, 431 (1957).

585. Schachner, H. Nowotny, H., Kudielka, H., Monatsh.., 85, 1140 (1954).

586. Moissan, H., Holl, A., Compt. rend., 135, 78 493 (1902).

587. Laraduc-Müller, L., Rev. Met., 7, 657 (1910).

588. Lebeau, P., Ann.chim. Phys., 8[1], 553 (1904).

589. Giebelhausen, H., Z. anorg. u. allgem. Chem., 91, 251 (1915).

590. Kieffer, R., Benesovsky, F., Schmid, H., Z. Metallkunde, 47, 247 (1956).

591. Wallbaum, H., Z. Metallkunde, 31, 363 (1939).

592. Parthe, E., Nowotny, H., Schmid, H., Monatsh., 86, 413 (1955).

593. Preller, H., VDJ-Zeitschrift, 98, 1611 (1956).

594. Novikov, I. I., Dautova, L. I.,Zhur.Neorg. Khim., 2, 2766 (1957).

595. Brauer, G., Scheele, W., "Anorganische Chemie" Weisbaden (1948).

596. Knapton, A., Nature, 175, 730 (1955).

597. Samsonov, G. V., Neshpor, V. S., Yermakova, V. A., Zhur. Neorg. Khim., 3, 868 (1958).

598. Schachner, H., Thesis, Univ. of Vienna (1953).

599. Parthe, E., Lux, B., Nowotny, H., Monatsh., 86, 859 (1955).

600. Nowotny, H., Schachner, H., Kieffer, R., Benesovsky, F., Monatsh., 84, 1 (1953).

601. Myers, C., Searcy, A., Jour. Amer. Ceram. Soc., 40, 526 (1957).

602. Metallurgica, 53, 175 (1956).

603. Moissan, H., Compt. rend., 120, 290 (1895).
     Moissan, H., Compt. rend., 121, 621 (1895).

604. Warren, H., Chem. News, 78, 318 (1898).

605. Frilley, R., Rev. Metall., 8, 457 (1911).

606. Boren, B., Archiv. Kem. Min. Geol., A 11, 2 (1933).

607. Andersson, L., Jette, E., Trans. Amer. Soc. Metals, 24, 375 (1956).

608. Kurnakov, N. N., Proc. Acad. Sci., USSR, 26, 362 (1940).

609. Same as Ref. 446.

610. Kieffer, R., Benesovsky, F., Schroth, H., Z. Metallkunde, 44, 437 (1953).

611. Chalmot, G., Amer. Chem. Jour., 19, 69 (1897).

612. Matignon, C., Trennoy, R., Compt. rend., 141, 190 (1905).

613. Zettel, C., Compt. rend., 126, 833 (1897).

614. Dauben, C., Templeton, D., Myers, C., Jour. Phys. Chem., 60, 443 (1956).

615. Vigouroux, E., Compt. rend., 129, 1238 (1899).

616. Watts, O., Bull. Univ. Wisc., 145, 255 (1906).

617. Defacqz, E., Compt. rend., 144, 1424 (1907).

618. Schönberg, N., See Ref. 626.

619. Schachner, H., Cerwenka, E., Nowotny, H., Monatsh., 85, 245 (1954).

620. Parthe', E., Nowotny, H., Schmid, H., Monatsh., 86, 385 (1955).

621. Fitzer, E., Schwab, J., Metall., 9, 1062 (1955).

622. King, E., Christensen, A., Jour. Phys. Chem., 62, 499 (1958).

623. Samsonov, G. V., Neshpor, V. S., "Problems of Powder Metallurgy and Strength of Materials," No. V, 3 (1957).

624. Materials and Methods, 43, 131 (1956).
      Metallurgica, 318, 175 (1956).

625. Vigouroux, E., Compt. rend., 127, 393 (1898).

626. Aronsson, B., Acta Chem. Scand., 9, 1107 (1955).

627. Ewald, P., Neumann, C., "Strukturberichte," Leipzig (1931).

628. Davydov, K. P., Gel'd, P. V., Fiz. Metal. Metallov., 2, 192 (1956).

629. McNees, R., Searcy, A., Jour. Amer. Chem. Soc., 77, 5290 (1955).

630. Pauling, L., Sodlate, A., Acta Cryst., 1, 212 (1948).

631. Akumov, Y. I., Jour. Appl. Chem. (USSR), 21, 227 (1948).

632. Haughton, J., Becker, M., Jour. Iron Steel Inst., 121, 315 (1930).

633. Batz, W., Mead, H., Birchenall, C., Jour. Metals, 4, 170 (1952).

634. Z. Phys. Chem., 29, 231 (1935).

635. Geller, S., Acta Cryst., 8, 83 (1955).

636. Schneider, A., Meyer-Jungnik, W., Angew. Chem., 67, 306 (1955).

637. Toman, K., Acta Cryst., 4, 462 (1951).

638. Lashko, N. F., Doklady Acad. Nauk SSSR , 81, 605 (1951).

639. Savitskiy, Y. M., Izd. vo. Acad. Nauk SSSR , (1957).

640. Reynor, G., Phil. Mag., 39, 318 (1948).

641. Bertaut, F., Blum, P., Compt. rend., 231, 626 (1950).

642. Schubert, K., Pfisterer, H., Z. Metallkunde, 41, 433 (1950).

643. Moissan, H., Manchot, W., Compt. rend., 137, 229 (1903).

644. Buddery, J., Welch, A., Nature, 167, 362 (1951).

645. Korst, W., Finnie, S., Searcy, A., Jour. Phys. Chem., 61, 1541 (1957).

646. Geller, S., Wood, E., Acta Cryst., 1, 441 (1954).

647. Voronov, M., Izv. Sekt. Plat. ION Kh., Izv. vo. Akad. Nauk SSSR, 13, 144 (1936).

648. Lebeau, P., Nowitzky, A., Compt. rend., 145, (1907).

649. Pfisterer, H., Schubert, H., Z. Metallkunde, 41, 359 (1950).

650. Bender, S. L., Dreikom, R. E., Einwohner, T. H., Ferber, R. C., Gannon, R. E., Hanst, P. L., Ihnat, M. E., Phaneuf, J. P., Schick, H. L., and Ward, C. H., "Thermodynamics of Certain Refractory Compounds," Vol. I, ASD-Tr-61-260 (May 1962).

651. Goldsmith, A., Waterman, T. E., "Thermophysical Properties of Solid Materials," WADC-TR-58-476 (1959); AD-207905.

652. Runnals, O., Boucher, R., Acta Cryst., 8, 592 (1955).

653. Zachariasen, W., Nat'l. Nucl. Energy, Ser. Div. IV, 14 B, 1451 (1949).

654. Same as Ref. 212.

655. Defacqz, E., Compt. rend., 147, 1050 (1908).

656. Same as Ref. 233.

657. Brauer, G., Haag, H., Z. anorg. u. allgem. Chem., 259, 197 (1949).

658. Parkinson, D. H., Simon, F. E., Spedding, F. H., "Atomic Heats of Rare Earth Elements," Proc. Roy. Soc., A-207, 137-155 (1951).

659. James, N. R., Legvold, S., Spedding, F. H., "The Resistivity of Lanthanum, Cerium, Praseodymium and Neodymium at Low Temperatures," Phys. Rev., 58[5], 1092-8 (1952).

660. Legvold, S., Spedding, F. H., "Magnetic and Electric Properties of Gadoliniur Dysprosium and Erbium Metals," Revs. Mod. Phys., 25, 129-30 (1953).

661. Hersh, H. N., "Vapor Pressure of Copper," Jour. Amer. Chem. Soc., 75, 1529-31 (1953).

662. Edwards, J. W., Johnston, H. L., "Vapor Pressure of Inorganic Substances, XI Titanium Between 1587 and 1764°K, Copper Between 1143 and 1292°K," Jour. Amer. Chem. Soc., 75, 2467-70 (1953).

- 672 -

663. Boosz, H. J., "The Average Specific Heat of Hard Metals Between Room Temperature and - 190°C," Metall., 11, 22 (1957).

664. Geglia, M. J., Hawkins, G. A., Deverall, J. E., "Determination of Thermal Conductivity of Copper and Deoxidized Copper - Iron Alloys: Apparatus and Technique," Anal. Chem., 24, 493 (1952).

665. White, G. K., Woods, S. B., "Thermal and Electrical Conductivities of Solids at Low Temperatures," Can. Jour. Phys., 33 58 (1955).

666. Sidles, P. H., Danielson, G. C., "Thermal Diffusivity of Metals at High Temperatures," Jour. Appl. Phys., 25, 58 (1954).

667. Rosenthal, D., Friedmann, N. E., "Thermal Diffusivity of Metals at High Temperature," Jour. Appl. Phys., 25[8], 1059 (1954).

668. Betz. H. T., Olson, O. H., "Determination of Emissivity and Reflectivity Data on Structural Materials," WADC-TR-56-222 Part II (1957).

669. Clements, J. F., "Specific Heat of Some Refractory Materials," Trans. Brit. Ceram. Soc., 61[8], 452-462 (1962).

670. Whittemore, O. J., Jr., Ault, N. N., "Thermal Expansion of Various Ceramic Materials to 1500°C," Jour. Amer. Ceram. Soc., 39[12], 443 (1960).

671. Matkovich, V. I., "Interstitial Compounds of Boron," Jour. Amer. Chem. Soc., 83, 1804-6 (1961); CA-55, 19569.

672. Epprecht, W., "Crystal Chemistry of Metal Carbides and Their Importance in Metallurgy," Chimia, 5, 49-60 (1951).

673. Moissan, H., Stock, A., Compt. rend., 131, 139 (1900).

674. Yevstrop'yev, K. S., Toropov, K. A., "Chemistry of Silicon and Physical Chemistry of Silicates," p. 223 (1950).

675. Tone, F., Ind. Eng. Chem., Ind. Edit., 30, 232 (1938).

676. Samsonov, G. V., Latysheva, V. P., Doklady Acad. Nauk SSSR, 105, 499 (1955).

677. Kotel'nikov, R. B., Samsonov, G. V., Zavod. Lab., 22, 375 (1955).

678. Zhuravlev, N. N., Kristallog., 1, 666 (1956).

679. Nowotny, H., Dimenopoulow, E., Kudielka, H.; See Ref. 444, no. 383.

680. Guryevich, M. A., Epel'baum, V. A., Ormont, B. F., Zhur. Neorg. Khim., 2, 206 (1957).

681. Mauer, F. A., Bolz, L. H., "Thermal Expansion of Cermet Components at High Temperature by X-ray Diffraction," NBS Report No. 3445 (1954).

682. DeSorbo, W., "Heat Capacity of Chromium Carbide from 13-300°K," Jour. Amer. Chem. Soc., 75, 1825 (1953).

683. Oriana, R. A., Murphy, W. K., "Heat Capacity of Chromium Carbide (Cr_3C_2)," Jour. Amer. Chem. Soc., 76, 343 (1954).

684. Kelley, K. K., "Specific Heats at Low Temperatures of Crystalline B_2O_3, B_4C and SiC," Jour. Amer. Chem. Soc., 63, 1137 (1941).

685. Olson, O. H., Morris, J. C., "Determination of Emissivity and Reflectivity Data on Aircraft Structural Materials," WADC-TR-56-222 Part III (1958).

686. Chang, H. C., Lindberg, O. H., Goldberg, C., "Power Rectifier Development Program," WADD-TR-59-11 (March 1960).

687. Dreger, L. H., Dadape, V. V., Margrave, J. L., "Sublimation and Decomposition Studies on Boron Nitride and Aluminum Nitride," Jour Phys. Chem., 66, 1556 (1962).

688. Coughlin, J. P., Orr, R. L., "High Temperature Heat Contents of Meta and Orthosilicate of Barium and Strontium," Jour. Amer. Chem. Soc., 75, 530 (1953).

689. Seltz, H., Dimkerley, F. J., DeWitt, B. J., "Heat Capacities and Eutropies of Molybdenum and Tungsten Trioxide," Jour. Amer. Chem. Soc., 65, 600 (1943).

690. Uyeno, K., "Determination of Vapor Pressure of Solids. IV Vapor Pressure of WO_3, MoO_3, CdO, TeO, and Their Thermodynamic Values," Jour. Chem. Soc. Japan, 62, 990 (1941).

691. Cosgrove, L. A., Snyder, P. E., "High Temperature Thermodynamic Properti of MoO_3," Jour. Amer. Chem. Soc., 75, 1227 (1953).

692. Sawada, S., Ande, R., Namura, S., "Thermal Expansion and Specific Heat of Tungsten Oxide at High Temperature," Phys. Rev., 84, 1054 (1951).

693. Sawada, S., "Thermal and Electrical Properties and Crystal Structure of Tungsten Oxide at High Temperatures," Phys. Rev., 91, 1010 (1953).

694. Deem, H. W., "Thermal Conductivity and Electrical Resistivity of Hafnium," USAEC-BMI-853 (1953).

695. Mauer, F. A., Bolz, L. H., "Thermal Expansion of Cermet Components by High Temperature X-ray Diffraction," NBS Report No. 4884 (1956).

696. Mauer, F. A., Bolz, L. H., "Thermal Expansion of Cermet Components by High Temperature X-ray Diffraction," NBS Report No. 3463 (1954).

697. Buessem, W. R., Bush, E. A., "Thermal Fracture of Ceramic Materials under Quasi-static Thermal Stresses," Jour. Amer. Ceram. Soc., 38, 27-32 (1955).

698. Long, G., Foster, L. M., "Aluminum Nitride Containers for Vacuum Evaporation of Aluminum," Amer. Ceram. Soc. Bull., 40[7], 423 (1961).

APPENDIX B

BIBLIOGRAPHY - ALPHABETICAL

BIBLIOGRAPHY
(Alphabetical)

Accountius, O. E.,	WADC-TR-53-287	91
Adams, M.,	Jour. Amer. Ceram. Soc., 37, 74 (1954)	534
Adenstedt, H. K.,	Trans. Amer. Soc. Metals, 44, 949 (1952)	228
Ageyev, N. V.,	Bull. Acad. Sci. USSR., Div. Chem. Sci., 1, 31 (1952)	494
	Doklady. Acad. Nauk., 112, 853 (1957)	578
Agte, C.,	Z. techn. Physik, 11[6], 182 (1930)	54
	Z. anorg. u. allgem. Chem., 198, 223 (1931)	52
	Thesis, Tech. Hochsch. Berlin (1931)	361
Akumov, Y. I.,	Jour. Appl. Phys. USSR , 21, 227 (1948)	631
Alaball, E. R.,	WADC-TR-55-155	101
Alexander, P.,	Metals and Alloys, 9, 179 (1938)	570
Alfred University	Prog. Report No. 237	10, 45
Allard, G.,	Compt. rend., 189, 108 (1929)	367
Allen, J.,	Trans. Roy. Soc. Canada, 25, 13 (1931)	468
Allen, R. D.,	Jour. Appl. Phys., 31[8], 1382 (1960)	539
Alterthum, H.,	Z. techn. Physik, 11[6], 182 (1930)	54

Ande, R.,	Phys. Rev., <u>84</u>, 1054 (1951)	692
Andersson, L. H.,	Acta Chem. Scand., <u>4</u>, 160, 209 (1950)	341
	Trans. Amer. Soc. Metals, <u>24</u>, 375 (1956)	607
Andrews, K. F.,	Jour. Metals, <u>187</u>, 746 (1949)	476
	Jour. Metals, <u>188</u>, 586 (1950)	256
Andrews, M. R.,	Jour. Amer. Chem. Soc., <u>54</u>, 1845 (1932)	230
Andrieux, J. L.,	Thesis, University of Paris (1929)	67
	Thesis, University of Grenoble (1948)	355
	Compt. rend., <u>189</u>, 1279 (1929)	70
	Compt. rend., <u>229</u>, 210 (1949)	371
	Ann. Chim., <u>12</u> [10] , 423 (1929)	69
	Rev. Met., <u>45</u>, 49 (1948)	339
	Bull. Soc. Chim. France, <u>15</u>, 598 (1948)	356
Anthony, F. M.,	WADC-TR-59-744 (1960)	529
Arbiter, W.,	WADC-TR-53-190 (1953)	13
Armstrong, C. T.,	Jour. Amer. Chem. Soc., <u>71</u>, 3583 (1949)	517
Armstrong, P. E.,	Trans. Amer. Soc. Metals, Reprint No. 48 (1955)	31
Aronson, E. V.,	Zhur. Tekh. Fiz., <u>24</u>, 1448 (1954)	425
Aronsson, B.,	Acta Chem. Scand., <u>9</u>, 1107 (1955)	626

Arvin, M.,	Jour. Appl. Pnys., 24, 498 (1953)	455
A.S.T.M.,	X-Ray Powder Data File	154
Ascherman, G.	Physik. Z., 42, 349 (1941)	514
Ault, N. N.,	Jour. Amer. Ceram. Soc., 39 [12], 443 (1960)	670
Ayton, M. W.,	WADC-TR-56-423 (1956)	251
Baensiger, N. C.,	Jour. Amer. Chem. Soc., 70, 99 (1948)	61
	USAEC-1-SC-53 (1949)	373
Baker, T. W.,	AERE-M/M-143 (1957)	231
Baldwin, W. M., Jr.,	Jour. Electrochem. Soc., 103, 429 (1956)	296
Baratta, F.,	A.R.S. Jour., 32[1], 53 (1962)	184
Bardgett, W. E.	Jour. Iron Steel Inst. London, 179, 211 (1955)	424
Barr, H. N.,	Amer. Ceram. Soc. Bull., 35, 47 (1956)	110, 122
Barr, R.,	Climax Molybdenum (1953)	305
Barrett, T.,	Ann. Physik., 77 (1925)	545
Basart, J. C. M.,	Z. anorg. u. allgem. Chem., 216, 209 (1934)	58
Batz, W.,	Jour. Metals, 4, 170 (1952)	633
Baur, J. P.,	Jour. Electrochem Soc., 103, 266 (1956)	232

Beall, R. A.,	RI-5951 (1962)	362
Beals, R. J.,	Amer. Ceram. Soc. Bull., 37 [11], 486 (1958)	555
Becker, G.,	Z. physik. Chem., 159, 1 (1932)	389
Becker, K.,	"Hochschmelzende Hartstoffe"	74
	Z. Physik, 31 368 (1925)	77
	Z. Physik, 34, 185 (1933)	181
	Z. Physik, 51, 481 (1928)	179
	Z. techn. Physik, 11, 216 (1930)	177
	Z. Physik, 32, 489 (1937)	505
Becker, M.,	Jour. Iron Steel Inst., 121, 315 (1930)	632
Beckman, C.,	Nature, 178, 1341 (1956)	233, 656
Belle, J.,	BMI-829 (1953)	279
Belliotti, J. V.,	Carborundum	14
Bender, S. L.,	ASD-TR-61-260 (1962)	650
Benesovsky, F.,	Z. Metallkunde, 43, 284 (1952)	30
	Z. Metallkunde, 44, 242 (1953)	23
	Z. anorg. u. allgem. Chem., 268, 191 (1952)	65
	"High Melting Hard Metals" (1953)	68

Benesovsky, F.,	Plansee Proc., 1955, 154-172 (1956)	12
	Monatsh., 88, 336 (1957)	64
	Z. Metallkunde, 45, 493 (1954)	582
	Z. Metallkunde, 47, 247 (1956)	590
	Monatsh., 84, 1 (1953)	600
	Z. Metallkunde, 44, 437 (1953)	610
Bennett, D. G.,	WADC-TR-52-111	116
Bertaut, F.,	Acta Cryst., 4, 72 (1951)	357
	Compt. rend., 234, 2621 (1952)	369
	Compt. rend., 229, 666 (1949)	372
	Acta Cryst., 3, 319 (1950)	569
	Compt. rend., 231, 626 (1950)	641
Betz, H. T.,	WADC-TR-56-222 Part II (1957)	668
Bidwell, C. C.,	Phys. Rev., 10, 756 (1917)	234
Biltz, W.,	Z. anorg. u. allgem. Chem., 148, 153 (1925)	382
Binet du Jassoneix, A.,	Compt. rend., 143, 169 (1906)	352
	Compt. rend., 143, 897 (1906)	349
	Compt. rend., 143, 1149 (1906)	349
Birchenall, C.,	Jour. Metals, 4, 170 (1952)	633

Birnbrauer, E.,	Z. anorg. u. allgem. Chem., 87, 167 (1914)	377
Blix, R.,	Z. Physik. Chem., 3 B, 229 (1929)	520
Blum, P.,	Acta Cryst., 3, 319 (1950)	569
	Acta Cryst., 4, 72 (1951)	357
	Ann. Soc. Chim. France, 14, 1077 (1947)	364
	Compt. rend., 229, 210 (1949)	371
	Compt. rend., 229, 666 (1949)	372
	Compt. rend., 231, 626 (1950)	641
	Compt. rend., 234, 2621 (1952)	369
Blum, S. L.,	Electronic Prog., 17 (1960)	186
Blumenthal, H.,	Trans. A.I.M.E., 185, 749 (1949)	89
	Jour. Amer. Chem. Soc., 74, 2942 (1952)	198
Blumenthal, W. B.,	"Chemical Behavior of Zirconium" (1958)	40
Boach, J. D.,	Jour. Electrochem. Soc., 98, 160 (1951)	92
Bokiy, G. B.,	"Introduction to Crystal Chemistry" (1954)	445
Bolef, D. J.,	Bull. Amer. Phys. Soc., 6, 76 (1961)	193
Bolz, L. H.,	N.B.S. Report No. 3445 (1954)	681

Bolz, L. H.,	N.B.S. Report No. 4884 (1956)	695
	N.B.S. Report No. 3463 (1954)	696
Booker, J.,	AF 33 (616)-6540	171
Boorse, H.,	Phys. Rev., 79, 7021 (1951)	470
Boosz, H. J.,	Metall., 11, 22 (1957)	663
Boren, B.,	Mineral. Geol., 11A [10], 28 (1933)	26
	Archiv. Kem. Min. Geol., A 11, 2 (1933)	606
Born, K.,	Stahl u. Eisen, 72, 1268 (1952)	422
Boucher, R.,	Acta Cryst., 8, 592 (1955)	652
Boulger, F. W.,	AECD-3382 (1949)	315
Bowman, M. G.,	Jour. Chem. Phys., 36,[2], 350 (1962)	207
Bradshaw, W. G.,	LMSD-2466 (1958)	227
Brager, A. K.,	Acta. physicochim., USSR , 13, 595, 600 (1940)	497
Brauer, G.,	Z. anorg. u. allgem. Chem., 259, 197 (1949)	657
	Z. anorg. u. allgem. Chem., 249, 325 (1942)	215
	Z. anorg. u. allgem. Chem., 267, 198 (1952)	567
	Z. Electrochem., 46, 397 (1940)	512
	Naturwiss., 37, 210 (1950)	568
	"Anorganische Chemie" (1948)	595

Brenner, A.,	NBS Report-4670 (1956)	241
Brewer, L.,	Jour. Amer. Ceram. Soc., 33, 273 (1950)	83
	Jour. Amer. Ceram. Soc., 33, 291 (1950)	11
	Jour. Amer. Ceram. Soc., 34, 173 (1951)	87
	UCRL-2544 (1954)	235
	Jour. Electrochem. Soc., 38, 103 (1956)	108
	"Thermodynamics" (1950)	393
Briar L. B.,	Jour. Amer. Ceram. Soc., 33, 27 (1950)	173
Bridges, D. W.,	Jour. Electrochem. Soc., 103, 266 (1956)	232
Bridgman, P. W.,	Proc. Amer. Acad. Sci., 66, 255 (1932)	387
Brixner, L. H.,	Jour. Amer. Chem. Soc., 80, 3214 (19 58)	556
Bromley, L. A.,	"Thermodynamics" (1950)	393
Brown, A. R. G.,	MET/PHYS. 325, AD-249, 685 (1960)	503
Browne, C. N.,	ONR-ACR-17 (1946)	301
Brunner, E.,	Z. anorg. u. allgem. Chem., 93, 91 (1915)	384
Buddery. J.,	Nature, 167, 362 (1951)	644
Buessem, W. R.,	Jour. Amer. Ceram. Soc., 38, 27 (1955)	697
Burdick, M. D.,	NACA-TN-1561 (1949)	236

Burgers, W. G.,	Z. Krist., 94, 299 (1933)	223
	Z. anorg. u. allgem. Chem., 216, 209 (1934)	58
Burke, J. E.,	Amer. Soc. Metals, Cleveland (1955)	319
Bush, E. A.,	Jour. Amer. Ceram. Soc., 38, 27 (1955)	697
Cadoff, I.,	Plansee Proc., 1955, 50-55 (1956)	135
	Jour. Metals, 5, 248 (1953)	93
Calhoun, H.,	NP-5539 (1955)	237
Calvert, E. D.,	RI-5951 (1962)	362
Campbell, D. E.,	Jour. Amer. Ceram. Soc., 45 [7], 310 (1962)	329
Campbell, I. E.,	"High Temperature Technology"	2
	"Vapor Plating" (1955)	298
	Proc. Electrochem. Soc., 93, 284 (1948)	226
	Jour. Electrochem. Soc., 96, 318 (1949)	334
Campbell, W. J.,	RI-5982 (1962)	324
Cape, J. A.,	WADD-TR-60-581 Part II	501
Carborundum Company Data		38
Carborundum Data Sheet		156
Carlson, O. N.,	Trans. Amer. Soc. Metals, Reprint No. 48 (1955)	31

Carniglia, S. C.,   "Dense Beryllium Oxide" (1961)   190

Catterall, J. A.,   "Thermochemical Data of Alloys" (1956)   273

Cerwenka, E.,   Thesis (1951)   50
                    Z. Metallkunde, 43, 101 (1952)   27
                    Monatsh., 85, 245 (1954)   619

Chaberski, A. Z.,   WADD-TR-60-749 Part I (1961)   183

Chalmot, G.,   Amer. Chem. Jour., 19, 69 (1897)   611

Chang, H. C.,   WADD-TR-59-11 (1960)   686

Chizhevskaya, S. N., "Physics and Physicochemical Analyses" (1957)   492

Chiotti, P.,   Jour. Amer. Ceram. Soc., 35 [5], 123 (1952)   100, 141

Christensen, A.,   Jour. Phys. Chem., 62, 499 (1958)   622

Chubb, W.,   Amer. Ceram. Soc. Bull., 41 [9], 564 (1962)   447
                    BMI-1488 (1960)   528

Chung, D. H.,   "Mechanical Properties" (1961)   191

Clark, H. K.,   Jour. Amer. Chem. Soc., 65, 2115 (1943)   379

Clausing, P.,   Z. anorg. u. allgem. Chem., 208, 401 (1932)   441

Cleek, G. W.,   Jour. Amer. Ceram. Soc., 45 [7], 219 (1962)   327

Cleland, B. B.,   BMI-829 (1953)   279

Clements, J. F.,	Trans. Brit. Ceram. Soc., 61 [8], 452 (1962)	669
Climax Molybdenum,	Bulletin cdb-6 (1956)	239
	Data Sheet (1953)	240
Coble, R. L.,	Jour. Amer. Ceram. Soc., 37, 107 (1954)	125
Coffin, L.,	WADC-TR-54-38 (1954)	229
Coffman, J. A.,	WADD-TR-60-646 (1961)	144
Cole, N. W.,	U.S. Pat. 2,088,838 (1937)	350
Comstock, A. A.,	Jour. Amer. Ceram. Soc., 45 [7], 310 (1962)	328
Cook, D.,	Phys. Rev., 79, 7021 (1951)	470
Coors Porcelain Co.,	Data Sheet No. 0001	157
Corak,	Cited from "Hartstoffe und Hartmetalle"	461
Cosgrove, L. A.,	Jour. Amer. Chem. Soc., 75, 1227 (1953)	691
Cotter, P. G.,	Amer. Mineral., 41, 355 (1956)	153
	Jour. Amer. Ceram. Soc., 39 [1], 11 (1956)	82
Couch, D.,	NBS Report 4670 (1956)	241
Coughlin, J. P.,	Jour. Amer. Chem. Soc., 75, 530 (1953)	688
	Jour. Amer. Chem. Soc., 72, 2262 (1950)	507

Cowie, P. A.,	ONR -ACR-17 (1946)	301
Crandall, W. B.,	"Mechanical Properties" (1961)	191
	NP-1779 (1950)	111
	NP-5796 (1955)	123
Crist, R.,	(June, 1960)	203
	(December, 1960)	204
	(June, 1961)	205
	(March, 1962)	206
Cronin, L. J.,	Amer. Ceram. Soc. Bull., 30 [7], 234 (1951)	39
Croxton, F. C.,	Jour. Amer. Chem. Soc., 70, 1718 (1948)	62
Currie, L. M.,	Geneva Conf. (1955)	242
Curtis, C. E.,	Jour. Amer. Ceram. Soc., 30, 180 (1947)	243
	Jour. Amer. Ceram. Soc., 37, 458 (1954)	76, 80
	TAM (1948)	551
Daane, A. H.,	USAEC-1SC-53 (1949)	373
Dadape, V. V.,	Jour. Phys. Chem., 66, 1556 (1962)	687
Dahl, O.,	Wiss. Veröff. Siemens-Konz., 8, 255 (1929)	489
Danielson, G. C.,	Jour. Appl. Phys., 25, 58 (1954)	666
Dauben, C. H.,	Jour. Amer. Ceram. Soc., 33, 291 (1950)	11

Dauben, C. H.,	Acta. Cryst., 3, 261 (1950)	28
	Jour. Phys. Chem., 60, 443 (1956)	614
Dautova, L. I.,	Zhur. Neorg. Khim., 2, 2766 (1957)	594
Davey, F. K.,	WADC-TR-55-155	101
Davydov, K. P.,	Fiz. Metal. Metalloved., 2, 192 (1956)	628
Dawihl, W.,	Z. anorg. u. allgem. Chem., 244, 191 (1940)	509
de Boer, J. H.,	Z. anorg. u. allgem. Chem., 187, 177 (1930)	75
	Z. anorg. u. allgem. Chem., 148, 345 (1925)	504
Deem, H. W.,	WADC-TR-55-496 (1956)	542
	DMIC-Memo 148 (1962)	208
	USAEC-BMI-853 (1953)	694
Defacqz, E.,	Compt. rend., 144, 1424 (1907)	617
	Compt. rend., 147, 1050 (1908)	655
de Klerk, J.,	Bull. Amer. Phys. Soc., 6, 76 (1961)	193
de Santis, V. J.,	AF 33(616)-6841 (March, 1962)	220
DeSorbo, W.,	Jour. Amer. Chem. Soc., 75, 1825 (1953)	682
Deverall, J. E.,	Anal. Chem., 24, 493 (1952)	664
DeWitt, B. J.,	Jour. Amer. Chem. Soc., 65, 600 (1943)	689
Dickerson, R. F.,	Jour. Amer. Ceram. Soc., 41 [9], 564 (1962)	447
	Nucleonics, 18 [3], 74 (1960)	448
Dimenopoulow, E.,	See Ref. 444 No. 383	679
Dodero, M.,	Thesis, Univ. of Grenoble	574

Dodero, M.,	Bull. Soc. Chim. France, 17, 545 (1950)	566
	Compt. rend., 199, 566 (1934)	564
Dolloff, R. T.,	A F 33(616)-6286 (December 31, 1959)	325
Doney, L. M.,	Jour. Amer. Ceram. Soc., 37 [10], 458 (1954)	76, 80
Dorfman, Y. G.,	Fiz. Metalloved. G.T.T.I., 405 (1934)	460
Dreger, L. H.,	Jour. Phys. Chem., 66, 1556 (1962)	687
Dreikorn, R. E.,	ASD-TR-61-260 (1962)	650
Druyvesteyn, M. J.,	Phillips Res. Reports, 2, 260 (1947)	419
Duckworth, W. H.,	WADC-TR-55-473	7
	BMI-1124 (1956)	277
	BMI-1223 (1957)	309
	BMI-1313 (1959)	142
Dunkerley, F. J.,	Jour. Amer. Chem. Soc., 65, 600 (1943)	689
Dushman, S.,	"Vacuum Technique"	244
Duwez, P.,	Jour. Electrochem. Soc., 97, 299 (1950)	60
	Jour. Metals, 3, 772 (1951)	16
Earhart, W. H.,	N. P. -4970 (1953)	118
Eberhart, J. L.,	Materials and Methods, (1954)	43

Ebert, F.,	Z. Physik, 31, 368 (1925)	77
	Z. techn. Physik, 11, 216 (1930)	177
Eckerlin, P.,	Z. anorg. u. allgem. Chem., 280, 3215 (1955)	479
Economos, G.,	NYO-3144 (1953)	291
Edmonds, W. H.,	U.S. Pat. 2,088,838 (1937)	350
Edwards, J. W.,	Jour. Amer. Chem. Soc., 75, 2467 (1953)	662
Edwards, R. K.,	AECD-3394; UCRL-1639 (1952)	245
Ehrlich, P.,	Angew. Chem., 59, 163 (1947)	196
	Z. anorg. u. allgem. Chem., 259, 1 (1949)	197
Einwohner, T. H.,	ASD-TR-61-260 (1962)	650
Elchardus, E.,	Rev. Ind. Minerals, 375, 861 (1936)	418
Ellinger, F. H.,	Trans. Amer. Soc. Metals, 31, 89 (1943)	57
Elliott, D. E.,	Jour. Phys. Chem., 62, 630 (1958)	176
Elliott, R. P.,	Metal Progress, 246 (October, 1960)	400
Ellis, W. C.,	Metals Technology, 15 (1948)	255
Ellison, F. O.,	ONR-ACR-17 (1946)	301
Ende, J. C.,	Jour. Iron Steel Inst. London, 175, 52 (1953)	427

Fawcett, S. L.,	Batt. Tech. Rev., 11 [9], 2-11 (1962)	557
Ferber, R. C.,	ASD-TR-61-260 (1962)	650
Ferguson, F. A.,	Chem. Eng., 237 (1957)	261
Fetzer, K.,	Ber. Dtsch. Chem. Ges., 40, 297 (1907)	348
Fichter, F.,	Z. anorg. u. allgem. Chem., 93, 91 (1915)	384
Fieldhouse, J. B.,	WADC-TR-55-495 (1956)	540
	WADC-TR-57-487 (1957)	246
Filbert, R. B.,	BMI-1165 (1957)	88
Fillmore, C. L.,	NBS Circ. 568 (1956)	275
Finlay, G. R.,	Chem. in Canada, 41 (1952)	105, 127
Finnie, S.,	Jour. Phys. Chem., 61, 1541 (1957)	645
Fishman, N.,	Chem. Eng., 237 (1957)	261
Fitzer, E.,	Metall., 9, 1062 (1955)	621
Fitzgerald,	Jour. Less Common Metals, 1, 132 (1959)	554
Fitzsimmons,	Jour. Amer. Ceram. Soc., 38, 214 (1955)	546
Flament, P.,	Compt. rend. 27th Cong. Intern. Belge., 20, 372 (1955)	429
Földex, G.,	Helv. Phys. Acta, 26, 199 (1954)	458

Frilley, R.,	Rev. Metall., 8, 457 (1911)	605
Fritz, O.,	Z. Elektrochem., 46, 42 (1940)	375
Frueh, A. J., Jr.,	Acta Cryst., 4, 66 (1951)	344
Fuks, M. Y.,	Zhur. Tekh. Fiz., 24, 1448 (1954)	425
Furusawa, K.,	Bull. Nagoya Inst. Tech., 5, 260 (1953)	428
Gallistl, E.,	Thesis, Tech. Hoch. Graz. (1951)	449
	Z. Metallkunde, 43, 284 (1952)	30
Gangler, J. J.,	NACA-TN-1911 (1949)	9
	Jour. Amer. Ceram. Soc., 33, 367 (1950)	94
Gannon, R. E.,	ASD-TR-61-260 (1962)	650
Garrett, A. B.,	Jour. Amer. Chem. Soc., 70, 1718 (1948)	62
Geglia, M. J.,	Anal. Chem., 24, 493 (1952)	664
Gel'd, P. V.,	Fiz. Metal. Metalloved., 2, 192 (1956)	628
Geller, R. F.,	NACA-TN-1561 (1949)	236
	Jour. Res. NBS, 35 [1], 87 (1945)	249
Geller, S.,	Acta Cryst., 1, 441 (1954)	646
	Acta Cryst., 8, 83 (1955)	635
Gemmill, M. G.,	Jour. Iron Steel Inst., London, 179, 211 (1955)	424
Gerds, A. F.,	Jour. Electrochem. Soc., 99, 197 (1952)	63

Giebelhausen, H.,   Z̲. anorg. u̲. allgem. C̲hem., 91, 251 (1915)   589

Gilbert, E.,   J̲our. C̲hem. P̲hys., 1, 753 (1933)   487

Gilles, P. W.,   J̲our. C̲hem. P̲hys., 36 [2], 350 (1962)   207

"Thermodynamics," (1950)   393

Glaser, F. W.,   J̲our. A̲ppl. P̲hys., 22, 103 (1951)   49

J̲our. C̲hem. P̲hys., 20, 1050 (1952)   330

J̲our. C̲hem. P̲hys., 22, 1264 (1954)   81

J̲our. M̲etals, 1, 475 (1949)   451

J̲our. M̲etals, 4, 387 (1952)   95

J̲our. M̲etals, 4, 391 (1952)   202

J̲our. M̲etals, 4, 631 (1952)   338

J̲our. M̲etals, 5, 1119 (1953)   44

P̲owd. M̲et. B̲ull., 6 [6], 178 (1953)   450

M̲etal. P̲rog.,67 [4], 77 (1955)   113

A̲cta M̲etal., 2, 20 (1954)   72

P̲lansee P̲roc., 1955, 173 (1956)   134

Z̲. M̲etallkunde, 44, 353 (1953)   478

Glasier, L. F.,   J̲our. A̲ppl. P̲hys., 31 [8], 1382 (1960)   539

Gleiser, M.,   ASTM, Cincinnati (1955)   250

Goldberg, C.,   WADD-TR-59-11 (1960)   686

Goldschmidt, H. J.,   J̲our. I̲ron S̲teel I̲nst., 160, 345 (1948)   51

Gruzin, P. L.,	Zav. lab., 21, 417 (1955)	475
Gulbransen, E. A.,	Jour. Metals, 188, 586 (1950)	256
	Jour. Metals, 187, 746 (1949)	476
Guryevich, M. A.,	Zhur. Neorg. Khim., 2, 206 (1957)	680
Guseva, L. N.,	Doklady Acad. Nauk SSSR , 112, 681 (1957)	452
Haag, H.,	Z. anorg. u. allgem. Chem., 267, 198 (1952)	567
	Z. anorg. u. allgem. Chem., 259, 197 (1947)	657
	Naturwiss., 37, 210 (1950)	568
Haddad, R. E ,	Jour. Appl. Phys., 20, 1130 (1949)	395
	Jour. Appl. Phys., 20, 886 (1949)	398
	Jour. Appl. Phys., 22, 70 (1951)	337
Haebler, H.,	Z. anorg. u. allgem. Chem., 204, 81 (1932)	415
	Z. anorg. u. allgem. Chem., 207, 145 (1932)	524
Haffitte, P.,	Rev. Ind. Minerals, 375, 861 (1936)	418
Hagg, G.,	Z. physik. Chem., 7B, 339 (1930)	522
	Z. physik. Chem., 11, 433 (1930)	222
	Nature, 170, 245 (1952)	96
Hahn, H.,	Z. anorg. u. allgem. Chem., 258, 58 (1949)	500
Hale, R. M.,	"Mechanical Properties," (1961)	189

Hamister, U. C.,	Geneva Conf., (1955)	242
Hamjian, H.,	NACA-TN-2731 (1952)	257
Hampel, C. A.,	"Rare Metals Handbook," (1954)	258
Handwerk, J. H.,	ANL-5053 (1956)	276
	Chem. Eng. Prog., 53 [2], 60F (1950)	259
Hansen, M.,	Trans. Amer. Soc. Metals, 44, 518 (1952)	15
	Trans. Amer. Soc. Metals, 45, 901 (1953)	18
	Amer. Soc. Metals, Preprint No. 4 (1951)	576
	Amer. Soc. Metals, Preprint No. 41 (1952)	581
	"Structures of Binary Alloys," (1941)	486
Hanst, P. L.,	ASD-TR-61-260 (1962)	650
Hardy, G. F.,	Phys. Rev., 89, 884 (1953)	216
Harman, C. G.,	BMI-748 (1952)	260
Hasselman, D. P. H.,	WADD-TR-60-749 Part I (1961)	183
	Carborundum Research Notebook 5793	132
Hauck, C. A.,	Jour. Amer. Ceram. Soc., 38, 450 (1955)	120
Haughton, J.,	Jour. Iron Steel Inst., 121, 315 (1930)	632
Hawkins, G. A.,	Anal. Chem., 24, 493 (1952)	664

Hedge, J. C.,	WADC-TR-57-487 (1957)	246
Hellner, E.,	Z. anorg. u. allgem. Chem., 261, 266 (1950)	483
	Angew. Chem., 62, 125 (1950)	484
Hersh, H. N.,	Jour. Amer. Chem. Soc., 75, 1529 (1953)	661
Hiester, N. K.,	Chem. Eng., 237 (1957)	261
Higson, G. I.,	Jour. Chem. Soc., 115, 215 (1919)	510
Hinnüber, J.,	ZVDI., 92, 111 (1950); Ref. 19	386
Hodge, J. C.,	AECU-3380	262
Hoffman, A.,	Naturwiss., 21, 676 (1933)	547
Hoffman, C.,	NACA-RM -E-53007 (1953)	263
Hoffman, G. A.,	RN-1868 (1957)	264
Hofman, O.,	"Metallurgy of Lead," (1918)	559
Holl, A.,	Compt. rend., 135, 78, 493 (1902)	586
Holland, L.,	"Vacuum Deposition of Thin Fibers"	265
Honak, E. R.,	Thesis	71
	Z. anorg. u. allgem. Chem., 268, 191 (1952)	65
Hönigschmid, O.,	"Karbid u. Silizid"	25
	Compt. rend., 142, 157, 280 (1906)	213

Hönigschmid, O.,	Compt. rend., 143, 224 (1906)	571
	Monatsh., 27, 205 (1906)	214
	Monatsh., 27, 1067 (1906)	572
	Monatsh., 28, 1017 (1907)	573
Hopkins, B. E.,	"Oxidation of Metals and Alloys," (1953)	274
Horn, F.,	Jour. Electrochem. Soc., 69, 2762 (1947)	469
Horst, C.,	Ber. Dtsch. Chem. Ges., 46, 1203 (1913)	363
Hove, J. E.,	Amer. Nucl. Soc., New York (1957)	266
Hower, L. D.,	Jour. Amer. Ceram. Soc., 34, 309 (1951)	102
Hulm, J. K.,	Phys. Rev., 82, 273 (1951)	340
Humenik, M.	NYO-3144 (1953)	291
Humphrey, G. L.,	Jour. Amer. Chem. Soc., 73, 2261 (1951)	392
	Jour. Amer. Chem. Soc., 75, 2806 (1953)	79
	Jour. Amer. Chem. Soc., 76, 978 (1954)	103
Ihnat, M. E.,	ASD-TR-61-260 (1962)	650
Inouye, J.,	ONRL-1565 (1953)	267
Ivanick, W.,	Jour. Metals, 4, 387 (1952)	95

Jacobs,	Z. Krist., 94, 229 (1936)	223
Jacobson, E. L.,	Jour. Amer. Chem. Soc., 78 [19], 4850 (1956)	212, 654
Jaeger,	Proc. Acad. Sci. Amsterdam., 39, 442 (1936)	225
Jaffee, R. I.,	Jour. Metals, 3, 335 (1951)	331
	BMI-1169 (1957)	307
	BMI-1170 (1957)	270
James, N. R.,	Phys. Rev., 58 [5], 1092 (1952)	659
Janeff, W.,	Z. Physik., 142, 619 (1955)	430
Jette, E.,	Trans. Amer. Soc. Metals, 24, 375 (1956)	607
	Jour. Chem. Phys., 1, 753 (1933)	487
Jellinek, E.,	Z. anorg. u. allgem. Chem., 97, 315 (1916)	374
Jenkins, J.,	Acta Metal., 3, 598 (1955)	472
Jochem, O.,	Ber. Dtsch. Chem. Ges., 46, 1205 (1913)	353
Johnson, A. F.,	U.S. Patent 2,480,473	411
	U.S. Patent 2,480,475	412
Johnson, J. R.,	Jour. Metals, 662 (1956)	6
	Jour. Amer. Ceram. Soc., 37 [10], 458 (1954)	76
Johnston, H. L.,	Jour. Amer. Chem. Soc., 75, 2467 (1953)	662

Joly, A.,	Compt. rend., 82, 1905 (1876)	401
	Acad. Sci. Ecole Norm., 6, 148 (1877)	402
Jones, G. A.,	Amer. Ceram. Soc. Bull., 30, 103 (1957)	99
Jones, W. H.,	Batt. Tech. Rev., 11 [9], 2-11 (1962)	557
Jordan, P. L.,	Jour. Appl. Phys., 31 [8], 1382 (1960)	539
Justi, E.,	Physik. Z., 42, 349 (1941)	514
Kakhovskii, N. I.,	Automat. Svarka, 7 [5], 24 (1954)	426
Kalita, P. T.,	Jour. Gen. Chem. USSR, 3, 872 (1933)	416
Kasatkin, B. S.,	Automat. Svarka, 7 [5], 24 (1954)	426
Katz, J. J.,	"Chemistry of Uranium"	140
Kebler, R.,	Linde Bulletin	269
Kelley, K. K.,	Jour. Amer. Chem. Soc., 62, 818 (1940)	404
	U.S. Bur. Mines Bull., 407 (1937)	390
	Ind. Eng. Chem., 36, 865 (1944)	391
	Jour. Amer. Chem. Soc., 63, 1137 (1941)	684
Kempter, C. P.,	Jour. Phys. Chem., 62, 630 (1958)	176
	Anal. Chem., 31 [1], 156 (1959)	138
Kerr, P. F.,	"Optical Mineralogy," (1942)	558

Kessler, H.,	Trans. Amer. Soc. Metals, <u>44</u>, 518 (1952)	15
	Amer. Soc. Metals, Preprint No. 4 (1951)	576
Kibler, G. R.,	WADD-TR-60-646 Part I (1961)	144
	DA-30-069-ORD-2787 (March 1962)	206
	AF33(616)-6841 (March 1962)	220
Kieffer, R.,	"Refractory Hard Metals"	19
	"High Melting Hard Metals," (1953)	68
	Z. Metallkunde, <u>43</u>, 101 (1952)	27
	Z. Metallkunde, <u>43</u>, 284 (1952)	30
	Z. Metallkunde, <u>44</u>, 242 (1953)	23
	Z. Metallkunde, <u>45</u>, 493 (1954)	582
	Monatsh., <u>88</u>, 336 (1957)	64
	Metallforschung, <u>2</u>, 257 (1947)	55
	Powd. Met. Bull., <u>4</u>, 4 (1949)	35, 172
	Z. anorg. u. allgem. Chem., <u>268</u>, 191 (1952)	65
	Monatsh., <u>84</u>, 1 (1953)	600
	Z. Metallkunde, <u>44</u>, 473 (1953)	610
	Z. Metallkunde, <u>47</u>, 247 (1956)	590
Kiessling, R.,	Acta Chem. Scand., <u>1</u>, 893 (1947)	90
	Acta Chem. Scand., <u>3</u>, 90 (1949)	335
	Acta Chem. Scand., <u>3</u>, 595 (1949)	343
	Acta Chem. Scand., <u>3</u>, 603 (1949)	342
	Acta Chem. Scand., <u>4</u>, 160, 209 (1950)	341
	Jour. Metals, <u>3</u>, 639 (1951)	351

Kiessling, R.,	Nature, 178, 1341 (1956)	233, 656
	Fortschr. Chem. Forschung., 3, 41 (1954)	462
Kikoin, I. K.,	Fiz. Metallov., GTTI 405 (1934)	460
King, E. G.,	Jour. Amer. Chem. Soc., 71, 316 (1949)	511
	Jour. Amer. Chem. Soc., 72, 2262 (1950)	507
	Jour. Phys. Chem., 62, 499 (1958)	622
Kingery, W. D ,	"Property Measurements at High Temperature"	158
	Jour. Amer. Ceram. Soc., 37, 107 (1954)	125
	NEPA-1446 (1950)	114
	NYO-599 (1951)	536
	NYO-601 (1952)	117
	NYO-3144 (1953)	291
Kirk, M. M.,	RI-5951 (1962)	362
Klemm, W.,	Z. anorg. u. allgem. Chem., 201, 24 (1931)	394
Klopp, W. D.,	BMI-1169 (1957)	307
	BMI-1170 (1957)	270
Knapton, A.,	Nature, 175, 730 (1955)	596
Knowles, P. R.,	IGR-RIC-190 (1957)	271
Knudsen, F. P.,	Jour. Amer. Ceram. Soc., 45 [2], 94 (1962)	219
	NBS Circular-568 (1956)	275

Koch, W.,   Stahl u. Eisen, 72, 1268 (1952)   422

Koenig, J. H.,   Ind. Eng. Chem., 43, 2008 (1951)   115

Kohn, J. A.,   Amer. Mineral, 41, 355 (1956)   153

Jour. Amer. Ceram. Soc., 39 [1], 11 (1956) 82

Kölbl, F.,   Powd. Met. Bull., 4, 4 (1949)   35, 172

Konrad, H. E.,   WADC-TR-53-287   91

Korst, W.,   Jour. Phys. Chem., 61, 1541 (1957)   645

Köster, W.,   Z. Metallkunde, 39, 11 (1948)   175

Kotel'nikov, R. B.,   Zavod. Lab., 22, 375 (1955)   677

Kovalskii, A. E.,   Acad. Nauk SSSR , 170-186 (1950)   560

Kramer, J.,   Physik. Z., 42, 349 (1941)   514

Krikorian, O. H.,   Jour. Electrochem. Soc., 103, 38 (1956)   108

UCRL-2544 (1954)   235

UCRL-2888 (1955)   272

UCRL-6132 (1960)   329

Kröger, C.,   Z. anorg. u. allgem. Chem., 204, 81 (1932)   415

Z. anorg. u. allgem. Chem., 207, 145 (1932) 524

Z. anorg. u. allgem. Chem., 218, 379 (1934) 437

Lange, N. A.,	"Handbook of Chemistry"	3
Larach, S.,	Phys. Rev., 102, 582 (1956)	164
	Phys. Rev., 104, 68 (1956)	165
Larrabee, R. D.,	Opt. Soc. Amer., 49 [6], 619 (1959)	148
Laraduc-Müller, L.,	Rev. Met., 7, 657 (1910)	587
Larsen, W. L.,	ASTM-Cincinnati (1955)	250
Lashko, N. F.,	Doklady Acad. Nauk SSSR , 81, 605 (1951)	638
Latysheva, V. P.,	Fiz. Metal. Metallov., 2, 303 (1956)	474
	Doklady Acad. Nauk SSSR , 105, 499 (1955)	676
Laube, E.,	Monatsh., 88, 336 (1957)	64
	Planseeber., 54 (April 1960)	169
Lauchner, J. H.,	Amer. Ceram. Soc. Bull., 37 [11], 486 (1958)	555
Laves, F.,	Z. Krist., 101A, 78 (1939)	17
Lawrence, W. G.,	N. P. -1779 (1950)	111
Lebeau, P.,	Compt. rend., 121, 496 (1895)	380
	Compt. rend., 145, (1907)	648
	Ann. Chim. Phys., 8 [1], 553 (1904)	588
	Ann. Chim. Phys., 6 [7], 476 (1899)	381

Legvold, S.,	Phys. Rev., 58 [5], 1092 (1952)	659
	Rev. Mod. Phys., 25, 129 (1953)	660
Lehman, G. W.,	WADD-TR-60-581 (1960)	182
Leitnaker, J. M.,	J. Chem. Phys., 36 [2], 350 (1962)	207
	LA-2402 (1960)	502
Lemmon, A. W.,	BMI-1165 (1957)	88
Lenie, C.,	Jour. Electrochem. Soc., 107 [4], 308 (1960)	150
Levy, L.,	Compt. rend., 121, 1148 (1895)	575
Lidman, W.,	NACA-TN-2731 (1952)	257
Liedeholm, C. A.,	CWR-481 (1957)	119
Lindberg, O. H.,	WADD-TR-59-11 (1960)	686
Litz, L. M.,	Jour. Amer. Chem. Soc., 70, 1718 (1948)	62
Liu, Y. H.,	Jour. Metals, 3, 639 (1951)	351
Livey, D. T.,	Jour. Nucl. Energy, 2, 202 (1956)	278
Loch, L. D.,	BMI-1124 (1956)	277
	USAEC-TID-7350 Part I (1957)	195
Lofgren, N. L.,	"Thermodynamics" (1950)	393
Loman, H.,	Jour. Inst. Metals, 49, 369 (1932)	488

Londeree, J. W.,	Jour. Amer. Ceram. Soc., 34, 309 (1951)	102
Long, G.,	Jour. Amer. Ceram. Soc., 42 [2], 53 (1959)	431
	Amer. Ceram. Soc. Bull., 40[7], 423 (1961)	698
Long, R. A.,	Metal Prog., 68 [9], 983 (1952)	85
Lorey, G. E.,	WADC-TR-55-155	101
Lowrie, R.,	DA-30-069-ORD-2787 (June, 1960)	203
	DA-30-069-ORD-2787 (December, 1960)	204
	DA-30-069-ORD-2787 (June, 1961)	205
	DA-30-069-ORD-2787 (March, 1962)	206
	DA-30-069-ORD-2787 (December, 1961)	192
Lozier, W. W.,	DA-30-069-ORD-2787 (June, 1960)	203
Lucks, C. F.,	DCIM Memo-148 (March, 1962)	208
	WADC-TR-55-496 (1956)	542
Lundin C.,	Trans. ASM, 45, 901 (1953)	18
	Amer. Soc. Metals, Preprint No. 41 (1952)	581
Lundin, L. E.,	ASM-Cleveland (1953)	210
Lux, B.,	Monatsh., 86, 859 (1955)	599
Lynch, J. F.,	WADC-TR-55-473 (1954)	7
Lyon, T. F.,	AF 33 (616) 6841 (March, 1962)	220

Mallett, M. W.,	Jour. Electrochem. Soc., 99, 197 (1952)	63
	BMI-829 (1953)	279
	AEC-TIS Reactor Handbook, 3 [1], Chap. 17 (1955)	97
Maier, C. G.,	U.S. Bur. Mines Bull., 436 (1942)	521
Manchot, W.,	Compt. rend., 137, 229 (1903)	643
Margrave, J. L.,	Private communication; see Mezaki, MSc Thesis (1961)	128
	Jour. Phys. Chem., 66, 1556 (1962)	687
Maschenschalk, R.,	Monatsh., 84, 677 (1953)	583
	Z. Metallkunde, 45, 493 (1954)	582
Mash, D. R.,	Amer. Soc. Metals, Monterey, California (1957)	280
Masing, G.,	Wiss. Veröff. Siemens-Konz., 8, 255 (1929)	489
Mathews, C. O.,	LMSD-2466 (1958)	227
Matignon, C.,	Compt. rend., 141, 190 (1905)	612
Matkovich, V. I.,	Jour. Amer. Chem. Soc., 83, 1804 (1961)	671
Matthias, B. T.,	Phys. Rev., 82, 273 (1951)	340
Mauer, F. A.,	N.B.S. Report No. 3445 (1954)	681
	N.B.S. Report No. 3463 (1954)	696
	N.B.S. Report No. 4884 (1956)	695

McBride, C.,	WADC-TR-53-287	91
McClelland, J. D.,	Jour. Amer. Ceram. Soc., 43, 54 (1960)	537
	Int. Jour. Phys. Chem. Solids, 15, 17 (1960)	538
McDonald, R. A.,	Jour. Amer. Chem. Soc., 70, 99 (1948)	61
McGuire, J. C.,	Anal. Chem., 31 [1], 156 (1959)	138
McKenna, P. M.,	Ind. Eng. Chem., 28, 767 (1936)	59
McLennan, J.,	Trans. Roy. Soc. Canada, 25, 13 (1931)	468
McMurtry, C. H.,	AT (40-1)-2558 (1961)	136
McNees, R.,	Jour. Amer. Chem. Soc., 77, 5290 (1955)	629
McNutt, J. E.,	NACA-TN-1911 (1949)	9
McPherson, D.,	Trans. Amer. Soc. Metals, 44, 518 (1952)	15
	Trans. Amer. Soc. Metals, 45, 901 (1953)	18
	Preprint No. 4 (1951)	576
	Preprint No. 41 (1952)	581
McQuarrie, M.,	Jour. Amer. Ceram. Soc., 37, 84 (1954)	533
Mead, H.,	Jour. Metals, 4, 170 (1952)	633
Megaw, H. D.,	Proc. Phys. Soc. London, 58, 133 (1946)	548
Meijering, J. L.,	Phillips Research Reports, 2, 260 (1947)	419

Meissner, W.,	Z. Physik, 65, 39 (1930)	283
	Z. Physik, 75, 521 (1932)	333
	Z. Physik, 63, 558 (1930)	465
	Erg. des Exakt. Naturwiss., XI (1932)	463
	Z. Ges. Kälteind., 39, 104 (1932)	466
	Ann. Phys., 17, 593 (1933)	467
Messerknecht, C.,	Z. anorg. u. allgem. Chem., 148, 153 (1925)	382
Meyer-Jungnik, W.,	Angew. Chem., 67, 306 (1955)	636
Michaelson, H.,	Materials and Methods (1953)	284
Mikol, E. P.,	ORNL-1131 No. 2 (1952)	544
Milford, F. J.,	Batt. Tech. Rev., 11 [9], 2 (1962)	557
Miller, A. R.,	"Thermodynamic Study of Metals," (1957)	285
Miller, E.,	Plansee Proc., 1955, 50-55 (1956)	135
Miller, G. L.,	"Zirconium"	211
	Materials and Methods, 131 (1957)	286
Milne, T. A.,	NAA-SR-2124 (1957)	322
Miner, C. F.,	U.S Patent 1,803,720	410
Mitchell, L.,	Ind. Eng. Chem., 48, 1702 (1956)	287
Mitius, A.,	Z. anorg. u. allgem. Chem., 249, 325 (1942)	215
Mixer, W. G., Jr.,	BMI-748 (1952)	260

Moch, M.,	Jour. Amer. Chem. Soc., 77, 304 (1955)	41
Moeller, T.,	"Inorganic Chemistry"	73
Moers, K.,	Z. anorg. u. allgem. Chem., 198, 233 (1931)	52
	Z. anorg. u. allgem. Chem., 198, 243 (1931)	52
	Z. anorg. u. allgem. Chem., 198, 262 (1931)	42
Moissan, H.,	Compt. rend., 118, 501 (1894)	385
	Compt. rend., 119, 185 (1894)	345
	Compt. rend., 120, 290 (1895)	603
	Compt. rend., 121, 621 (1895)	603
	Compt. rend., 123, 15 (1896)	865
	Compt. rend., 131, 139 (1900)	673
	Compt. rend., 135, 78 (1902)	586
	Compt. rend., 135, 493 (1902)	586
	Compt. rend., 137, 229 (1903)	643
	Ann. Chim. Phys., 8, 565 (1896)	346
Monack, A. J.,	Materials and Methods, 158 (1957)	288
Mong, L. E.,	N.B.S. Report No. 2427	34
Moody, A. R.,	Jour. Chem. Soc., 81, 14 (1902)	347
	Proc. Chim. Soc., 17, 129 (1901)	354
Moody, L. S.,	Jour. Amer. Ceram. Soc., 33, 27 (1950)	173

Mordike, B. L.,	Jour. Less Common Metals, 1, 132 (1959)	554
Moreland, R. E.,	NACA-TN-1561 (1949)	236
Morgan, F. H.,	Jour. Appl. Phys., 20, 886 (1949)	398
	Jour. Appl. Phys., 20, 1130 (1949)	395
	Jour. Appl. Phys., 22, 108 (1951)	336
Morris, J. C.,	WADC-TR-56-222 (1956)	530
	WADC-TR-56-222 Part III (1958)	685
Moskowitz, D.,	Powd. Met. Bull., 6 [6], 178 (1953)	450
	Acta Met. 2, 20 (1954)	72
	Jour. Chem. Phys. 22, 1264 (1954)	81
	Plansee Proc., 1955, 173 (1956)	134
Mott, B. W.,	"Micro Indentation Hardness Testing"	8
Mott, W. R.,	Trans. Amer. Electrochem. Soc., 35, 255 (1919)	388
Mowry, A. L.,	Trans. AIME., 185, 133 (1949)	56
Mulford, R.,	Chem. Eng. News, 56 (July 30, 1962)	435
Murphy, W. K.,	Jour. Amer. Chem. Soc., 76, 343 (1954)	683
Murray, P.,	Jour. Nucl. Energy, 2, 202 (1956)	278
Muthman, W.,	Liebigs Ann., 355, 92 (1907)	513

Myers, C.,	Jour. Amer Ceram. Soc., 40, 526 (1957)	601
	Jour. Phys. Chem., 60, 443 (1956)	614
Myers, D. K.,	"Aluminum Nitride Literature Search," (1957)	155
NACA-TN-1911 (1949)		9
NACA-TN-1918 (1949)		84
Nadler, M. R.,	Anal. Chem., 31 [1], 156 (1959)	138
	Jour. Amer. Ceram. Soc., 38, 214 (1955)	546
Nakata, M. M.,	AF 33 (657) 7136 (January, 1962)	146
	AF 33 (657) 7136 (April, 1962)	209
Namura, S.,	Phys. Rev., 84, 1054 (1951)	692
Naray-Szabo, S. V.,	Z. Krist., 97, 223 (1937)	21
	Jour. Amer. Chem. Soc., 71, 1882 (1949)	358
Naylor, B. F.,	Jour. Amer. Chem. Soc., 68, 370 (1946)	145
Neeley, J. J.,	Jour. Amer. Ceram. Soc., 33, 363 (1950)	289
	NEPA-818 (1948)	313
Nelson, H. R.,	Jour. Electrochem. Soc., 99, 197 (1952)	63
Nelson, J. A.,	WADC-TR-52-111	116
Neshpor, V. S.,	Fiz. metal. metalloved., 4 [1], 181 (1957)	166

Norton, J. T.,	Trans. AIME, 185, 749 (1949)	89
	Jour. Electrochem. Soc., 103, 107 (1936)	316
Novikov, I. I.,	Zhur. Neorg. Khim., 2, 2766 (1957)	594
Nowicki, D. H.,	Jour. Electrochem. Soc., 96, 318 (1949)	334
Nowitzky, A.,	Compt. rend., 145 (1907)	648
Nowotny, H.,	Monatsh., 84, 1 (1953)	600
	Monatsh., 84, 677 (1953)	583
	Monatsh., 85, 245 (1954)	619
	Monatsh., 86, 385 (1955)	620
	Monatsh., 86, 413 (1955)	592
	Monatsh., 86, 859 (1955)	599
	Monatsh., 88, 336 (1957)	64
	Metallforschung, 2, 257 (1947)	55
	Planseeber, 9 [1/2], 54 (1960)	169
	Z. Metallkunde, 44, 242 (1953)	23
	See Ref. 444 No. 383	679
Null, M. R.,	DA-30-069-ORD-2787 (June, 1960)	203
Oak Ridge	ORNL-1952 (1955)	294
	ORNL-1945 (1955)	294
O'Dell, F.,	Jour. Electrochem. Soc., 97, 299 (1950)	60

Ogden, H. R.,	Jour. Metals, 3, 335 (1951)	331
Olson, O. H.,	WADC-TR-56-222 Part I (1956)	530, 668
	WADC-TR-56-222 Part III (1958)	685
Olson, W.,	Chem. Eng. News, 56, (July 30, 1962)	435
Oriana, R. A.,	Jour. Amer. Chem. Soc., 76, 343 (1954)	683
Ormont, B. F.,	Zhur. Fiz. Khim., 20, 459 (1946)	498
	Zhur. Fiz. Khim., 21, 3 (1947)	499
	Zhur. Fiz. Khim., 34, 2329 (1960)	221
	Zhur. Neorg. Khim., 2, 206 (1957)	680
	Zav. Lab., 14, 104 (1948)	496
Orr, R. L.,	Jour. Amer. Chem. Soc., 75, 530 (1953)	688
Osborne, R. H.,	Opt. Soc. Amer. Jour., 31, 428 (1941)	543
Ott, H.,	Z. Physik, 22, 201 (1924)	151
Ovechkim, B. I.,	Doklady Acad. Nauk SSSR , 112, 681 (1957)	452
Paine, R. M.,	AF 33 (616)-6540	171
Pallmer, P.G.,	AT (45-1)-1351 (February, 1962)	432
Pappis, J.,	Electronic Prog., 17 (June, 1 960)	186
Paprocki, S. J.,	BMI-1143 (1956)	295

Parché, M. C.,	"Facts About Fused Alumina"	36
Parkinson, D. H.,	Proc. Roy. Soc., A-207, 137 (1951)	658
Parthe, E.,	Powd. Met. Bull., 8 [1/2], (1957)	491
	Monatsh., 86, 385 (1955)	620
	Monatsh., 86, 413 (1955)	592
	Monatsh., 86, 859 (1955)	599
Pauling, L.,	Acta Cryst., 1, 212 (1948)	630
Pearl, H. A.,	WADC-TR-59-744 (1960)	529
Pearson, J.,	Jour. Iron Steel Inst. London, 175, 52 (1953)	427
Pease, R. S.,	Nature, 165, 722 (1950)	433
	Acta Cryst., 5, 356 (1952)	162
Pechman, A.,	Ceramics (1954)	200
Petrova, L. A.,	Acad. Nauk SSSR , 170 (1950)	560
Pfisterer, H.,	Z. Metallkunde, 41, 359 (1950)	649
	Z. Metallkunde, 41, 433 (1950)	642
Phalniker, C. A.,	Jour. Electrochem. Soc., 103, 429 (1956)	296
Phaneuf, J. P.,	ASD-TR-61-260 (1962)	650
Pietrobansky, P.,	Jour. Metals, 3, 772 (1951)	16
Pike, J. N.,	DA-30-069-ORD-2787 (December 1960)	204

Plotnikov, V. A.,	Jour. Gen. Chem. USSR , 3, 872 (1933)	416
Plummer, W. A.,	Jour. Amer. Ceram. Soc., 45 [7], 310-316 (1962)	328
Polikarpov, Y. A.,	Zav. lab., 21, 417 (1955)	475
Pollard, F. H.,	Trans. Far. Soc., 46, 190 (1950)	143
	Jour. Chem. Soc., 2444 (1952)	508
Pollock, B. D.,	NAA-SR-5439 (1961)	525
Popper, P.,	Trans. Brit. Ceram. Soc., 60, 603 (1961)	526
Popper, P.,	"Special Ceramics," (1960)	160
Porembka, P.,	BMI-1003 (1955)	303
Porter, H. B.,	NOTS-1191 (1955)	297
Post, B.,	Acta Met., 2, 20 (1954)	72
	Plansee Proc., 1955, 173 (1956)	134
	J. Chem. Phys., 22, 1264 (1954)	81
	J. Chem. Phys. 20, 1050 (1952)	330
	Jour. Metals, 4, 631 (1952)	338
Potter, R. A.,	Jour. Amer. Ceram. Soc., 39 [1], 11 (1956)	82
	Amer. Mineral, 41, 355 (1956)	153
Powell, C. F.,	"Vapor Plating," (1955)	298
	Jour. Electrochem. Soc., 96, 318 (1949)	334

Powell, R. W.,	Philosophical Mag., 44 [7], 645 (1953)	299
Preller, H.,	V.D.J. Zeitschrift, 98, 1611 (1956)	593
Prescott, C. H.,	Jour. Amer. Chem. Soc., 48, 2534 (1926)	397
Quirk, J. F.,	BMI-1165 (1957)	88
	AEC-TIS Reactor Handbook, 3 [1], Chap 15 (1955)	98
Rabinowitch, E.,	"Chemistry of Uranium"	140
Rasor, N. S.,	Int. Jour. Phys. Chem. Solids, 15, 17 (1960)	538
Rauscher, W.,	Z. Metallkunde, 39, 111 (1948)	175
Reed, E. R.,	AECU-110, NEPA-465 (1948)	300
Renaux, L.,	Thesis (1900)	131
Reynor, G.,	Phil. Mag., 39, 318 (1948)	
Rice, H. H.,	Amer. Ceram. Soc. Bull., 35, 47 (1956)	110, 122
Rice, W. H.,	NP-4970 (1953)	118
Riedelbauch, R.,	Liebigs Ann., 355, 92 (1907)	513
Riethof, T. R.,	WADC-TR-60-646 (1961)	144
Rix, W.,	Z. anorg. u. allgem. Chem., 244, 191 (1940)	509
Robards, C. F.,	NACA-TN-1911 (1949)	9

Roberts, E. L.,   "Inorganic Chemistry" (1954)   405

Robins, D.,   Acta Metall., 3, 598 (1955)   472

                       Phil. Mag., 3, 313 (1958)   453

Rogers, A. F.,   "Optical Mineralogy" (1942)   558

Rossinni, F. P.,   ONR-ACR-17 (1946)   301

Roth, R. S.,   NBS Circular-568 (1956)   275

Roth, W. A.,   Z. Elektrochem., 46, 42 (1940)   375

                       Z. physik. Chem., 159, 1 (1932)   389

Rough, F. A.,   BMI-1000 (1955)   139

                       BMI-1488 (1960)   528

                       Nucleonics, 18 [3], 74 (1960)   448

Rosenthal, D.,   Jour. Appl. Phys., 25, 1059 (1954)   667

Ruddlesden, S. N.,   Trans. Brit. Ceram. Soc., 60, 603 (1961)   526

Ruff, O.,   Z. anorg. u. allgem. Chem., 97, 315 (1916)   374

                       Z. anorg. u. allgem. Chem., 128, 96 (1923)   396

Rundle, R. E.,   Jour. Amer. Chem. Soc., 70, 99 (1948)   61

                       Acta Cryst., 1, 180 (1949)   518

Runnals, O.,   Acta Cryst., 8, 592 (1955)   652

Ryshkewitch, E.,	Jour. Amer. Ceram. Soc., 36, 65 (1953)	37
	WADC-TR-50-633 (1950)	107
	AF-TR-6330 (1950)	302
Saller, H. A.,	BMI-1000 (1955)	139
	BMI-1003 (1955)	303
Samsonov, G. V.,	Fiz. metal. metalloved., 4 [1], 181 (1957)	166
	FTD-TT-61-409 (January 29, 1962)	444
	"Physics and Physico-chemical Analysis," (1957)	454
	Fiz. Metal. Metallov., 5, 565 (1957)	473
	Fiz. Metal. Metallov., 2, 303 (1956)	474
	Zhur. Neorg. Khim., 3, 868 (1958)	597
	"Problems of Powder Metallurgy and Strength of Materials," No. V, 3 (1957)	623
	Doklady Acad. Nauk SSSR , 105, 499 (1955)	676
	Zavod. Lab., 22, 375 (1955)	677
Samsonov, V. P.,	Doklady Acad. Nauk SSSR , 112, 853 (1957)	578
Sandler, E.,	Thesis, Tech. Hoch., Munchen (1911)	563
Sato, S.,	Bull. Inst. Phys. Chem. Res. (Tokyo), 14, 862 (1938)	417
	Sci. Pap. Inst. Phys. Chem. Res. (Tokyo), 34, 888(1938)	438
Savitskiy, Y. M.,	Izd. vo. acad. Nauk SSSR , (1957)	639

Sawada, S.,	Phys. Rev., 84, 1054 (1951)	692
	Phys. Rev., 91, 1010 (1953)	693
Schachner, H.,	Monatsh., 84, 1 (1953)	600
	Monatsh., 84, 677 (1953)	583
	Monatsh., 85, 245 (1954)	619
	Monatsh., 85, 1140 (1954)	585
	Thesis, Univ. of Vienna (1953)	598
	Z. Metallkunde, 44, 242 (1953)	23
Scheele, W.,	"Anorganische Chemie," (1948)	595
Schick, H. L.,	ASD-TR-61-260 (1962)	650
Schissel, P. O.,	DA-30-069-ORD-2787 (February, 1962)	199
Schliegphake, O.,	Z. anorg. u. allgem. Chem., 11, 1951 (1926)	481
Schmid, H.,	Monatsh., 86, 385 (1955)	620
	Monatsh., 86, 413 (1955)	592
	Z. Metallkunde, 47, 247 (1956)	590
Schmidt, J. M.,	Bull. Soc. Chim., 43 [4], 49 (1928)	383
Schneider, A.,	Angew. Chem., 67, 306 (1955)	636
Schnorrenberg, E.,	Z. physik. Chem., B-27, 37 (1934)	376
Schomaker, V.,	DA-30-069-ORD-2787 (June, 1960)	203
	DA-30-069-ORD-2787 (December, 1960)	204

Schomaker, V.,	DA-30-069-ORD-2787 (June, 1961)	205
	DA-30-069-ORD-2787 (March, 1962)	206
Schönberg, N.,	See Ref. 626	618
Schrader, R. E.,	Phys. Rev., 102, 582 (1956)	164
	Phys. Rev., 104, 68 (1956)	165
Schroth, H.,	Thesis, Univ. Graz (1952)	577
	Z. Metallkunde, 44, 437 (1953)	610
Schubert, K.,	Z. Metallkunde, 41, 359 (1950)	649
	Z. Metallkunde, 41, 433 (1950)	642
Schuff, K.,	Z. anorg. u. allgem. Chem., 209, 33 (1932)	485
Schüth, W.,	Z. anorg. u. allgem. Chem., 201, 24 (1931)	394
Schwab, J.,	Metall., 9, 1062 (1955)	621
Schwartzkopf, P.,	"Refractory Hard Metals"	19
	Electrochem. Soc., Cleveland (1950)	46, 174
	Z. Metallkunde, 44, 353 (1953)	478
Scuderi, T. G	Jour. Amer. Ceram. Soc., 45 [7], 319 (1962)	327
Searcy, A. W.,	Jour. Amer. Ceram. Soc., 33, 291 (1950)	11
	Jour. Amer. Ceram. Soc., 40, 431 (1957)	584
	Jour. Amer. Ceram. Soc., 40, 526 (1957)	601

Searcy, A. W.,	Jour. Amer. Chem. Soc., 77, 5290 (1955)	629
	Jour. Amer. Chem. Soc., 78 [19], 4850 (1956)	654, 212
	Jour. Phys. Chem., 61, 1541 (1957)	645
Seibel, R. D.,	WAL-821/9 (1954)	304
Seltz, H.,	Jour. Amer. Chem. Soc., 65, 600 (1943)	689
Semchyshen, M.,	Climax Molybdenum (1953)	305
Serpek, O.,	U.S. Patent 888,044	407
	U.S. Patent 987,408	408
	U.S. Patent 996,032	409
Settig, L.,	Z. anorg. u. allgem. Chem., 144, 169 (1925)	53, 178
	Z. anorg. u. allgem. Chem., 143, 293 (1925)	516
Sevast'yanov, N. G.,	Compt. rend. acad. sci. URSS , 32, 432 (1941)	378
Seyforth, H.,	Z. Krist., 67, 295 (1928)	20
Shaffer, P. T. B.,	WADD-60-749 Part I (1961); Part II (1962)	183
	Carborundum Research Notebook 5625	149, 170
Shapiro, H.,	NBS Report-4670 (1956)	241
Shaw, H. L.,	AECD-3382 (1949)	315
Sheipline, V. M.,	AEC-TIS Handbook, 3, Chap. 17 (1955)	97
Sherwood, E. M.,	Ind. Eng. Chem., 48, 1735 (1956)	306

Shevlin, T. S.,	WADC-TR-54-173 Part I	121
	Jour. Amer. Ceram. Soc., 38, 450 (1955)	120
Shirane, G.,	Phys. Rev., 84, 854 (1951)	549
Shomate, C. H.,	Jour. Amer. Chem. Soc., 68, 310 (1946)	439
Shukow, J.,	Jour. Russ. Phys. Chem., 42, 40 (1910)	440
Shumilov, A. M.,	Zav. lab., 21, 417 (1955)	475
Sidgwick, N. V.,	"The Chemical Elements and Their Compounds"	1
Sidles, P. H.,	Jour. Appl. Phys., 25, 58 (1954)	666
Simon, F. E.,	Proc. Roy. Soc., A-207, 137 (1951)	658
Sims, C. T.,	BMI-1169 (1957)	307
	BMI-1170 (1957)	270
Sindeband, J. J.,	Trans. AIME, 185, 198 (1949)	66
	Trans. AIME , 185, 749 (1949)	89
	97th Electrochem. Soc. Meeting, Cleveland (1950)	46, 174
Sinke, G. C.,	"Thermodynamic Properties of the Elements" (1956)	311
Sirota, N. N.,	"Physics and Physico-chemical Analysis" (1957)	454, 492
Skinner, B. J.,	Amer. Mineral., 42, 39 (1957)	553
Slade, R. E.,	Jour. Chem. Soc., 115, 215 (1919)	510

Slyh, J. A.,	WADC-TR-55-473	7
Smagina, E. I.,	Zhur. Fiz. Khim., 34, 2329 (1960)	221
Smith, W. H.,	Jour. Amer. Ceram. Soc., 33, 27 (1950)	173
Smithells, C. J.,	"Metals Reference Book" (1955)	308
Snyder, M. J.,	BMI-1124 (1956)	277
	USAEC-TID-7530 Part I (1957)	195
	BMI-1313 (1959)	142
	BMI-1223 (1957)	309
Snyder, N. H.,	Ind. Eng. Chem., 43, 2008 (1951)	115
Snyder, P. E.,	Jour. Amer. Chem. Soc., 75, 1227 (1953)	691
Sodlate, A.,	Acta Cryst., 1, 212 (1948)	630
Solonnikova, L. A.,	Fiz. Metal. Metallov., 5, 565 (1957)	473
Spedding, F. H.,	Proc. Roy. Soc., A-207, 137 (1951)	658
	Phys. Rev., 58 [5], 1092 (1952)	659
	Revs. Mod. Phys., 25, 129 (1953)	660
Spiess, K. F.,	Z. phys. Chem., A-175, 140 (1935)	152
Spindler, W. E.,	AERE-M/M-143 (1957)	231
Spinner, S.,	Jour. Res. NBS, 65C, 89 (1961)	194

Strel'nikova, N. S.,   "Physics and Physico-chemical Analysis" (1957)   454

Stull, D. R.,   "Thermodynamic Properties of the Elements" (1956)   311

Styri, H.,   Metals and Alloys, 3, 273 (1932)   180

Swanson, H. E.,   NBS Circular-539 (1953)   527

Swartz, E. L.,   NP-5796 (1955)   123

Tamaru, S.,   Z. anorg. u. allgem. Chem., 62, 81 (1909)   480

Tatke, E.,   NBS Circular-539 (1953)   527

Taylor, K. M.   Ind. Eng. Chem., 47, 2506 (1955)   163

Materials and Methods (January, 1956)   159

Jour. Electrochem. Soc., 107 [4], 308 (1960)   150

AT (40-1) 2558 (February, 1961)   136

Taylor, R. E.,   WADD-TR-60-581 Part II (1962)   501

AF 33 (657)-7136 (January, 1962)   146

AF 33 (657)-7136 (April, 1962)   209

Jour. Amer. Ceram. Soc., 45 [7], 353 (1962)   326

ASD-TR-62-34 (1962)   531

NAA-SR-4905 (1960)   532

Jour. Amer. Ceram. Soc., 44 [10], 525 (1961)   535

Jour. Amer. Ceram. Soc., 45 [2], 74 (1962)   550

Technical Information Service, Reactor Handbook (1955)   312

Teeter, C. E., Jr.,	Jour. Amer. Ceram. Soc., 33, 363 (1950)	289
	NEPA-818 (1948)	313
Tefft, W. E.,	Jour. Res. N.B.S., 64A, 212 (1960)	188
	"Mechanical Properties" (1961)	187
Templeton, D. H.,	Jour. Amer. Ceram. Soc., 33, 291 (1950)	11
	Acta Cryst., 3, 261 (1950)	28
	Jour. Chem. Phys., 18, 391 (1950)	366
Templeton, D.,	Jour. Phys. Chem., 60, 443 (1956)	614
Tharp, A. G.,	Jour. Amer. Chem. Soc., 78 [19], 4850 (1956)	212, 654
Thielke, N. R.,	Jour. Amer. Ceram. Soc., 33, 304 (1950)	78
	N.P.-4970 (1953)	118
Thomas, E. A.,	T.A.M. (1948)	551
Thorne, P. S. L.,	"Inorganic Chemistry" (1954)	405
Thorpe, J. F.,	"Thorpe's Dictionary" (1943)	406
Tiede, E.,	Z. anorg. u. allgem. Chem., 87, 167 (1914)	377
Tinklepaugh, J. R.,	Amer. Ceram. Soc. Bull., 30, 103 (1957)	99
	WADC-TR-54-38 (1954)	229
Tobias, C. W.,	Jour. Amer. Chem. Soc., 71, 1882 (1949)	358

Todd, S. S.,	Jour. Amer. Chem. Soc., 72, 2914 (1950)	506
Toman, K.,	Acta Cryst., 4, 462 (1951)	637
Tone, F.,	Ind. Eng. Chem., 30, 232 (1938)	675
Toropov, K. A.,	"Chemistry of Silicon" (1950)	674
Trennoy, R.,	Compt. rend., 141, 190 (1905)	612
Trice, J. B.,	NEPA-818 (1948)	313
Tripler, A. B.,	BMI-1313 (January, 1959)	142
Trulson, O. C.,	DA-30-069-ORD-2787 (February, 1962)	199
Tsumura, Y.,	Jap. Patent 179,269	420
Tucker, S. A.,	Jour. Chem. Soc., 81, 14 (1902)	347
	Proc. Chem. Soc., 17, 129 (1901)	354
Tyle, R. P.,	Jour. Less Common Metals, 3, 13 (1961)	541
Udy, M. C.,	"Chromium" (1956)	314
	AECD-3382 (1949)	315
Ulik, F.,	Ber. Wien. Acad. (II) 52, 115 (1865)	562
Umanski, J. S.,	Jour. Phys. Chem. USSR , 14, 332 (1940)	515
Urbain, E.,	Fr. Patent 677, 330	414

Urbain, M.,	Rev. Met., 50, 617 (1953)	423
Uyeno, K.,	Jour. Chem. Soc. Japan, 62, 990 (1941)	690
Van Arkel, A. E.,	Physica, 4, 286 (1924)	436
	Z. anorg. u. allgem. Chem., 148, 345 (1925)	504
Vasilos, T.,	Jour. Amer. Ceram. Soc., 37, 107 (1954)	125
Vig, G.,	Carborundum Research Notebook 5843	133
Vigouroux, E.,	Compt. rend., 129, 1238 (1899)	615
	Compt. rend., 127, 393 (1898)	625
Villa, H.,	Z. Electrochem., 53, 32 (1949)	482
Vogel, R.,	Z. anorg. u. allgem. Chem., 61, 46 (1909)	490
	Z. anorg. u. allgem. Chem., 84, 323 (1914)	561
Voronov, M.,	Izv. vo. Akad. Nauk SSSR, 13, 144 (1936)	647
Wachtman, J. B.,	"Mechanical Properties" (1961)	187
	Jour. Res. N.B.S., 64A, 212 (1960)	188
	Jour. Amer. Ceram. Soc., 45 [7], 319 (1962)	327
Wagner, C,	Jour. Electrochem. Soc., 103, 107 (1936)	316

Wallbaum, H.,	Z. Krist., 101, 78 (1939)	17
	Z. Metallkunde, 33, 378 (1941)	22
	Z. Metallkunde, 31, 363 (1939)	591
Waller, C. E.,	"Mechanical Properties" (1961)	189
Wallstein, R.,	Z. anorg. u. allgem. Chem., 128, 96 (1923)	396
Ward, C. H.,	ASD-TR-61-260 (1962)	650
Warde, J. M.,	Bulletin 94, Refractories Institute	5
Warren, H.,	Chem. News, 78, 318 (1898)	604
Waterman, T. E.,	WADC-TR-57-487 (1957)	246
	WADC-TR-58-476 (1959)	651
Watts, A. A.,	WADD-TR-60-646 Part I (1961)	144
Watts, O.,	Bull. Univ. Wisc., 145, 255 (1906)	616
Webb, W. T.,	Jour. Electrochem. Soc., 103, 107 (1936)	316
Weber, W. P.,	BMI-1165	88
Wedekind, E.,	Chem. Ztg., 29, 1032 (1905)	218
	Ber. Dtsch. Chem. Ges., 40, 297 (1907)	348
	Ber. Dtsch. Chem. Ges., 46, 1205 (1913)	353
	Ber. Dtsch. Chem. Ges., 46, 1203 (1913)	363
	Ber., 35, 3929 (1902)	579

Wedekind, E.,	Chem. Ind. Koll., 7, 249 (1900)	580
Weiss, G.,	Bull. soc. Chim. France, 14, 1077 (1947)	364
	Ann. chim., 1, 446 (1946)	201
	Thesis, Univ. Grenoble (1946)	355
	Bull. Soc. chim. France, 15, 598 (1948)	356
Weiss, L.,	Liebigs Ann., 355, 92 (1907)	513
Welch, A.,	Nature, 167, 362 (1951)	644
Wells, A. F.,	"Structural Inorganic Chemistry" (1962)	248
Welty, H. F.,	Jour. Amer. Ceram. Soc., 34, 309 (1951)	102
Wentorf, R. H.,	Jour. Chem. Phys., 26, 956 (1957)	434
Westerhoff, H.,	Z. Physik, 75, 521 (1932)	333
	Ann. Phys., 17, 593 (1933)	467
Westinghouse	A-2173 (1956)	317
Westphal, R. C.,	AECD-3864 (1954)	318
Wheelock, N. R.,	CWR-481 (1957)	119
White, D. W.,	ASM Cleveland (1955)	319
White, G. K.,	Can. Jour. Phys., 33, 58 (1955)	665
Whiteley, M. A.,	"Thorpe's Dictionary" (1943)	406

Whittemore, O. J.Jr., Jour. Amer. Ceram. Soc., 39 [12], 443 (1960)   670

Ind. Eng. Chem., 47, 2510 (1955)   32

Jour. Can. Ceram. Soc., 28, 43 (1959)   129

Wilhelm, H. A.,   Trans. ASM, Reprint 48 (1955)   31

Wilhelm, J.,   Trans. Roy. Soc. Canada, 25, 13 (1931)   468

Wilkinson, D.,   AERE-M/M-143 (1957)   231

Willmore, T. A.,   WADC-TR-52-111   116

Wilson, A. S.,   Jour. Amer. Chem. Soc., 70, 99 (1948)   61

Wilson, R. E.,   Amer. Ceram. Soc. Bull., 30, 103 (1957)   99

Wilson, W. B.,   Jour. Amer. Ceram. Soc., 43, 77 (1960)   137

Winter, R. M.,   Ann. Physik, 77 (1925)   545

Witsett, T.,   Iowa State College, Jour. Sci., 31, 541 (1957)   493, 495

Wöhler, L.,   Z. anorg. u. allgem. Chem., 11, 1951 (1926)   481

Z. anorg. u. allgem. Chem., 209, 33 (1932)   485

Wolf, U.,   Z. Elektrochem., 46, 42 (1940)   375

Wölfel, E.,   Z. anorg. u. allgem. Chem., 280, 3215 (1955)   479

Wood, E.,   Acta Cryst., 1, 441 (1954)   646

Wood, W. D.,   DMIC-148 (1962)   208

Woods, S. B.,	Can. Jour. Phys., 33, 58 (1955)	665
Woodward, P.,	Trans. Far. Soc., 46, 199 (1950)	143
Wygant, J.,	Jour. Amer. Ceram. Soc., [12] (1951)	321
Yajima, E.,	Bull. Nagoya Inst. Technol., 5, 260 (1953)	428
Yavorsky, P. J.,	Jour. Res. N.B.S., 35 [1], 87 (1945)	249
Yermakova, V. A.,	Zhur. Neorg. Khim., 3, 868 (1958)	597
Yevstrop'yev, K. S.,	"Chemistry of Silicon" (1950)	674
Yosim, S. J.,	NAA-SR-2124 (1957)	322
Young, R. A.,	Oxford Conf. (1951)	323
Zachariasen, W.,	Nat'l. Nucl. Energy, 14B, 1451 (1949)	653
Zachariasen, W. H.,	Z. physik. Chem., 128, 39 (1927)	48
	Acta Cryst., 2, 94 (1949)	332
	Acta Cryst., 2, 388 (1949)	523
Zalabak, C. F.,	NASA-TN-D-761 (March, 1961)	403
Zalkin, A.,	Jour. Chem. Phys., 18, 391 (1960)	366
Zehms, E. H.,	Jour. Amer. Ceram. Soc., 43, 54 (1960)	537
Zemansky, M.,	Phys. Rev., 79, 7021 (1951)	470

Zettel, C.,   Compt. rend., 126, 833 (1897)   613

Zeytts, F.,   Fiz. metallov., G.I.T.T.L., 205 (1947)   477

Zhadanov, G. S.,   Compt. rend. acad. sci. U.R.S.S., 32, 432 (1941)   378

Zhuravlev, N. N.,   Kristallog., 1, 666 (1956)   678

Ziegler, W. T.,   Oxford Conf. (1951)   323

                     Jour. Amer. Electorchem. Soc., 69, 2762 (1947)   469

APPENDIX C

UNITS AND CONVERSION FACTORS

APPENDIX C

UNITS AND CONVERSION FACTORS[3]

Property	Units Used	Convert to:	Multiply by:
Formula weight	g/mole	lb/mole	0.0022
Formula volume	cc/mole	-	-
Melting point	$^{\circ}$C	$^{\circ}$F $^{\circ}$K	(1.8 x $^{\circ}$C) + 32 Add 273
Boiling point	$^{\circ}$C	$^{\circ}$F $^{\circ}$K	(1.8 x $^{\circ}$C) + 32 Add 273
Vapor pressure	mm of Hg	atmospheres psi	0.00131 0.0193
Evaporation rate	g/cm^2/sec	lbs/ft^2/sec	2.048
Density (ρ)	g/cc	lb/cu ft	62.43

CHEMICAL

Theoretical analysis	percent	-	-

ELECTRICAL

Resistivity	ohm-cm	ohm-in	0.394
Critical temperature* (T_c)	$^{\circ}$K	$^{\circ}$C	Subtract 273
Temperature coefficient of resistivity	per $^{\circ}$C	per $^{\circ}$F	0.555
Thermal EMF	μv/$^{\circ}$C	μv/$^{\circ}$F	0.555
Dielectric constant	none	-	-
Dissipation factor	none	-	-
Thermionic work function	ev	-	-

*Temperature at which a material becomes a superconductor.

Property	Units Used	Convert to:	Multiply by:
MAGNETIC			
Susceptibility	Henry's/meter	-	-
Critical magnetic field** (H_c)	Oersteds	-	-
MECHANICAL			
Strength	psi	kg/mm^2	7.03×10^{-4}
Hardness: Mohs	none	-	-
Vickers	kg/mm^2	-	-
Knoop	kg/mm^2	-	-
Rockwell	none	-	-
Young's modulus	psi	kg/mm^2	2.926×10^2
Torsion modulus	psi	kg/mm^2	2.926×10^2
Shear modulus	psi	kg/mm^2	2.926×10^2
Bulk modulus	psi	kg/mm^2	2.926×10^2
Poisson's ratio	none	-	-
Creep rate	in/in/hr	percent/hr	100
NUCLEAR			
Cross section	barns/atom or	$cm^2/atom$	10^{-24}
	cm^2/cm^3	-	-
OPTICAL			
Self explanatory			
STRUCTURE			
Self explanatory			

**Field at which a superconductor becomes a "normal" conductor (557)

Property	Units Used	Convert to:	Multiply by:
THERMAL			
Conductivity	cal/sec-cm-$^{\circ}$C (CGS)	btu/hr-ft-$^{\circ}$F	244
Diffusivity	cm^2/sec	in^2/sec	0.155
Expansion	per $^{\circ}$C	per $^{\circ}$F	0.556
Specific heat	cal/gm/$^{\circ}$C	btu/lb/$^{\circ}$F	1
Heat content	cal/mole $^{\circ}$K	cal/mole $^{\circ}$C	1
Heat capacity	cal/mole $^{\circ}$K	cal/mole $^{\circ}$C	1
Thermodynamic properties	kcal/mole	–	–
	eu (entropy units)	–	–

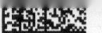